The Surgery of Henri De Mondeville, Volume II

The Surgery of Henri De Mondeville, Volume II

Surgeon of Philip the Fair, King of France

Written from 1306 to 1320

A French Translation, with Notes, an Introduction and a Biography
by
E. Nicaise
Professor in the Faculty of Medicine of Paris
Surgeon at the Hospital Laennec
Former President of the Society of Surgery of Paris
Former Member of the Council for Oversight of Public Assistance
with the collaboration of Dr. Saint-Lager and F. Chavannes

This monument of French Surgery deserves its place among the
predecessors of Guy de Chauliac
Littré, Hist.litt. vol. XXVIII, p.351

PARIS
Ancienne Librairie Germer Bailliére et Cie
1893

Translated into English In Two Volumes
With Annotations, Comments and Appendices
by
Leonard D. Rosenman, M.D.
1996
Copyright © 1998

VOLUME TWO

To order additional copies of this book, contact:
Xlibris Corporation
1-888-795-4274
www.Xlibris.com
Orders@Xlibris.com
19536

TRANSLATOR'S PREFACE TO VOLUME II

For reasons given by Mondeville in his own Introduction on page 1, we have divided Treatise II of Mondeville's *SURGERY*. Its First Doctrine concluded Volume I, and Doctrine II opens Volume II.

This Volume contains the four chapters of Doctrine II and the first two Doctrines of Treatise III. The Third Doctrine was not written; a sad and sick Henri left for us only his detailed Rubrics for the forty-three chapters he had hoped to write and his remorseful comments which supplement the Rubrics. His terminal illness also explains his failure to write Treatise IV on Fractures and Dislocations as he had planned it. Treatise V is Mondeville's Antidotary; it comes at the end of his great work. It is followed by Appendix I, the Pharmacopeia which I have expanded to include all of the medications used by the seven great medieval surgeons who wrote the texts which served to reawaken surgery in Europe during the epoch 1170 to 1320. Those surgeons were Mondeville's predecessors, his teachers and his contemporaries.

LDR

NOTE RE RUBRICS FOR VOLUME II

For the convenience of the Reader, a Master Table Of Contents is placed at the beginning of both volumes of this edition. The many items to be found there in detail to some extent may substitute for an Index at the end of the Work, for which I have had neither time nor opportunity. I offer The Table with apologies.

As noted in the Preface, this Volume II opens with Doctrine II of Treatise II: An Introduction and Four Chapters: 1. Ulcers; 2. Bites and Stings Of Venomous Animals; 3. Fistulas. 4. Cancerous Ulcers. The many sections of those chapter as well as the contents of Doctrines I and II of Treatise III are described in the Master Table. The latter consist of Mondeville's brief Introduction, the 24 Chapters in Doctrine I and the 23 Chapters in Doctrine II.

The Rubrics for Doctrine III of Treatise III are set apart. They follow the text of Doctrine II, and are limited to Mondeville's Introduction and to the extensive list of what he had planned to write. Together they represent the only extant portions of his planned text.

The Rubrics of Treatise V, the Antidotary, are given in the Master Table.

Volume II concludes with an Appendix, which is the Compendium Pharmacopeia.

CONTENTS

VOLUME I   PART I

DEDICATION

MONDEVILLE'S TEXT

VOLUME II   PART I

GENERAL TABLE OF CONTENTS FOR VOLUMES I AND II

TREATISE II-DOCTRINE II TEXT

VOLUME II PART II

TREATISE III DOCTRINE II.

TREATISE II

DOCTRINE II

Introduction

Nicaise's Note: After completing Doctrine I of Treatise II, Mondeville stopped work on his Surgery, which he had begun in 1306. He added Doctrine II which follows here in 1312 and he altered his original plan for the chapters, as he restated it in the Introduction below.

Here begins the Particular Introduction to Doctrine II of Treatise II of the Chirurgia of Master Henri de Mondeville, Surgeon to our Illustrious Leader, The King Of France

Having completed Treatise I of this *Surgery*, which contains the Anatomy limited to what is useful for a surgeon, and having completed Doctrine I of Treatise II on the treatment of wounds, I now undertake with the support of God to write Doctrine II to be followed directly by Treatise III. My plan is to devote an entire chapter to an ailment, to describe it, to offer treatments for it, and to follow with explanations to clarify the more obscure items.

I shall deal with each ailment in four ways: 1. Define it and describe it. 2.Describe in detail only those varieties which come within the purview of a surgeon, and the various sources of certain difficulties that he may encounter. 3. The causes of the ailments themselves. 4. The symptoms (ie signs) in common for all the varieties, and those which distinguish each particular case.

Then I shall provide up to three methods for treating each ailment, but only if they have rational bases. First will come preventive measures, followed by curative treatments and then by palliation.

I. Preventive Treatments are the medical and surgical measures that will deter the progress of an illness. Here I quote Galen (*Commentaries on the Aphorisms,* 22), "If an impending cause is aborted, that is an anticipation, not a cure." He said the same in *Techni, Tr. of Causes,* Ch. 27. Haly agreed (*Commentaries*), "This treatment (ie prevention) can be used in some but not all surgical conditions. For example, a person may suffer from chilblains in his feet all winter long, whereas he could have avoided the suffering with proper treatment." The same is true for many conditions; many persons can be spared illness with preventive measures. However, such are not applicable in cases of wounds, contusions, fractured bones, etc., conditions which are not predictable. As Aristotle wrote (*On Harmony,* Bk. I and *Physics,* Bk. II), "There is no science that can foretell and prevent all chance events; one cannot tell in advance, but once they begin, one may get some inkling of what will happen."

I. The preventive treatments are used by some medical practitioners in certain cases when they are indicated, as one may read in Galen's *De Ingenio,* Bk.XII. Ch. 3, where he tells us to treat syncope in anticipation of it happening. "You who are wise physicians will be put to shame if you allow your patient to faint."[1]

II. Second, I will describe Curative Treatments and include references to the recommendations of the medical and surgical authorities, especially to Galen, the citations where he describes his own treatments. I shall deal with the treatments first by giving the General Precepts that I have garnered both from the older authorities and from the moderns. Those are the principles and canons from which one may abstract and infer nearly all the procedures designed for particular cases and for the special operations.

I shall refer to authorities for their treatments in all except three types of cases in which surgeons, simply to earn fees, will perform

---

[1]   I cannot trace this reference. I assume that Galen rebukes the physician who bleeds his patient until he faints (LDR).

operations really designed for palliation. 1. When the patient is absolutely incurable by surgery, as in internal cancers, when fistulas communicate with the urinary system or with the marrow-cavities of bones and in other similarly evil conditions that you cannot cure with surgical procedures unless you can radically extirpate the entire lesion, roots and all; when it is clear that the patient's vital forces cannot withstand such procedures which are depleting or lethal. 2. Some curable illnesses do not warrant surgical procedures. Avicenna (Bk. IV. Fol. 4, Tr. III, Ch. on unhealed fistulas) wrote this about the treatment of chronic fistulas, "The treatment consists of a total ablation with the knife or the cautery or with caustic medications, procedures that cause such profound suffering that a patient will choose to live with his disorder rather than submit to the treatment." 3.When a disorder is curable by a surgical procedure that both the surgeon and the patient agree to undertake, but the treatment will produce another disorder that is worse than the original one. In that category are operations for a fistula in ano that is a complication of chronic hemorrhoids, and when the internal opening of the fistula is above the anal muscles.[2] Another example is a case of extensive mal-mort.[3] The surgical procedures in such cases lead to disorders and to suffering that are worse than the original. Hippocrates (Aphorisms, Part 6) said the same. Later on in this Treatise if you are attentive, you will read how common sense will demonstrate the truth of the foregoing.

III. Palliative Surgical Treatments aim to ease and to diminish the effects of a disease, but do not attempt to cure the patient. The surgeon can blunt the severity of the illness and render it more tolerable to the patient. All the medical authors palliate with dietary regimens for incurable patients and for cases when there is no advantage to treat for cure when a condition is incurable. Hippocrates), in many places (*Aphorisms*, Part 6) and Galen and Haly (*Techni, Treatise on Causes*, Ch. 33) all advised that we fall back on palliative treatment when curative measures avail little or cannot be used, saying, "a curative treatment

---

[2]   An incision through the sphincters may lead to permanent incontinence of feces (LDR).

[3]   A chronic ulcer with necrotic slough; often a deep stasis ulcer on the leg or a necrotizing ulcer anywhere (LDR).

should get rid of a disease. If you cannot accomplish that, palliate it by changing it to a less serious form." These treatments are not the same as in the three cases where curative treatment are not at all suitable. There the patient is absolutely incurable, or if he is curable, the patient will be unable to withstand the radical treatment, or when the treatment causes a more serious disorder. In all three kinds of cases, curative treatments are contraindicated

However, when the patient wishes to obtain relief and he can afford to pay for his services, which may include palliative treatments,[4] the surgeon will provide a suitable diet and will take care of all the patient's needs, and will fulfill all the orders which are the concerns of a supervising physician, all of it to the extent of the patient's generosity in payments already received and those to be anticipated. The palliative treatments in the last two cases often come when the patient has refused to accept the recommendations to undergo curative treatments. In my own experience in Paris, I have seen that happen many times when my colleagues have disagreed with my judgment.[5]

CHAPTER 1

THE TREATMENT OF ULCERS

Three General Comments

A. Description of Ulcer

Definition : An ulcer is a wound that has suppurated or has been caused to suppurate (ie by medications), or it discharges other impure matter for a long time or in great quantity, or simply one which has not begun to heal within the expected usual time. According to the Salernitans that was about two days.[6] However, Galen

---

[4]   The patient is wealthy enough to maintain a personal resident-surgeon to attend him (LDR).

[5]   When he accepted the patients for palliation (LDR).

[6]   The expected time for a properly treated wound to show signs of healing (LDR).

(*De Ingenio,* Bk. IV, Ch.4) fixed the time at seven days, stating "one may blame our All-Powerful God and cast blame for prolonging the time for healing up to a year, when it should take place within a week."

The Varieties:: One type of ulcer is called apparent, it can be flat or concave. There are five types of the latter: the virulent, dirty (ie sordid), corrosive, putrid and simply slow-to-heal. We shall discuss the treatments of each variety later on in this chapter. Now we shall go on to describe their similarities and differences.

Other than the principal varieties there are ulcers that develop after certain complications or dispositions, and they do not fall into the categories that are listed in surgical handbooks (ie Practicas). The medical authors, especially Avicenna (Bk. IV, Fol. 4, Tr. 3, Ch. 1) said "one ulcer derives from an internal source, another from an external; one is painful, another is not; some are complicated by an abscess or tumor, others not; some are accompanied by heat, others by cold; some with fever, others not; some are caused by humors that cause burning pain, in others the humors do not burn; some are venomous or cause poisoning, others do not; some have indurated margins or beds; some are livid or black or green; some are granular and shed hair, and others do not. Some provoke spasms, some develop in fleshy tissues, others in nerve-bearing regions. Some produce superfluous granulation tissues, whereas none appears in others. In some the ulcer-bed is a gangrenous bone. Some appear in persons or in parts of their bodies endowed with healthy complexions, others are to the contrary. Some merely are damp, some are very wet, others are moderately so. Some are recent, others very old and others are in between. Some are round, some are irregular. Some have lost much tissue, some none, some only a small amount, some a moderate amount. Some bleed. Some bulge with superfluous humors (ie edema), others have none."

In addition to the apparent and the deep ulcers, other types exhibit more differences than we have already listed. Some of those differences include cancers, esthiomene, fistulas, mal-mort etc., as we will see in the chapters that define them. All these varieties, differences, conditions of ulcers and other matters should be known by a surgeon who wants his treatments to succeed. Indeed, one must treat according to what he encounters in any ulcer. These generalities will apply in all subsequent chapters.

The Causes And The Varieties Of Causes: Some are proximate (ie material) and others are predispositions. That is what Avicenna wrote (Bk. I, Fol. 2, Ch. 27 "Causes of Ulcers"). The materials which cause ulcers are blood, bile, phlegm, melancholy, wind and water.

The predisposing causes are the innate tendency of a part of the body to impel the bad humors to the ulcer and the weakness of the ulcerated part which receives them; the two cooperate in forming the ulcer. Another predisposition is the looseness of the tissue, its sponginess or its thinness, such as found in glands or emunctories. Another predisposition derives from the diameters and the lengths of the blood vessels and other channels that are directed toward the ulcer. Another cause is the location of the parts that are ulcerated. For example, the legs and the feet are lower than the rest of the body.[7] Another cause is an excess of nutritive matter delivered to a sick region which cannot convert it, and another is superfluous heat attracted to the part.[8] Other causes are venomous bites and stings and the lack of resistance to them. Excessive exercise of a limb may attract humors, as well as a contusion, a fracture, a dislocation and a tight constriction. Another cause is pain that attracts heat and the vital spirits.

Furthermore, we see certain ulcers come from external causes and others from internal ones. The ignorant mob speaks of and believes in chance or of Divine intervention as the internal cause of these disorders, sent to us by our glorious and sublime God. The common folk, say that it is He who sends them, and we should not interfere and use our man-made treatments. The outcome of the treatment is as it pleases the Lord. They support their opinions by insisting that God alone cures us of our chronic debility; he wounds

---

[7]   To explain why the legs are common sites for what we today call stasis ulcers (LDR).

[8]   To explain dependent edema and the frequent company of low-grade cellulitis (LDR).

and he provides the remedies. What he has given he can take away. Others think differently and they defend God by saying that diseases with internal causes do not come from him but from the bad luck of the patient.

The surgeon must understand all of this; the causes and their effects as listed here, and others, too.

The Symptoms: There are two kinds, those which are common to all ulcers and those that belong to each variety. The symptoms in common are described in all the surgical handbooks and by the medical writers. You can infer others from the name of the ulcers in our list. Every good surgeon knows them. The particular symptoms will be described when we get to their treatments.

B. The Treatment of Ulcers

We divide them into Preventive, Curative and Palliative.

I. Preventive Measures consist of defending against and eliminating all the causes which we have described, which favor the development of these maladies, especially those which are proximate or impending. Indeed, even if we do intervene, an ulcer may be inevitable, yet often we are effective and no ulcer will form. That is what we mean by the preventive (ie preservative) treatment of wounds. The prevention of inflammation (aposthems) keeps wounds from forming pus and from bursting open (ie to become ulcers). By prevention we avoid ulcers.

II. Treatments For Cure: 39 General Guides:

1. When a specific dyscrasia complicates an ulcer, such as inflammation or swelling (ie edema), treat it for cure before you attack the ulcer. That is Avicenna's rule (Bk. I, Fol. 4, Ch. 21; and Bk. IV, Fol. 4, Tr. 3, Ch. 1). The medications for ulcers have drying actions, and they interfere with the wet medicines needed to dissolve and cleanse the complicating dyscrasia, which must be treated first.

2. Whenever the ulcer is extremely painful, you try to relieve it

before acting on the preceding Rule. So said Avicenna (ibid., Ch. "On Relief of Pain"). And Galen said it, too, (*Acute Regimens,* Bk. III).

3. Your medication for healing (ie closing) the ulcer will fail until you get rid of the dry crusts and the filthy matter. (Galen, *Megatechni,*Ch. 3, and Avicenna, Ch. on Ulcers).

4. When the ulcer or the rest of the body suffers a dyscrasia, treat that with evacuants, proper diets and topicals directed against the dyscrasia. Then treat the ulcer. (Avicenna, ibid.)

5. If the liver, the spleen or the stomach suffers dyscrasias, restore their health first. Those organs are necessary for the health of the entire body; they make healthy blood. When they are sick, the blood from them is not strong enough to heal ulcers.

6. When the humors of the entire body or of the ulcerated member alone or of an organ that usually provides normal humors for the body show signs of deterioration or diminution or both, you should get rid of them with evacuations, or temper them with phlebotomies or laxatives or both and with a healthy dietary regimen that acts contrary to the impaired humors. Those humors are not suitable for replacing ulcerated tissue or for keeping the region healthy.

7. Inasmuch as the humors which cause ulcers come from other parts of the body, we try to intercept or deflect (ie repercuss) them (Avicenna, ibid.)

8. All ulcers are treated by foods, by internal medicines and by suitable topicals which are desiccative and by a regimen that is proper in its six non-natural elements and is contrary to the causes of the ulcer, (Avicenna, ibid.).

9. When you treat ulcers, avoid using suppuratives (Avicenna, ibid.), both internally and in your topicals.

10. Avoid using naturally humid medicines, because, as Avicenna explained, we should emphasize desiccatives.

11. When ulcers are exposed during the mid-day to warm moist air, the cure is retarded, nor can we find any favorable air that does not cause ulcers to suppurate and thereby to interfere with the healing.

12. Do not use active cold medicines. Hippocrates (*Aphorisms*, Part 5) said that cold irritates ulcers.

13. You may apply cool medicines on ulcers that tend to be warm, and use the warm topicals when the ulcers tend to be cool. Galen (*Commentaries on the Aphorisms*, Part 2) said we should always treat with contraries, and Hippocrates agreed (*Aphorisms*, at the end of Part 2).

14. Do not apply corrosive topicals on internal ulcers and on those that penetrate, such as those that enter the chest cavity, because the medicaments then cannot be extracted (ie before they damage vital organs).

15. On deeper and more moist ulcers use drier topicals. As it works its way into the depths of the ulcer the medicine loses its strength. Therefore, when the distance is great you must resort to the most potent compounds.

16. In some cases you must take care not to allow cohesion of adjacent thin layers of tissue, for example the eyelid must not adhere to the eye, and fingers should not be allowed to adhere to each other. This principle really is self-evident.

17. Ulcers which come after other illnesses are difficult to treat because in curing the illness Nature traps the residues of the corrupt superfluidities in the ulcer.

18. The skin surrounding some ulcers may lose its hair. That is a sign that the bad humors have corrupted the region. That ulcer is difficult to cure. So said Galen (*Commentaries on the Aphorisms*, Part 6).

19. Ulcers at the coccyx are not easy to cure (Avicenna, Bk. I, ibid.) for four reasons. a. Many nerves end there, and they separate the parts.[9] b. It is near the immodest openings. c. It is a low part. d. When the patient strains to expel feces he distends

---

[9]   Nerve-tendon-ligament: here is another example of the confusion. The 'nerves' were contractile structures. They form clusters and plexuses which separate the fleshy tissues. By definition, a wound is a separation! (LDR).

the region and prevents the ulcer from closing. Galen (*Megatechni*, Bk. V, Ch. 3, in the part where he discusses wounds of the lung) said that immobility is necessary (ie for ealing) in a wounded or an ulcerated part

20. Ulcers at the knee and at the tip of the elbow lie over joints which move and stretch or contract the edges of the lesion. The same is true for the lips, the eyelids, the penis et al. Treatments are difficult (Avicenna, ibid.). Again, the reason is that the regions cannot be put at rest.

21. Ulcers in nerve-bearing regions are difficult to cure because they are very painful. Galen (*Commentaries on the Aphorisms*, Part 5) said that pain impairs the flow of humors, and, since nerves are spermatic structures, their tissues are not easily repaired.

22. Round ulcers are difficult to cure. a. They have lost more tissues and need more time to regenerate them. b. Nature cannot decide where to begin the repair; a circle has no sides, no beginning and no ending. c. The margins are equally separated, whereas in oblong ulcers the opposing sides can be brought together by traction at the ends.

23. Round ulcers can be very painful and babies may suffer unto death from pain. That must be relieved. Avicenna (Bk. IV, ibid.) said that the vital force is depleted and overcome by strong pain.

24. Ulcers over or near the dorsal spine, and those in the front of the lower leg are difficult to treat and because they are near large nerve-bearing muscles they are disposed to spasm and they cause much trouble (Avicenna, ibid.)

25. If the ulcer is the only ailment, apply the topicals as the only treatment. (Theodoric, Bk.I, Ch. 8.)

26. Laxatives and emetics help in the treatment of old and painful ulcers. (ibid).

27. Round and excavated ulcers will persist until you can elongate them by incising opposite margins to convert them to ovals. (ibid.)

28. When the surfaces of the ulcer are indurated, cut them away

with a knife or a cautery, to allow the granulation tissue to adhere.[10]

29. Purge away and obstruct the return of humors that have accumulated in an ulcer. Use good purges to deflect them. If the influx has ceased but the the stench of the ulcer remains, evacuate the region by cupping and bleeding, and use desiccatives in the ulcer. If once is not enough, purge again.

30. A resistant chronic ulcer may be treated by making a fresh wound alongside it, and keeping it open while the primary ulcer heals. That will intercept and deflect the humors which caused the first ulcer. When that heals, the second ulcer will be easy to cure. (ibid. Ch. 10).

31. Whoever treats ulcers should have available plasters and other medications that attract bony fragments and any other abnormal stuff that may be lodged in ulcers. (Avicenna, Bk I, ibid.).

32. When you first encounter an ulcer do not bathe it or the patient with hot water. However, when it is well along and is healing, and when it drains healthy pus, you may bathe it or irrigate it as we have instructed. When the ulcer is new and you have to bathe the patient, cover the ulcer with a sparadrap [11] or wrap it. Galen taught us how to wash an ulcer (*Megatechni,* Ch. 1). But, if you wash it at the outset you attract more humors and bring on inflammation (Avicenna, Bk.I, ibid.).

33. A deep or an obscure ulcer needs stronger medications than does the exposed kind. The deeper and the less exposed they are the more potent the medications must be, because as they work into the depths they lose some of their power, (Avicenna, ibid.)

34. Whenever some of the bad humor is attracted (ie from nearby) to an ulcer you have to deflect it before reaches the ulcer, but

---

[10] Mondeville means that healthy blood will adhere to the surfaces and be the source of the granulation tissue (LDR)

[11] A section of cloth made sticky with egg-white. It will seal the ulcer during the immersion (LDR).

you should not be too aggressive (ie with potent medicines) and make matters worse by attracting even more (ie by injuring the member).

35. Also, when the humors come from elsewhere in the body, you should use evacuants in a distant part. On the other hand, if the all the inflow is from nearby, evacuate from nearby. The first method requires more potent measures because of its distance, but the same is not true for the nearby source. If we are too violent there we will bring in more of the humors than we pull out. If there are no distant sources, we need only to evacuate locally.

36. If the ulcer is complicated by another ailment or an abscess or by severe pain, use a topical that will treat the ulcer as well as the complication. A complicated lesion requires a complicated compound (ie not just a Simple).

37. The topicals must take into account the delicacy of the site and the sensitivity of the region and of the noble and principal organs nearby. Consider the needs of the whole region and all the other things. Therefore, we should apply topicals which are suitable in strength: weak, medium-strong or very energetic.

38. If worms[12] appear in an ulcer, kill them with the juice of river-calamint, persicaria, leaves of parsley, absinthe, centaury and wild mint and the like. (Avicenna, Bk. IV, ibid.).[13]

39. Ulcers that have internal causes or are due to repletion of the body (ie dropsy) or others that occur in doddering old folks, and ulcers that provoke much pain in muscles, or cause fevers, or those that derive from burned tissues or from poisonous matter, or those which are indurated, dark, livid or sickly green, or are accompanied by spasm, or which form excessive amounts of proud flesh, or those that hide a patch of gangrenous bone or a foreign body, or are very wet or very old (ie chronic), or have lost substance, especially if the loss is extensive, and occurred with hemorrhage, and those which received their offensive matter from other sites, all of them are more difficult

---

[12]   Maggots (LDR).

[13]   All of the medications are described in the pharmacopeia of the Antidotary, Treatise V (LDR).

to treat, all other things being equal, than are ulcers that occur in conditions opposite to the foregoing. I have already given a partial explanation for that. A curious student should pay attention to the following sections where we will deal with what already has not been specifically discussed.

A practicing surgeon should be well-prepared to offer prognoses as to which ulcers are curable, and which of those are easier to cure if he has learned our definitions and the explanations, the causes, clinical signs, and the guiding principles for treatments as we have disclosed them. And he can infer what nearly all the treatments will consist of, both in general and in what will apply in specific types of ulcers. All the principles given above are the bases for the particular treatments which follow.

Special Treatments For Seven Types Of Ulcer[14]

The apparent (ie exposed) ulcers: In the flat or filled ulcer the skin alone is missing, such as are excoriations caused by friction, etc. Also, all formerly deeper ulcers that have filled with

---

[14]   Nicaise summarizes: " Henri de M. defined seven principal types of ulcers that constitute two categories, the deep and the exposed (ie apparent). The former has only one member, the latter has six. The six are divided in two: flat and concave. The first has only one, the second has five categories: virulent, sordid, corrosive, putrid and difficult-to-heal by cicatrization." Here is Nicaise's outline (LDR).


```
          PROFOUND _____          _____ PROFOUND

ULCERS_____          FLAT_____          _____ FLAT

                                                _____ VIRULENT
          APPARENT_____                   _____ SORDID
                           CONCAVE_____         _____ CORROSIVE
                                                _____ PUTRID
                                                _____ DIFFICULT
                                                       TO HEAL
```

new tissue up to but not including the skin are flat ulcers. We treat them with non-erosive desiccatives. The corrosives will impede cicatrization by eating away the materials that regenerate skin. We will devote a chapter in the Antidotary to this discussion of simple and compound medicines designed for ulcers of every type. We will describe how to prepare the compounds for each and how to apply them, when and how often, and when to change the medications.

Five Types of Concave Ulcers

Concave ulcers have lost skin and some deeper tissues. The treatments have General and Particular features. The General includes two parts:

1. In all types of concave ulcers, after the ulcer has been filled with pads containing the ointments etc. suited to each type, we lay atop them in all cases and at every change of dressing, a cleansing medicament consisting of honey et al. spread on a piece of linen of a suitable size. The mondificatives work continuously, bit-by-bit and not too energetically. After the ulcer has filled with good granulation tissue and it no longer is a concave ulcer, eliminate the detergent and treat it as a flat ulcer.

2. Treat all concave ulcers from the beginning until the granulations fill them by smearing the surfaces of the ulcer with a defensive ointment to repel the influx of more of the bad humors. Especially coat the side most suspect as the portal of entry. An example of the defensive ointment: Take 1 oz. of bol d'armenie, 4 oz of terra sigillata and use 2 parts of oil of roses and 1 part of vinegar to mix and make an ointment as thick as honey. Stir the powdered solids as you gradually add the liquids. You will find many defensives described in our Antidotary, as well as the instructions for making them. You can modify them to suit your case by following the instructions given here.

The Particulars for the five types of concave apparent ulcers:

1. *Bandaging any ulcer.* Let it begin directly over the ulcer and
    pass over the entire region, make it moderately tight, to press
    more firmly directly over the ulcer. Loosen the wrap
    progressively as you wind it farther away from the ulcer.

2. *Virulent ulcers,* cold and warm: A virulent ulcer exudes a thin or
    semi-liquid matter, the virus. That substance is of two sorts, hot or
    cold. The hot virus is formed in overheated ulcers with these
    properties: The ulcer and the surroundings are reddish and the
    virus is red like the color of a lotion after it washes granulation
    tissue. The virus is acidic and it irritates the ulcer and erodes tiny
    pockets in the surface. The cool virus is formed in cool ulcers,
    which are white or off-white. The surrounding skin has its natural
    color or perhaps is slightly paler. The virus does not bite or erode.
    Both the viruses derive from unnatural heat which is hotter than
    the normal body-heat, but the process differs in the two cases. In
    the hot virus the unnatural heat exceeds the natural more than
    does that of the cool virus, and its action is stronger.

    The treatment is suitable for both kinds. Use cool drying
    topicals for the hot ulcer. An example is made with rose-water,
    antherum, barley, balaustium, psidia, alum, lentils, leaves of
    medlars, peas and plantain, all of which have similar cool
    virtues. A better lotion for some cases is made from the juice of
    plantain in which you boil rose petals and anthers. After a
    good wash, use some to wet the pads and bandages, enough to
    soak through to the bottom of the ulcer.

    Treat a cool ulcer with warm drying lotions. Here is one
    composed of wine made from a decoction of myrrh, absinthe,
    marrubium, oregano, calamint, pouliot, ambroise, centaury,
    lavender and tansy. Take all or some of them and add some
    honey. If the two lotions are successful and the ulcers and
    their viruses are dried, then you may apply the incarnatives
    and cicatrizers which are described in the Antidotary.[15]

---

[15] The frequent references to "My Antidotary" explain the enthusiasm of
Mondeville's students who pleaded with him to complete his Antidotary

If, on the other hand, the lotions do not enlarge or shrink the ulcer, neither to improve nor worsen the effects of the virus, and if they do not dry the ulcer or the affected limb or the body as a whole, then you should add some abstersives like honey and astringents like oak gall or alum, et al. If the drying effect occurs too rapidly or too drastically you should dilute the abstersives, desiccatives and detergents by mixing in some medicines with contrary actions.

Whoever ignores the principle of using contrary medicines will never cure an ulcer. Avicenna said it (Bk. IV, Fol. 4, Ch. 3 on treating virulent ulcers), "If such an ulcer happens to be cured, it will be by sheer luck rather than by doing what is needed."

3. *Sordid Ulcers* have thick scales, resembling those of fishes, or some flakes of dry salty phlegm. They may have some spots of gangrene (mal-mort) or a type of scrofula. They may present with globules of granulation tissue or lumps of scaly scrofulas formed from salt-phlegm or black bile or other humors as well.

    When you treat sordid ulcers, remove the crusts from the surfaces with oily moisteners and softeners, such as Rhazes' white ointment. After that, use corrosive detergents until the entire surface is clean. Begin with your most potent medicaments, as prescribed by Avicenna (Bk. IV, Fol. 4, Ch. 'On The Treatment of Sordid Ulcers') and Theodoric. Then gradually decrease the strength until you come to the gentler compounds. Once clean, promote the growth of granulation tissue with regeneratives followed by cicatrizers, which you will find in our Antidotary. Some already have been described in Chapter 10 of Doctrine I of this Treatise II.

4. A *Corrosive Ulcer* grows large by eroding insidiously and destructively. It eats away without causing pain, and you can see it going deep as it corrodes intensively. The blood and the humors are destructive with their cutting and burning properties.

---

before he wrote Treatise IV. Mondeville complied and he did not live to write Doctrine III of this Treatise and Treatise IV (LDR)

You treat the ulcer by suppressing the intensity of the humors by providing refreshing beverages and food, by purging and by applying defensives around the ulcer as well as inside it. Use cool consolidatives such as Rhazes' white ointment or others like it described in our Antidotary. Theodoric warned us to persist and not to change our treatments unless they seem to make things worse. Occasionally, the corrosive ulcer is so destructive that you must extirpate it to spare the rest of the body. Avicenna said the same (Bk. IV, ibid,). Theodoric said that some of the corrosive ulcers contain much more pus than the others. The latter may be treated with coolants and desiccatives such as myrtle-water, rose-water, rain-water, lead-water, juice of plantains, vinegar and similar. Those that produce a lot of pus should receive water-of cinders (lye), sea-water, brine, et al. The formulas can be found in books by other authors, as well as in our Antidotary.

5. *Putrid Ulcers* stink and are filled with pus and are hot. The stench is horrible, beyond words, different from any other malodorous sources of pus and inflammation. Any experienced surgeon can make the diagnosis as he enters the sickroom, having been told nothing about the patient. The malignant miasma is poisonous and penetrating, as if it comes from a dead body. This is an erysipelous corrosive ulcer, and it often attacks the penis.[16] Its stench permeates the entire abode. In France it is called St. Mary's disease; in Burgundy it is St. Anthony's[17] and in Normandy it is St. Laurent's fire. And there are many more names.

In the treatments, first try to suppress the stench and then the suppuration. For the odor use lotions of hydromel, myrrh, sandalwood, roses, violets, water-lilies, camomille, melilot, camphor and other deodorants. For the pus use the most potent detergents listed above, such as lye, sea-water, brine, the capitel (ie potash lye) et al.

---

[16] Probably these are pyoderma gangrenosa rather than a venereal sore. (LDR)

[17] St Anthony's Fire was ergotism causing gangrenous digits (LDR).

Avicenna (Bk. IV, ibid.) classified corrosive and putrid ulcers as wandering and non-wandering, and at various times they can be both. The wanderers invade and undermine at the margins and spread in any or all directions into adjacent tissues, but they do not erode deeply. They are limited to the skin and subcutaneous tissue as superficial ulcers.[18] Theodoric did not classify these superficial ulcers as distinct from the others, and I believe that he was correct in thinking that any ulcer could wander as well as the corrosive and the putrid types. If an ulcer wanders (ie undermines) you add a medicament that arrests that process to the medications you use for the others. Some authors have devoted special chapters to the wanderers, or what they call such, where they prescribe special phlebotomies and purges to evacuate the fiery and irritating bilious humors. One type of evacuation for the local region includes scarification and cupping and the application of defensives and cold mondificatives.

The cautery is used as a last resort, and if even that fails, and you know how to carry it out skillfully, you must amputate the ulcerated limb to prevent the spread of the gangrene (corrosion).

6. The *Difficult-to-Cicatrize Ulcer*, according to Avicenna (Bk. IV, ibid.) is neither a fistula nor a cancer nor localized gangrene (ie mal-mort). It does not fit in any one of the categories. Simply put, it is a chronic wound that will not close. That defect stems from some inherent abnormality in the ulcer itself or perhaps acquired from a defective body as a whole. If the source cannot be diagnosed by its appearance or by touch, you can use your common sense to determine that the cause is not something that we have described in the foregoing sections; it is something that impairs the treatment of the original wound or the ulcer.

---

[18]   To make sense of Avicenna's classes: His ambulant putrid ulcers may represent what we have called Meleney's ulcer and necrotizing fasciitis, and what he called nonambulant was ischemic gangrene (LDR).

In brief, the treatments are as follows: Try to discover the cause for the persistence of the ulcer and get rid of it. As I explained in the next-to-last chapter of Doctrine I, you cannot cure an ulcer unless you suppress its causes. Having succeeded in that, apply the medicines used for all ulcers: the regeneratives and cicatrizers. When the cause cannot be eliminated and you see no improvement, then will know that the causes are within the body, and you must purge them with laxatives and clysters. Also, you purge the member that bears the ulcer with scarification, cupping and leeches. Then apply the topicals for ulcers which you will find in the handbooks and in our Antidotary, and follow our directions when you use them.

The General Chapter deals with when and how to purge, and the details are important as to what comes first and what should follow. They are explained in Chapter I of Doctrine II in this Treatise II. More details are to be found in the Medical texts.

How To Dress An Apparent Ulcer

One Method: A tightly bound bandage, wrapped only over the ulcer with its plies sewn in place is an old method which today I reject for three reasons: a. It is a belt, tight only at one level. The neighboring unbound regions can continue to deliver humors, and the pain persists and inflammation ensues. b. The neighboring tissues are not helped at all, although they share the suffering. c. The bandage tends to slip down from the ulcer to the slimmer part of the limb.

Second Method: You add pledgets under the bandage, the better to hold it in place. Better but not good enough; the wrap over the calf is not tight enough, and, although the pads offer firm pressure, the gaps between them suffer from looseness.

Third Method: Add a bandage snugly applied at the knee, and fasten it to the lower one to prevent it from slipping. It avoids some of the problems of the first two methods.

Fourth Method: Use one bandage with a large number of pledgets, extending from the bend of the knee to below the ulcer. Do not remove all of it when you change the dressings; simply replace the pledgets that fill the ulcer-bed.

Fifth Method: Use a single long bandage that covers the entire leg, adjusting the tightness in relation to the site of the ulcer, as we have described.

Sixth Method. Use a double-headed bandage, centered at the ulcer and unrolled with properly adjusted tension.

Seventh Method: Use a single bandage with two 'heads', fixed in place at its center and unrolled in two directions, or a single bandage, or two bandages unrolled in the opposite directions, and cover the entire leg. Leave the ulcer itself uncovered or cut an opening centered over it. When you change dressings you can easily expose the ulcer by unwrapping some of the cover.

The fifth and sixth methods were used by old-time surgeons. The seventh is a recent development and is better than all of them, as confirmed by any experienced and skilful surgeon.

C. Deep (ie Profound), Cavernous and Hidden Ulcers

These terms all refer to the same group of ulcers, alike in that you cannot see the full extent of the ulcer from its outer opening. Sometimes the cavity is too large, or the opening is too small, or the tract is too sinuous, or it may have several openings. The common folk and some ignorant surgeons mistakenly call it a fistula. I will explain later.

The treatment consists of purges, a proper diet, special topicals and surgical procedures.

The purges are of two sorts: the general purges affect the entire body, the particular purges affect only the ulcer and the neighboring region. The general evacuations are by phlebotomy and laxatives. The particular evacuations are scarification, lotions, detergents, massage, leeches, cups, et al.

The dietary regimen uses only easily digestible foods that promote the formation of healthy blood. They have drying qualities contrary to the causes of the ulcers. If the ulcer is cool use warm foods, and the reverse for warm ulcers. Use the simple topicals in two ways: a. Just as they are found, not mixing them or changing them. b. Combine them but not in the ways you make compounds.[19] The compounds are plasters,

---

[19]   Do not cook, add oils and waxes, etc.(LDR).

ointments, powders, lotions, all of which should be desiccative, and that holds for all the simples alone or in the compounds. They should be warm or cool in their actions, and be contrary to the causes of the ulcers and opposite to the complexions of the ulcer and the region. Besides those above, we use detergents, regeneratives, and cicatrizers, ranging from strong to medium to weak, and used according to the needs. All of this has been well covered in the surgical handbooks and by medical authors, and we briefly touched on them early in this chapter. We will list the Simples, and the Compounds, their uses, when and how often to apply them, when we should change from one to another and how to make the compounds from the simples, all this in our Antidotary.

A general rule for treating this type of ulcer: For a deep ulcer of recent origin with a suitably large opening in respect of a moderately large cavity, it will be sufficient to use enough mondificatives to fill the cavity. If the cavity angles off and lies in tissues at a distance from nerves, bones and joints and from vital (ie noble) organs, and if it does not communicate with any of the body-cavities, so far as you can tell, such as ulcers in the ears or in the chest-wall, all other things being equal, it will recover sooner and more easily if you use medicines that have properties contrary to the condition.

The treatment has four parts:

1. How to apply the topicals: Most surgeons are familiar with the method, and you can learn it with a little practice. However, more skill is called for in the use of the lotions (ie ablutions). You inject one of the lotions described here for treating the virulent ulcer, cold or warm, taking the one best suited to the lesion and to the affected member and insuring that it gets to the bed of the ulcer. You may use a clyster-like device, or some other like it, or a so-called injector, which the common folk in France call an esclice which children use to shoot water (ie in games).

2. The technique for using incisions: We shall discuss the use of the knife in all its aspects in Treatise III, in the Doctrine on Evacuations. You will find a general discussion dealing with

the use of cauteries as therapeutic evacuants as well as routine hygienic exercises, that is, as scheduled purges for maintaining the health of one's body or of special organs.

3.   That is different from the use of a cautery in treating local lesions—ulcers, abscesses, excrescences—as it is in the following, where we use cauteries in two ways. a. In treating the deep ulcer itself, we try to remove the indurated surface, in an effort to rid the ulcer and the member of the dyscrasia, when no other measures have succeeded. This is called the method of excision and the surgeon should follow all the rules given here and in the chapter on cauteries. b. We have two other therapeutic uses for the cautery: In one we apply it in places at a distance from the ulcer to intercept the humors that flow to the ulcer and to dispel humors which have collected there. For example you apply a cautery in the hollow behind the knee of the same leg that harbors an ulcer lower down. That will intercept the humors in transit and it will attract them from the ulcer, drying it and assisting the local treatment. If that is not enough, use a double diversion, one close to the ulcer and another at a distance. You may choose one of two sites for the distant application, either the fontanelle of the arm (ie the subclavicular hollow) on the same side as the ulcerated leg, or the popliteal space of the opposite leg. The second use of cauterization is an application alongside the ulcer, choosing a place most apt to attract humors from it. This nearby cautery-wound should be kept open until the primary ulcer has healed.

4.   General Considerations For Bandaging Cavernous Ulcers. Here we have three General Instructions: 1. Dress a cavernous ulcer much as you do an aposthem, until it seems clean. 2. After the detergents and the ablutions with desiccative lotions, wrap a bandage which is very different from that used for aposthems; here it must be very snug. Begin over the shin just below the orifice of the ulcer and wrap upward almost to the ulcer, where you knot it. Begin the second bandage above the ulcer and wrap it downward almost to the orifice, and tie it. The center of the ulcer remains exposed. You need not unwrap the bandages

when you change the dressings; they remain in place for a long time, until the ulcer is filled with granulation tissue. That makes it easier for the surgeon, and the undisturbed ulcer 'ferments' more rapidly. 3. Now you apply an old-style bandage, a third one, to cover the wound and its easily changed dressings.

Special Considerations For Bandaging Here

Many methods have been promoted; that for an ulcerated leg as given above. You can infer from that and from what follows what your needs will be for any other ulcer.

First Method: Insert a drain into the opening and cover the region with a piece of linen cloth covered with lard. Then lay on some lint pads and oakum tampons, as we used in the first method described for apparent ulcers. But here the bandage should be different, for the same three reasons we gave there.

Second Method: The drain, the pads and the ointments are the same as the preceding, but the wrapping is different. We use a wide strip and many small pads. We fasten it to another strip which we wrap over the bulge of the calf.

Third Method: You dress with the same procedures but you attach the bandage to another snugly wound around the knee. That will keep the larger bandage from slipping down the narrower leg, as often happens when you use a single bandage. Although this is better than the others, I still reject it because there are gaps between the bandages.

Fourth Method: You saturate each of the bandages with the same topicals used in the pads, hoping to attract the bad matter from within the ulcer. I deem that a useless maneuver because you leave in the drain for too long a time, and it dams back pus.

Fifth Method: You use all the foregoing materials and you add lavages. I condemn that as useless, because you must expose the ulcer to air every time you perform the lavage.

Sixth Method: To the foregoing, you add compresses suited to the ulcer and a smoothly wrapped bandage. This method is of little value because the pus remains in the ulcer for long periods, defeating your treatment.

Seventh Method: This is similar excepting that it uses no drains. It, too, fails, because the small outer opening soon closes and traps the pus, a source of added complications.

Eighth Method: You incise from the opening to the bottom of the cavity. I object to it for reasons I will give in the chapter on incisions.

Ninth Method: You use a method to enlarge the wound, other than cutting it. This is a measure only of the last resort.

Tenth Method: Here we use one of the above-listed methods, but we replace the greasy drain with a hollow tube. This procedure fails because you need frequent changes of dressing, and that requires frequent loosening of the bandages, and that delays the formation of scar in the depths of the ulcer.

Eleventh Method: After an adequate mondification, you insert a drainage tube and wrap two bandages avoiding the ulcer, anchoring them with tampons and by other maneuvers we described in the General Considerations. This dressing fulfills all the requirements for perfect healing. The tampons, the large pads and the (ie third) bandage are kept moist with fluids, warmed with wine or cooled with rose-water or the juice of plantains, as indicated by the case; all of them are soaked and then squeezed to remove the excess.

D. Palliative Treatments For All Kinds Of Ulcers

These measures are very similar to the other kinds of treatments.

A. The preventive treatments and the palliative treatments of an easily-cured ulcer, when combined are the recommended treatment for a difficult-to-cure ulcer. In the same way, the preventive treatment against a non-ulcerating cancer is the palliative treatment of an ulcerating cancer, and the palliative treatment of an ulcerating cancer is the preventive treatment against a fistula. Repeated liberal applications of topicals and a diet regimen fulfill the needs for each patient.

B. The palliative treatments resemble the curative in the frequent applications of liberal amounts of topicals and the prescription of a special dietary regimen for each patient in accord with the six non-natural items. With these measures alone an ulcer sometimes unexpectedly is completely cured. So it is that the palliative treatment of a disorder may cure it. I often have seen cancerous ulcers and other seemingly hopeless cases that have been turned away by other surgeons. They simply had advised the patients to wash and to mondify their ulcers, and to dry them with oakum pads and linen lint. The patient who had seemed to the surgeon to be completely incurable had recovered with the help of Nature and the support of God, in spite of the desperate prognoses.

The General Palliative measures consist of applying defensive ointments on the skin around the ulcers. I described the ointment in the second General Rule for concave ulcers. Also, use detergents, abstersives and desiccatives and apply them once or twice a day for as long it seems useful to do so.

The Special Palliative measures: Three items for apparent ulcers: a. You may palliate the hot virulent, the sordid, the corrosive ulcer that is hidden, rampant and invasive, and other difficult-to-cure ulcers with Rhazes' white ointment or others of that type that can attenuate the biting humors and restrict their corrosive and spreading tendencies. b. You may palliate a cool virulent ulcer by following our protocols that make use of the same warm and dry topicals that we use in the curative treatments. c. Palliate a putrid ulcer and other ambulant ulcers with the treatments used for cancerous ulcers, given in Chapter 3, above, and following our General Principles.

We palliate deep (ie profound) ulcers with measures we described in Chapter 2, on Cicatrizing, and in our regiment for curable fistulas.

E. Twelve Comments on the Foregoing and on Related Matters

A. In Chapter 1 of Doctrine I, in Notable III, to which we refer the Reader, we said that all wounds, including those caused by weapons; ulcers, both the profound and the apparent; fistulas and cancers; gangrene, salt-phlegm sores, etc.; all of them are classed as solutions of continuity. If you add any of the last four in the list to an ulcer, you create an entity with its own characteristics, as you will see in the special chapters devoted to them.

In these comments we speak of them as solutions of continuity—a wound, an injury, an ulcer. You can see the resemblances and the differnces. Also, we refer you to Chapter 5 of Doctrine I of this Treatise II in which we discuss the features that can interfere with topicals, even when the correct ones are properly applied.

B. In regard of our definitions of ulcers, it is better to say that it is a wound rather than simply a solution of continuity, even though it belongs in that category, too. The ulcer is more immediate to a wound in our sense of that word. And now we must define the word 'immediate' as Aristotle (*Topics*, Bk.I) used it. By setting ulcers in the same subset as wounds, we distance them from fractures of bones, undrained abscesses et al. all of which are solutions of continuity, yet the interruption cannot be seen and therefore they do not merit the term wound as commonly used. I really believe that all solutions of continuity, whether internal or on the surface should be called wounds. When we say, "it is a wound that has pus or in which we provoke pus" we exclude simple wounds by definition and they are not ulcers (ie healed within a defined time). And when we say, "they have been discharging pus for a long time", we exclude the inflamed wounds which have drained pus for only a day or two, and not for a long time." Those definitions determine our choice of treatments.

C. We call it a putrid ulcer which begins to disharge a moist and unlaudable kind of pus different from what we drain from a mature abscess. We say that an ulcer has putrified when, after it had begun to emit laudable pus, it changed and began to discharge a foul sort. An ulcer really is not putrid unless it began as a wound. When an old unhealed wound discharges an unlaudable fluid, and forms crusts and scales and other matters which have none of the properties of laudable pus, from that time on we no longer call it a wound.

D. Now we shall define pus, virus, decay, dirty contamination, scales and crusts, their properties, their contents and how each forms.

Pus is a wet substance with both natural (ie derived from the body) and unnatural (ie from external sources) contents. It is not a pure superfluidity (ie edema) that forms in old wounds as part of the third digestion. *Laudable pus* is very white, sweet, smooth and soft to the touch. It is homogeneous and uniform. It holds together and does not layer out. It does not form lumps. It is odorless, that is, it has not soured. Hippocrates described all of this in the first part of his *Prognostics*. Galen (*Commentaries on the Aphorisms*) covered the same subjects with added explanations.

Pus forms from the blood that runs in veins (ie not arterial) and Nature has no use for the blood after it leaves the vessels. In effect, pus forms partly through the actions of Natural heat and partly from the weakness of the unnatural heat, It is generated when the Natural heat tries to save the innate properties of the shed blood and to correct the deterioration that already has begun. At the same time, the unnatural heat tries to corrupt the blood. Thus the conflict between those two forces causes the pain in a wound and generates the pus. It is as Hippocrates wrote (*Aphorisms*, Part. 2). Once suppuration occurs, the conflict is at an end and the pain subsides, especially after the pus drains from the wound. Then the heat subsides and the fever leaves. Galen described it, too. (ibid.)

A *Virus* is a thin pus of two sorts, warm and cool. We described it in the previous section on the treatment of virulent ulcers. It is a thin malodorous liquid, reddish if its warm, the color of lotion that has irrigated a fresh wound. If it is cool, it is whitish like whey (ie the serum of cheese). It is formed from watery humors, as Theodoric stated (Bk. I, Ch. 9) really the result of actions of the unnatural heat, an opposite result from the sort of conflict that generated pus. Decayed matter (*Putrefaction*) is an unnatural excretion: fetid beyond measure, as we have described it in the section on putrid ulcers. It is thick, lumpy, whitish or reddish in color, never homogeneous. It derives from the thick parts of humors acted on by the two sources of heat, as in the formation of a virus. *Sordid Matter* is a thick, unnatural, fetid mixture of bits of rotted and gangrenous tissues. Its appearance is more or less as described in the preceding paragraph. It is generated in the same way as the virus and most often occurs in contusions complicated by gangrene.[20]

---

[20] A description of a large subcutaneous hematoma with infarction of the covering skin (LDR)

Crusts (ie scabs) and scales resemble tree-bark and each other except that crusts are larger. They are seen as lids over ulcers.

Scales (ie desquammation of flakes of epidermis) are very small and numerous. They cover the skin surrounding ulcers and resemble fish-scales. They appear in salt-phlegm conditions without ulceration. They are brittle, solid and they range from dark to light in color. They are produced by condensation and desiccation of humors, brought about by the two sources of heat. Their humor is thin, dry, sticky and viscous, not a thin liquid. It dries and condenses itself and thickens as does soot adherent to the walls of a chimney.

E. The Salernitans and their followers, as well as the illiterate and the lay surgical practitioners, considered fistulas and cancers as wounds and ulcers that had not healed after six weeks to two months or more of treatments. They are wrong, as we shall show in our chapters on cancer and fistula. They claim to have cured those indolent ulcers, as described in their books that describe medicines for ulcers and fistulas. That has convinced some of our present-day surgeons to use the ulcer medicines for treating fistulas, although they are too weak to be effective; they neither destroy nor cure, and time runs out as the ailments worsen while the medications fail. Besides, when some modern surgeons treat true ulcers, they accept the opinions of the older generations in the belief that they are fistulas or cancers and they use extremely potent corrosives. The true ulcers which are easy to cure (ie as wounds) are converted into aggressive ulcers, which are the serious and the incurable kinds. The latter then attract harmful humors from elsewhere in the body. Galen said (*Commentaries on the Aphorisms*, and in *De ingenio* Ch. 4) that pain is the cause for the attraction of the worst parts of humors from the other regions of the body.

F. One may argue against the recommendation that we use a warm variety of alum in the lotions that wash warm ulcers, because it contradicts Galen's dictum (ibid.), and Hippocrates said the same (*Aphorisms*, Part 2). I must cite Avicenna (Bk. IV, Fol. 4, Doct.3, Ch. on the treatment of virulent ulcers) that alum-water washes, repercusses and dries. Perhaps he and others refer to an alum that we here do not have, which could have been cool. Although the alum that we use is warm and dry, we can dilute it with enough water and other cool medicaments to overcome its heat and excessive dryness.

G. One may question the precept given in the section on treating sordid ulcers that we should begin with our most potent medicines. Constantine (The African) *(On the Eyes,* Ch. on phlegm) wrote that one should begin with the most gentle measures. I have to comment that his recommendation befits conditions that are not violent, where the surgeon has the advantage of trying out a succession of medicines. But when it comes to treating sordid ulcers, you must attack them with the most potent kinds, as both Avicenna and Theodoric have taught.

H. Avicenna (Bk. IV, ibid.) compared fistulas with burrowing and cavernous ulcers. He said that the lumen of a fistula is lined by indurated tissue and it resembles a feather's quill, The burrowing ulcer penetrates for some distance between the skin and subcutaneous tissues but not deeply. A cavernous ulcer goes deep but has a larger external opening than a fistula.

I. Avicenna (Bk. I, Ch. on the treatment of deep ulcers) wrote that ulcers should be filled with granulation tissue and that the virus should not be allowed to accumulate. Often ulcers have not been treated with medications that were liquid enough to seep into the bottom. And there are some ulcers where the medications have not been sufficiently viscous (ie too liquid) to adhere to the surfaces for a long enough time to work well. Furthermore, he said (ibid.) that the topical should remain in place for three days after it has been applied.

J. The sparadrap which we use for ulcers and other ailments is a cloth thoroughly saturated with a viscous plaster of various sorts to suit the case. Here is a plaster in common use: 2 oz. wax, 2 lbs. resin, 2 lbs. black pepper, 2 drams each of frankincense and galbanum and mutton fat ad lib.

K. The reason we have not discussed the techniques for incisions (ie with a knife) in the chapter on ulcers is we dealt with them in the General Chapter, whereas sufficient instructions for using the cautery in treating ulcers are given here. The general instruction for the use of cauteries has been set apart from that for knives. The various uses of knives have more in common with each other than with cauteries. Indeed, knife-incisions are made only in cases of malignant indications in particular cases, never for treating a systemic illness. That is why the techniques for incision using knife are in one place in this book, On the other hand, there are more differences among the various uses of cauteries, for ulcers, fistulas etc. and the instructions are given in the chapters dealing with them. While the cautery can be used for treating certain systemic disorders, as preventives as well for cures, the techniques for those applications have very little in common with those for the local lesions, and we will describe them in their own special chapter.

L. Notwithstanding the principle stated above that we should begin our treatments with mild medications, there are several good reasons why we deal with palliative treatments after we have discussed curative treatments. First, whenever there are options, treat to cure. If that fails or if it is not tolerated by the patient, you can switch to palliative measures. That seems obvious. When you succeed in curing you will not need to palliate, whereas a failure to palliate leaves you trying to cure something you should have wanted to cure in the first place. Second: most patients want to be cured and not simply palliated. Third: Many surgeons who are trained to cure are not familiar with palliative treatments, and they either cure or kill their patients, or they desert a sick person, leaving no support because they are uncertain how to offer palliation. Those are the completely ignorant surgeons.

CHAPTER 2.

TREATMENT OF BITES AND STINGS

By Large Animals, Reptiles And Serpents And Birds: Venomous[21] And Not Venomous

Here we have two categories: 1. Animal bites and stings which are not venomous but cause wounds and ulcers that are more painful than the ordinary and so are suspect for poison. 2. Truly venomous wounds and stings as from rabid dogs or poisonous snakes, wounds

---

[21]   Mondeville had two types of venomous beast: those those who were venomous by nature and those who were envenomed by accident, such as a rabid dog, meaning those who became venomous as the result of contact, perhaps a bite from a naturally venomous animal. All the effects that followed a bite were ascribed to a venom. There was no concept of rabies viruses or other infectious agents (LDR)

which themselves are poisonous and which intoxicate the victim. We shall consider three aspects of both categories.

I. Descriptions. Four Items

A. Definitions: Some animal bites which are not venomous may arouse suspicions of carrying venom. Such are the bites by jack-rabbits, hares, non-rabid dogs, mosquitoes, fleas and horse-flies. In the other category are bites and stings which are envenomed when they are inflicted and continue as a source for poisoning the victim.

B. Varieties: Some lesions are not venomous and never become so and never cause maddening, such as bites of pigs, jack-rabbits and hares which never bite unless they are trapped or excited. The stingers include fleas, ticks, bed-bugs, horse-flies, etc. Some bites are by animals who may not be rabid, They are dogs, hares, mules, horses, wolves, foxes, cats, weasels, et al. Although the animals are not rabid, one may be suspicious that their bites are rabid at the time because the animal may have been trapped and angry and was maddened to a degree and appeared to be rabid. Finally, we have the lesions caused by animals who by nature are venomous, not as a result of an accident (ie as rabies). These animals are venomous both when they are at rest and when they are stirred to anger. They are the serpents, bees, wasps, Spanish flies (cantharides) et al. However, their bites are more toxic when they are inflicted by an angered animal or when they are in rut and heat.

Some lesions are larger than others, some are more deadly, some are more painful, some gape and others are constricted and some are fresh and others are old. The victims range from robust to delicate, their natural heat ranges from warm to cool. Some have warmer hearts than do others; some bites are nearer vital organs than are others; some bites are in regions connected to vital organs such as a bite near the great vessels that come from the heart; some bites are made by young and hungry male animals living in dry regions during the summer months, and other bites are the opposite. Some bites are treated with special medicines others by any and all: all at one time or one after another. All these facts must be known and be useful to an operating surgeon.

C. Causes: Frequently, we do not know why the animal had bitten the patient we are treating, although we may be able to find the answer. Sometimes we are aware that the rabid animal was hungry and that the encounter was by chance; the rabid dog had bitten an unsuspecting man who crossed his path. Sometimes they are predictable as when a hare bites the person who grabs him too tightly or a dog bites someone who teases it or a snake bites an unwary person who enters his cave, etc.

D. Symptoms (ie signs): *The General Signs:* a. The indications that a person has received a venomous wound are: he seems to be half dead or is comatose, as if drunk or asleep. The wound is painful at once, burning, biting and, bursting and the region may be swollen. These are not like small wounds made by a dagger or a strike by club or a fall or other contusion. The lesion itself, a bite or a sting, if venomous adds its own special symptoms to match the intensity of those of the worst kind of injuries.

Particular Diagnostic Signs: In most cases it is not important to know if the wounds are venomous, because all will be treated as we do for other wounds, excepting the most common case of all, the bite by a rabid dog. Was the dog rabid? The treatment will be different if we know that. Was the patient asleep when he was bitten? Was it during the day or the night? He may not have felt the bite or he may have thought that he had dreamed it. He may have been in bed in a dark room, or the animal may have run away unseen even during the day. Was it a bee or a serpent? Sometimes the patient may have had a good long look at the animal even in dim light, perhaps he may have been able to catch it. On the other hand, he may not have recognized the kind of animal and been able to name it or know the kind of venom it carried.

In all such cases and many others, when we are unable to determine any distinguishing characteristics of the wound and in the story of the event, then we must be most attentive to what happens after the incident. We need a third type of sign in addition to the two we have mentioned, some special distinguishing feature for each type of venom.

First, consider a bite by a *Rabid Dog.* There are two sorts of symptoms. a. The marks of madness in the dog: If we offer a hen a bit of bread or crumbs wet with the blood from the bite she will refuse it, but if she is very hungry and eats some of it she will die within ten days. The dog's ears droop; its back is arched; its tail hangs between its legs; it rarely barks and that will be hoarse; when it walks, it totters as if it was drunk; it grinds its jaws silently with only slight movements; it slinks along walls; it keeps to itself and no longer recognizes its master or its home; its eyes are red and it drools saliva; its nostrils exude disgusting matter; it barks at its own shadow; it laps its tongue as if it is drinking. b. Symptoms that appear in somebody bitten by a rabid dog: He has terrible nightmares; he acts foolishly and feels bites and stings all over his body; he has hiccups and complains of a dry mouth for no known reason; he is very fearful of water; he dies after a short illness. In certain cases, however, the venom of a similar bite seems to lie indolent for a long time until the victim suddenly remembers the events of the injury that happened fifteen days or a month or seven years before, depending on the kind of venom and the dispositions of the patient and style of life. Usually the symptoms appear within eight hours after the bite.

The signs of a *Scorpion's Sting*: the edges of the sting-mark are stony hard, but there is little swelling or redness; the pain appears at once and then may fade away; the patient sweats and his legs may be stiff and extended.

The signs of a bite from a *Poisonous Viper*: instantaneous pain at the bite which then spreads over the body; the male's bite shows only two punctures whereas the female's shows many more; some blood and greasy matter oozes from the punctures; the patient's skin turns green, at least near the bite.

The signs of a *Serpent's Bite*: pain and swelling at the site; the patient's face quickly turns pale when the skin cools as his heat retreats inward from the surface of the body. Then the pallor turns from greenish to blue or black when his vital spirits (heat) return to the surface and the bite itself becomes very hot. The heat is accompanied by vomiting, urination is painful and he suffers from colic.

The signs of a *Bee's Sting* and stings of other insects: the bee leaves behind its stinger in the puncture, other insects do not.

The signs of stings by the *Spanish Fly*: much more pain and terror; the victim feels a great urge to urinate.

The signs of stings by *Wasps and Flying Ants and many Other Insects* have little to distinguish them, one from another.

II The Treatments For Bites And Stings

A. Bites by non-venomous animals which introduce no venom. We shall not discuss preventive and palliative measures, which are not used in these cases.

General Measures

The treatments are the same as those for simple wounds designed to preserve the natural complexion of the wounded part and the rest of the body. However, if the wounding animal is scrawny the situation is worse, even more if the animal is young, and if it was made by an animal with a bad complexion, as is a human bite, or by a doddering old animal who eats foul food and is full of bad humors. If the bitten member is itself weak (ie lacks resistance), that is worst of all. In all of these cases apply this plaster: mix wine, onions, salt, oil of roses or olive oil, in amounts dictated by the various circumstances, and add honey. Apply it warm until the wound is clean. You also may use crushed or chewed raw fava-beans,[22] or alternate it with warm oil until the pain eases and then apply the chewed fava-beans or wheaten flour. The effect is better if the chewer is a teenager or a child. Another remedy is boiled onions with bee-hive honey or rose—honey (ie rosamel) mixed with chewed bread-crumbs. The best way is to treat all animal bites as if the animal is rabid. Follow the preceding medications with incarnatives and consolidatives, adding Black Ointment: 1/2 dr. each of wax, lard, resin and 3 dr. of galbanum. This ointment is excellent for bites and for punctures made by a hawk's beak or talons.

---

[22]   Maceration with saliva was a favored method for producing some pastes (LDR).

Special Procedures for treating lesions caused by particular animals: a. The bite of a weasel is very painful and is marked by its brown color. Prescribe onions and garlic in a good wine and apply a plaster made with the same; also good is a plaster of ripe figs with flour of vetch, You may use this weasel-bite plaster on dog-bites and obtain immediate relief. b. Bites by people, especially by children: Apply pine resin or crushed onions mixed with honey or the roots of the 'flammula' iris mashed in vinegar, or crushed roots of fennel with honey or fava-bean flour mixed with water or vinegar. c. Bites of monkeys' and of non-rabid dogs are treated with crushed salt and onions with honey or with fresh lanolin and oil and warm vinegar. You may use this ointment: 5 dr. each of wax and chicken-fat, 10 dr. of oil of lotus. If the bite is inflamed add litharge and some water to the ointment. d. Bites of bees, wasps and other venomous flying insects: We know that bee-bites causes the smallest wounds, and next is the winged ant. The sting of the wasp is much worse, but the sting of the Spanish fly (cantharides) is the worst. Treat all stings empirically with traditional methods. The empiricals include placing a cold piece of metal on the site; some practitioners catch an offender, crush it and apply it on the bite; some immerse the bitten limb in warm water for a full day and then apply brine and vinegar. The established surgical measures for treating all sorts of insect bites include potions and topicals. The potions are made with 5 dr. of althea seeds boiled in 1/2 lb. of water and 1 lb of wine. Another good potion: 1 1/2 dr. of althea with 2 oz. of a syrup of hemp seeds or with 1 1/2 dr. each of powdered dry coriander and sugar. Boil them with cold water or with the juice of some cool herb, such as lettuce or with a syrup of sour grapes. The topicals are powdered limestone or lentils with vinegar or bread soaked in vinegar and rose-water or green coriander with vinegar.

B. Treatment of Bites by Naturally Venomous Large Animals and by Reptiles, and of Bites by Animals that Have been Envenomed by chance.

1. Preventive Measures: Whoever wants to avoid these injuries should keep his distance and keep out of the way of possibly venomous beasts, especially when they are in heat or have been angered and in hot weather. Do not excite or injure them or stand in the path of their

escape. A person may be forced to fight with a dog or a wolf, or to travel in a region in which snakes congregate. In such cases he will do well to take some theriac of nuts as a preventive; it works well. In those who use it regularly the action of the venom in a bite is delayed for some time. Make it with 2 parts each of dried rue and almonds, 5 parts of salt, 1 part of dried figs. Mix and knead it by hand. In some cases the traditional Great Theriac is good. If the victim had taken some of the medicinal liquors before he was bitten, such as diamargariton, etc, which Avicenna and other authors preferred (see his *Book on Cordials*), the spread of all kinds of venoms into the victim's body is strongly resisted, especially if you use the topicals that I describe.

Fumigants put to flight serpents and other venomous beasts and are useful preventives for those who know how to use them and who go where they abound. Here are some: the smoke of a burning deer's antler will dispel venomous snakes. Other fumigants are made of goats' hooves, sulfur, mustard, burning human hair, galbanum or serapinum. Crushed mustard and water chestnuts put in the serpents' tunnels may kill them. If you burn some scorpions in a house, all the other scorpions will depart.

2. Curative Treatments: General Measures (three parts):

A. In General: 1. Rabbi Moses (ie Maimonides) (*Treatment of Poisons*, Ch.3, Div. 1) advised as follows, "In cases of this sort we prescribe oral medications. Whenever we do not know what the offending animal was, we must carefully assess the patient's symptoms. When he is very warm, as after serpent-bites, we prescribe medicines made with milk and vinegar. If he is very cool, as in cases of scorpion-stings and viper-bites, I prescribe medicines made with wine, and I give a decoction of anise to those who refuse wine. Indeed, the classical authorities recommend anise in all cases." 2. All the recommended doses for the oral medicines which follow are for persons between the ages of thirteen and forty. For those who are between ten and thirteen and for the very old reduce the amounts on a scale to match their ages, while keeping in mind such variables as the potency of the venom, the ages and the complexions of the victims, the geographic region, the season, and, most important, the vitality of the person. Indeed, you should give a very strong dose to someone of middle age who is vigorous and has a good complexion, who has received a potent venom during a cool season in a cool region. For contrary conditions reduce the dosage to fit the case. If the poisoned person is a child younger than six years or is a decrepit oldster, their escape from death is uncertain. Nevertheless, give them a quarter dose of simple medicines. Dosing the theriac to use between a quarter dram and a half may create a problem in measurement. In any event, the physician, having evaluated all the factors, will have his say.

B. Two Categories of Oral Medicines:

1. The Simples are: the pips of lemons and oranges—a lovely round yellowish fruit that grows along the Ligurian coast, that is called the Apple of Orange in France.[23] Its pips reduce the lethality of the venom and act as an antidote, both when used as a topical and when swallowed. Take 1/2 to 2 dr. and mash them, or use them as did Avicenna: 3 dr. with warm wine. Emeralds, the brilliant green gems, act in the same way, as Avenzoar used them: grind 9 gr. in cold water or wine. They will cause a discharge of venom in the emesis that follows that dose. Another Simple is spikenard, 1 dr. in wine. Equally good is crushed peeled garlic, 1 1/2 to 3 dr. in wine. Both of the latter act like a Theriac. Galen (*Megatechni*, Bk. XII) strongly recommended garlic against all cold venoms. Against the hot venoms he used the crushed roots of mandragore, 2 dr. with 1 oz. of honey. He also used long aristolochium, crushed and strained, 1 1/2 dr. with wine. Also, iris roots or wild celery, 3 dr. with wine; another: powdered cumin, 4 dr. in wine or water. Another: anise, 4 dr. in a broth of crayfish along with 3 dr. of comb-honey in cold water. Another: seeds of rue, either wild or garden-grown, 2 1/2 dr. in wine. Another: clotted hare's blood, 1 1/2 dr. in a small amount of vinegar. In France we often use the roots of tormentilla, chewed or powdered and swallowed, and taken to the limits of tolerance. It treats all bites and stings that affect people as well as quadripeds, and it is an antidote against swallowed poisons.

---

[23] The ancient territory of Orange in southeastern France, as part of the Savoy, was extended by Mondeville to include the Ligurian coast (LDR).

Hunters give it to their dogs when they have been bitten by snakes and they are promptly cured.

2. The most widely used and best known compounds to be taken by mouth are the following: the Grand Theriac, taken in 1/4 to 1/2 doses as recommended by Rabbi Moses Maimonides, or in a dose of a little more than 2 dr. in a small amount of watered wine as prescribed by Avicenna (*On The Theriac*). *The Antidotary of Nicolas* also recommends this type of theriac. He also gave the Theriac of Four Kinds, dosing 1 to 2 dr. of each, although each alone is a theriac: a. Take equal parts of Myrrh, shelled bayberries and the roots of gentian and aristolochium. Mix with 3 parts of whipped honey. This was the traditional classic theriac used as an antidote against poisons. Later came b. The theriac of asafoetida which is easier to take and is more effective against cold poisons. Prescribe it in warm climes in doses of 1 to 2 dr. In cool regions use 2 to 4 dr.: Take 1 oz. each of dried leaves of rue, costus, dry mint, black pepper and pyrethrum, and 1 1/2 dr. of asafoetida. Dissolve the last named in wine and grind the others and mix them with the fluid before adding boiled whipped honey.

The Compound Topicals:

Fourteen Precepts as to the methods:

a. Your purpose with the topicals is to extract and expel the venom using potent and penetrating attractives, to supplement the oral medications. (Galen, *Megatechni*, Bk. XIII, Ch. 4).

b. We should try to alter whatever venom remains after the extraction, to render it better. These two precepts are the bases for the next twelve. (Galen, ibid.).

c. When the lesion is in a limb distal to the shoulder or hip, tie a tight cord placed between the bite and the rest of the body, and place another ligature distal to the bite. These will isolate the venom and prevent its spread into the rest of the body, However, if the lesion is in the trunk such a ligature cannot be used.

d. Apply a theriac or a similar medication as an inunction near the upper ligature.

e.   Do as follows, even when you may not be able to follow through in a case. Use strong oral suction over the bite. The person who sucks should prepare himself so: rinse his mouth with oil or with oil and warm wine; oil his lips with oil of violets or similar; keep his stomach filled with garlic, nuts, rue, figs and wine. Repeat the performance frequently.

f.   If no one is willing or available to suck, remove the feathers from around the anus of a rooster and set it against the bite and hold the animal in place until it dies. Then apply another and a third in the same way.

g.   After all the preceding, scarify the entire region and apply cups with a tight seal, or use leeches if you can obtain them.

h.   If, praises be, you succeed with one or all these measures and you see improvement, do not rest easy and do not let the victim fall asleep until the victim's pulse is strong and his natural heat has improved and the pain subsides. Then treat the lesion with the proper topicals.

i.   If the pain persists after everything else is as you had wished, slit open young pigeons bellies and lay them open over the bite, replacing one with another when the first cools. You may use young chickens or wash with warm vinegar or wheaten flour in warm oil. All of these are good analgesics.

j.   When everything so far has failed to relieve the persisting or increasing pain and other bad symptoms and the patient collapses (ie syncope), you must change course and seek a physician capable of performing what we call the 'long art" (ie an operative procedure) as it is designed specially for the patient at hand and for the special animal at fault.

k.   If, in spite of all, the pain persists, cauterize the region, both superficially and deeply, as the situation of the wound will allow.

l.   All evacuations except for sweats and baths are harmful if used before all the venom has been extracted and all the topicals have had a chance to penetrate to reach their destination. You can observe the effects on the venom and on the patient to determine if the benefits are sufficient.

m. If by neglect no one has removed any of the venom and the
    surgeon has been summoned late, after much of the venom
    has drifted into the tissues, he immediately should apply the
    binding cord, massage the victim's hands and feet and
    administer strong evacuants—oral medicines, clysters,
    leeches—and do everything that should have been done at
    the time of the bite.[24]

n. If neither the physician nor the surgeon has been able to
    accomplish anything with their prescriptions and maneuvers,
    including the cautery, then you must amputate the limb: a
    finger, a hand, a forearm at the elbow, a foot, or leg at the knee.
    Go no higher.

If the perforations of the bite are small, enlarge them, and if they
have closed, reopen them. Once opened, keep them open with drains
which are saturated with ox-bile or another suitable liquid, until all
your concerns for retained venom are allayed.

The best way to enlarge a small opening is with the knife. Let the
wound bleed, and after a sufficient flow, apply the actual cautery. If
the victim cannot tolerate those measures, insert a narrow strip soaked
with a caustic ointment. Keep one in place, replacing it with a larger
one at each change. If the opening has closed, use a knife or cautery.
Again, if the patient will accept neither neither the knife nor the
fire,[25] use a caustic as described in our Antidotary (ie an eruptor).
Cantharides with yeast is good, or this plaster: opoponax and Greek
tar, melted before adding almond paste. Apply it for three hours.
After a blister appears open it and insert the series of ever-larger

---

[24] This recommendation for evacuants seems to contradict the previous precept
     (l.).However there is no contradiction for the medieval surgeon: The
     evacuation in 'l' would attract the venom into the body toward the gut, or
     toward the bleeding vein. The bath and the sweat attract the venom out
     through the skin. In 'm' the venom already has entered the body. Therefore,
     it must be evacuated through the 'general' route. See below *Dog-bite* Items 2
     and 3 (LDR).

[25] "ni fer ni feu" is Nicaise's elegant metaphor. (LDR).

drains. After the opening is large enough, follow with topicals of Simples and of Compounds, as follows:

Simple Topicals

River-mint (ie menthastrum); basil grown on stone walls; droppings of pigeons, ducks or goats; sulfur; asafoetida; bdellium; salt; garlic. Apply one or more after the venom has been expelled. The same results can be obtained with ox-bile and the crushed and powdered pips of citrons, as already cited. The innate properties of these Simples can rescue an envenomed victim from the gates of death.

Compound Topicals

a. Take equal parts of pigeon-droppings, garlic and salt. Crush and apply.

b. Crushed seeds of althea (green or dry) mixed with vinegar and oil.

c. Wild mint boiled in vinegar.

d. Ashes of fig-tree twigs or of grape-vines, salt and niter. Grind all together and mix with some vinegar and dog-bile.

e. Equal parts of serapenum, castoreum, asafoetida, sulfur, pigeon-droppings, mint and calamint. Mix with warm aged oil and shake vigorously, and stir. This compound is especially good for extracting venom and relieving pain in bites and stings

f. Cover the area with a plaster of theriac (major).

g. Equal parts of galbanum, serapinum, myrrh, asafoetida, peppers, sulfur. Grind all together and add some wine

h. Pomegranate and fig leaves, ground and applied

Special Treatments

A. Dog-bites. In Normandy the most esteemed treatment for rabid dog-bites is known to everybody, no matter how uneducated they may be, known for its easy use. It consists of nine immersions in the sea by every person or animal that has been bitten by a dog, whether or not it is known to be rabid and the victim need not be certain about the danger. All the surgeon has to do is treat a simple wound. Theriac is of no use. I have often seen people and animals taken to the seashore who already exhibit the bad signs, at peace and docile as they are led. In such cases I venture to give some tormentilla as a suitable intervention.[26]

However, more rational treatments are recommended by our authorities and I will add a few comments to those already given.

1. The surgeon should pay equal attention to all dog-bites, rabid or not. If he is not completely certain about the dog he should treat its bite as if it is rabid. In that way he avoids the error of a missed diagnosis. Indeed, even when the bite has healed before the fortieth day and the victim has returned to his usual activities, he may be taken with grave symptoms, such as fear of water, and he will go on to die. However, if the ulcer had been kept open, those symptoms would not have reappeared. I emphasize the need for great prudence—the poison of rabies is never moderate; the disease always is furious.

2. When a patient first comes to the surgeon a week or so after the bite, a sea-water bath is useless, although it may be a good early measure. If the wound has sealed, do not reopen it or use strong attractives. Indeed, the venom has already deepened and we would subject the patient to useless pain.

3. If the patient appears three days after the bite, do not prescribe purges, clysters, emetics or phlebotomy. If he arrives after the third day use them in a proper sequence: first a laxative, then an emetic, then a phlebotomy. Thereafter administer a daily gentle clyster. If the patient is plethoric, bleed him again and bathe him with strongly fortified waters.

4. Do not interrupt the evacuations, the reopening of the bites, the potions, the topicals and the dietary regimens before the fortieth day.

---

[26] Is this another Mondeville pun: 'tormentilla' for a 'tranquil' patient? (LDR).

How To Prepare The Potions

Soon after the bite the patient should take a theriac or something like it daily for forty days. Make it with 1 1/2 dr. of high grade lycium in cold water. Another: Roman nigella, crushed and crumbled, dosed as the first. Another: 5 dr. incinerated riverine crayfish with water. It is a well tested potion. Another: 5 dr. gentian, 1 dr. frankincense, 10 dr. crayfish ashes, all ground and sifted. Dose 1 dr. daily in cold water.

A special theriac for these cases: 1 part olibanum, 5 parts gentian, grind both and sift. Drink that daily for forty days, beginning with 2 dr. and increasing by 1/2 dr. daily until you reach the maximum daily dose of 6 dr.

The Topicals

Apply plasters of flour of vetch and honey or of crushed almonds with bee-hive honey or fresh mint leaves and crushed salt. Introduce asafoetida in wine into the dilated opening of the bite, or crush together some salt, a walnut, an onion and bee-hive honey, and apply it. A plaster of crushed verbena also is good.

B. Scorpion Stings. The scorpion is about the size of a scarab beetle with an added tail. It is well known in Italy, in Avignon and in other countries. Its small bite is barely visble, as described by Galen (*de Interioribus*, Part III).

Potions for Scorpion Stings

Galen's special treatment (*Megatechni*, Bk. XIII, Ch. 4) consists of garlic and asafoetida in wine, or a theriac. They can be swallowed as well as applied as topicals. It warms and attracts the venom and expels it. It supports the victim and impairs the toxicity of the venom.

The Mithradaticon, the Diatessaron and the Socrugene theriacs[27] have similar effects, as does the theriac of the four spices, which is

---

[27] I have not identified the Socrugene. The Diatesseron and Mithradaticon are in the Antidotary (LDR).

made so: 4 dr. of aristolochia, 3 dr. of pepper, 1 1/2 dr. of wild celery, 4 1/2 dr. of pyrethrum. Add honey and make pills about the size of Egyptian fava beans. Give two of them with 3 dr. of strong wine. Very good in these cases is a potion of 1 1/2 dr. of olibanum or garlic crushed in a small amount of wine.

Another effective and easily compounded remedy: Grind some absinthe and mix it with butter and bee-hive honey. Dose 3 dr. for a quick relief of pain. Another treatment for the same purpose is to gorge the patient with figs, walnuts, garlic, rue and enough wine to make him drunk.

Treat the pain that follows all bites and stings when they cause shivering chills or unbearable heat: Give 2 dr. of (ie ground) citron pips. Please note that as a rule all the oral medications prescribed above are to be taken with good wine. If, for any reason, the patient cannot take wine, use a decoction of anise leaves.

Topicals for Scorpion Stings

The well-tested and easily prepared special topicals are:
1.Anoint the sting with oil in which you boiled some scorpions; it is an unusual but widely used medicine. It rapidly eases the pain. 2. An inunction with terebinth. 3. Crushed citron pips cooked in vinegar an honey. 4. Boil 1 dr. of each of the above substances in 2 liters of water until their principal essences are extracted. Then use the water as a lotion. 5. Crushed garlic.

Adder and Viper Bites

These are two names for the same animals. They differ from other serpents in having larger heads and longer fangs, some as long as a hare's teeth. They bite with their side teeth, not those in front. The venom of the male is the worse.

The Special Treatments in addition to the General Measures a. Potions: 1. 1 to 1 1/2 dr. of a freshly made theriac, of clotted hare's blood; 2 dr. of asafoetida; 1 dr. dried and shredded deer's penis; 1 dr. doses of hemlock or citron-pips or orange pips or round aristolochia or old butter, or clotted deer's blood and many others. b. Topicals. All are strongly atttractive, especially good are crumbled aged cheese in cold water and leaves of orange boiled in water and then mashed.

Treatment of Other Kinds of Serpent-Bites[28]

Many kinds of serpents are more common in the warmer countries and we can read about them in the medical texts. We have mentioned vipers, scorpions, crawling and flying insects, dragons, asps, basilisks and many others that we never see in France. They are of no interest to us here. We have common grass snakes, lizards which have names like stellion (ie gecko), morons, leopard-lizards, all of which inhabit forests, meadows, lakes, etc. I have seen them in burrows in village stables. I have seen grass snakes hiding in tunnels until someone pulls them out and puts them into a sack. They are common in houses and they do not harm people unless they are challenged or injured. Their rare bites are superficial when compared with the others. The grass-snakes cause wounds with their tongues and bite with their teeth. When we use the word serpent here we usually refer to the grass snake.[29]

When lizards and stellions wound and bite they leave their teeth in the wounds. That causes a lot of pain which continues until they are extracted. The General Treatments usually are all that are needed to cure the bites of most of the animals; you seldom need more. The serpents of France really are not as dangerous or as venomous as those in the warmer regions.

However, we should describe what to do when the teeth of lizards are left in wounds. Simply rub the wound with oil until the teeth come out. If that alone does not succeed, lay on some oil and ashes. And, if yet more is needed to ease the pain, use other measures which we have described, including baths and doses of theriacs, as Galen instructed (ibid.)

---

[28]   Mondeville lumps biting insects with snakes as problems for surgeons (LDR).

[29]   The common European grass snake *Natrix natrix* (LDR).

Bites of Spiders

The common spiders are barely or not at all venomous. They live in houses and gardens and they fashion webs to catch flies. There are other kinds of spiders that are very poisonous, the ruteles, but I am not familiar with them; we do not find them in France.

The treatment of a common spider bite consists first of bathing the patient in warm water and applying some myrrh ground with salt and water. Or you may apply some chewed bread or flour cooked with oil and salt. Have the patient drink some powdered wild rue (ie pegamum) mixed with wine. The treatment of a bite by a scarab is the same as for scorpion-bites, adding 5 dr. of asparagus boiled in 6 oz. of good wine as a potion. Or give the victim to drink 2 to 6 dr. of crushed tamarisk fruit or 1 1/2 dr. of absinthe in fresh water. You may bathe the wound with water and wine containing myrrh.

III Palliative Treatments

The General Measures are not worth much in these cases because the sufferers either recover promptly with treatments or they die soon after they are poisoned. The latter is the usual outcome of a mad-dog bite, although in some cases the treatments may dely the fatal outcome.

IV Dietary Regimens For Envenomed Patients

A. The General Regimen of foods and beverages: bread-crusts fried in oil or butter; many kinds of figs, walnuts, pistachios, hazel-nuts, garlic, onions and rue—all together or separately, alone or with bread, honey and butter: juices of sour apples, marinated seeds (ie in vinegar) and cool legumes. Raw or cooked acorns are good for envenomed patients, both taken by mouth and applied as topicals. The diet improves the mental functions and the memory, and the same holds for pigeon-blood. Chewing lemon-peel is good, and so is a tea made by boiling the lemon-leaves. Add salt to everything because it consumes the venom and add wine for the same reason. Avoid all fresh meats, especially that of birds. Drink fresh milk to which you may add wine, if the patient can down it.

B. Special diets for victims of dog-bites, as recommended by William of Saliceto[30]: Three days of austerity, limited to cold foods, as barley-water with barley-flour, plums in cold water with bread-crumbs. Follow that with young hens cooked with wild grape-juice, soft eggs (in the shell) for forty days. Along with the food take this powder: 3 oz. cinnamon, 1/2 oz. cardamon, 1 oz. safron and white wine diluted two to one with water. That diet seems to me to be very rational. Rabbi Moses (Maimonides)[31] and others recommended that the diet not change from the beginning to the end. It consisted of all the foods in the General Regimen except that salt should be used only in moderation. Soups of small birds, excluding pigeons, are good, as well as cabbage and meat soups. Omit crayfish. Many of the following can be used to liven the diet as well as be used as topicals: garlic and onions, raw or cooked, frequently but not every day; the same for salt-fish. Serve these things as wet as is feasible: chicken-soup, for example. You may delude a hydrophobic victim of a rabid dog-bite by having him drink water in which you have quenched a red-hot iron and mixed some wine, half and half. Offer plain water, too. If you must restrain him in his madness, place a plaster of cold medications on his belly. There are no other treatments for hydrophobia.

V Twenty-one Explanatory Comments

A. In respect of our comment at the beginning of this chapter, stating our disagreement with those who believe that bites and stings should be classed as simple wounds, I add that the definition of a wound given in Notable IV of the General Precepts preceding Treatise II does not pertain here. Here is what I wrote: A bite or sting is a recent wound in an otherwise healthy part of the body. However, in those lesions, at the instant one occurs or is caused, some abnormal item is introduced and it no longer is just a simple wound; it has been corrupted.

---

[30] Williams' Chirurgia was published in 1275, and contained no chapter on dog-bites. Mondeville probably refers to William's earlier (ca 1270) medical treatise, the *Summa Conservationis et Curationis* (LDR).

[31] Maimonides. 1135-1204 (LDR).

Sometimes it is intermediate between a simple wound and a poisoning, sometimes it is a pure envenomization, as in the bite of a rabid dog or the sting of a scorpion. That indicates that it is not just a simple wound, even though there has been a solution of continuity, a break in the skin. Therefore, it is to be considered as ulcer, and that is why I have put this chapter in Doctrine II which deals with the treatment of ulcers. Indeed, I follow Aristotle in believing that the logical process that moves us from the General to the Particular is an inborn human trait. We first learn about Man in General and then about a particular person. This chapter precedes the chapters that deal with fistulas and cancers, because, if we treat these lesions (ie bites and ulcers) badly they degenerate into fistulas and cancers—not the reverse—and the victims may die. Since that is the case, when our treatments for bites and stings fail, we must rely on the treatments for fistulas and cancers. And since the reverse is not the case, our chapters should keep the order we have given.

B. Five ways to deal with the venom: 1. Detoxify it. 2. Resolve it or extract it. 3. Expel it. 4, Use antidotes, such as garlic, wine and asafoetida for scorpion bites. 5. Block its spread toward the heart with theriacs etc.

C. A poisoned person is more rapidly intoxicated and will die more quickly if his heart is warm than when it is cool, because the veins and arteries are larger when the heart is warm, allowing a more rapid spread of the venom to the heart.[32]

D. Warm venom is worse than cold because it spreads more rapidly. However, once it reaches the heart where it is cooled, the cold venom becomes more dangerous, because it bonds more tenaciously to the warm vital spirit, according to the principle of contraries. I took this explanation from Avicenna (Bk. IV, Fol. 5, Doct. 1, Ch.2).

E. According to Averrhoes (*On Theriac*) the medicines that give

[32] Be reminded: The heart provided the vital spirit (heat) for the body, and distributed it through the arteries. The air breathed by the lungs cooled the heart and prevented it from being overheated (LDR).

relief from venoms are intermediaries between other medicines, the body and the venoms. Also, a theriac is not suitable for well-adjusted bodies, nor for all sick bodies. It is suited only for large bodies, but not for all of them; it suits only those that have been altered (ie harmed) by phlegmatic and melancholic humors. Those who have been damaged by sanguinous or bilious (yellow) humors, at least those who are not too far from normal, are not helped by a theriac. Always be careful and use common sense when prescribing a theriac, and never use it in cases of fever.

F.   In re preventive measures note that fumigations from burning deer-horns, goat-horns, hooves of goats, sulfur, human hair, galbanum or serapinum will chase away serpents that live in houses. Crushed mustard or water-chestnuts also are effective when stuffed in the serpents' nests.

Rats will flee if you hang a small bell from the neck of one of them and let him run back to the group. If you sew the anus closed of one of them it will become enraged and bite some of the others, who in turn will be maddened killers. Also, you can kill them by adding crushed seeds of euphorbia to soft cheese, or another similar pâte that they will eat. Mice will leave a house which you fumigate with calamint. They say that mice will flee if you tie one of them by its foot in the center of the house, or char the face of one of them, or hang a bell from the neck of one, or cut off the tail or castrate one. To kill them, fill their nests with ashes of oak wood which causes lethal scabs to form on those who touch it. The weasel hates the odor of rue. A mole will die if you plug all but one of its exits, that which faces the prevailing winds, and at that one you burn some walnut shells filled with sulfur and cedar resin. Ants will flee if you place some cedar resin or sulfur or asafoetida or ox-bile in their nests, or place atop the nests some sulfur or oregano or ashes of sea-shells or ordinary wood cinders. Fleas will leave places where you put cantharides and polycaria. They will flee persicaria which in France is called culridge, and from leaves of the medlar tree. If you bathe a dog in water containing the juices of the leaves of costus he will be rid of fleas by the next day. The following anti-flea remedies have been well-tested and will kill them on contact: decoctions of rue, water-chestnuts, rose-laurel, absinthe, colocynth or nigella.

Stink bugs, the round, red and malodorous insects that live in old houses and in old fences, are killed by the smoke of burning straw or cow-dung or roseau or nigella. Bees will avoid persons who eat garlic or who have garlic breath or who have handled marjolain. They will not sting those who rub their hands and face with the juice of nettles, mauve or melissa (our peasants call it piment). Wasps flee the smoke of sulfur, yellow orpiment or olibanum. The water of or a decoction of black hellebore or yellow orpiment or their powders mixed in milk will kill them, and gnats, too. They will not sting what is coated with a juice of althea or mauve in oil.

G.   The ligature tied around a limb to block the spread of the venom should not be loose, a worthless gesture. Yet, it should not be so tight that it will be unbearable. Make it tight enough to arrest the spread and not attract the humors by causing pain.[33]

H.   Use a theriac as an inunction around the ligature to share the work of blocking the passage of the venom. The venom always yields to the theriac.

I.   Washing your mouth and oiling your lips allows you to use oral suction, and will keep you safe from any vestige of venom. An additional precaution is to avoid sucking if you have dental caries or ulcerated gums.

J.   Whoever will suck the bite should first fill his stomach with garlic and other similar foods to act as antidotes if some of the venom will be swallowed, and to keep it from passing further and to prevent it from remaining in the stomach. The antidote will easily cause emesis.[34] A young person sucks more vigorously

---

[33]   The overly tight tourniquet causes pain, and pain attracts humors to the source. That is the surgeon's concern here; not ischemia! (LDR).

[34]   A full stomach easily delivers its contents, unlike the retching of an empty stomach (LDR).

on behalf of the patient, but with more risk to the person who sucks. That is avoided if the sucker's stomach is full.

K. One way to make leeches adhere is to wash the place with warm water and then scarify it. Then rewash it and rub on a small amount of blood: human, lamb or pigeon.

L. After the initial excitement has quieted, keep the patient awake. In sleep the vital spirits and the venoms are attracted from the site and can move toward the vital organs. That is just the opposite of what our rules call for, and the patient may die (ie in his sleep).

M. At the outset, sweats and baths are effective because they attract humors from the inside to the surface, whereas phlebotomy, purgation and similar actions do harm by attracting the venom inward, causing it to penetrate toward the vital organs.

After they have functioned well, the sweats and baths become harmful, except as the final measures. But, between the beginning and the finale, the bleeds and purges are useful.

N. Enlarge the punctures with a knife and follow with the cautery. The bleeding from the incision will carry off some of the venom. Then the cautery will destroy some more of it and help restore the normal complexion.

O. They say that the basilisc kills people just by appearing in the scene, by poisoning the air, not by actual physical contact. Furthermore, a young man's saliva will kill a basilisc, and it is said that a basilisc will die if sees itself in a mirror.[35]

P. Some people doubt the efficacy of immersion in the sea as a cure for rabid dog-bites and they ask if it is effective against bites of serpents and scorpion-stings.

They ask if man-made brine is as effective as sea-water? Is the water of the Black Sea and the Mediterranean Sea which

---

[35] The fabulous basilisc was a hissing dragon whose foul emanations and terrifying appearance were fatal to humans. The name suggests that it was the ruler (basileus) of the serpent kingdom, and as such accounted for its appearance in this chapter (LDR).

have no tides as effective as that of the Atlantic? The people who live along the rivers that feed the seas know nothing of the practice. All of these questions and others are matters for investigation by the physicians.[36]

Q. The victim of a rabid dog-bite should not examine his own urine, because he may see what seem to be shreds of tissue. When you bleed him, he should not look at his own blood, because he will imagine that he sees bits of the dog's viscera; that is a commonly held belief.

R. The victims of rabid dog-bites have a fear of water, a symptom that we call hydrophobia, and when they reach the stage when the very sound of water is fearsome they are nearly beyond cure. When they complain of thirst, try to disguise the water by having them drink through tubes. Also, they should not see themselves in a painting or a mirror or a window-glass. They must be managed as if they are maniacal or deeply depressed.

S. Rabid persons dread water because of the internal corruption and contamination. When they see water their imaginations stray and they believe that they see water inside themselves. When asked why they dread water, they reply that it is full of the intestines and the viscera of the dog, a good enough reason for their horror. What little reason remains of their minds is enough to be horrified by the products of their own deranged imaginations.

T. You may be amazed that someone may become hydrophobic, who has not suffered a bite or an injury or any external cause. I saw it happen at Paris. The grocer of the Archbishop of Narbonne became hydrophobic and died within a span of eight hours; he had no marks on him. Much amazed, I took it upon myself to question all the medical authorities whom I knew to have lectured on the subject, and I found no one who could answer my questions. Although I was quite worn by my

---

[36] This is another of Mondeville's tongue-in-cheek remarks that made him unpopular to the Schoolmen of Paris (LDR).

searches, I happened to find this in one of the medical texts, a handbook by Bartholomew.[37] In his chapter on hydrophobia he wrote that the disorder may come from air that has been contaminated by foul emanations from cadavers. If one inhales that air when passing near nearby, he may be disposed to hydrophobia. In the absence of any other cause, I concluded that such was the case of my patient.

U. The authors of the medical books offer more complete discussions of these topics. I cite Avicenna (Bk. IV, Fol. 6, Ch.3), Rhazes (*Almansor*, Bk. VIII), Rabbi Mose Maimonides (*Treatise on Venoms*), Haly (*On Royal Dispositions*, Discourse IV, Part 2).

CHAPTER 3.

THE TREATMENT OF FISTULAS

As it is for all chapters in this Doctrine II, this chapter has three elements.

I. The Descriptions: Four Items

A. Definitions: A Fistula gets its name from the musical instrument used by a piper to accompany the songs of the country-folk. The Arabs (see Avicenna) call it "assucati", that is, a quill from a bird's feather; also in Arabic it is "garab" or "algarab", which is a hollow cane or reed.

It is a deep ulcer with a small external opening, and an opening at the bottom which may be large or small. In time the tract is lined with a hard callosity, and that is what makes it resemble a feather-quill or a hollow reed.

B. There are many Varieties, only some of which are curable and worthy of treatment; the incurable ones do not warrant such treatments. Some fistulas are recent and others are chronic; some are superficial, even subcutaneous, others go deep; some enter vital structures or pass nearby while others go nowhere near; some affect very sensitive

---

[37]   Bartholomew , a physician of Salerno, fl. 1125 (LDR).

structures while others range from moderately sensitive to not at all; some enter or come near joint-cavities; the same is true for muscles, nerves and bones; some affect robust persons, while other victims are more delicate; some involve structures or persons with normal complexions, others are abnormal; some have more than one external opening and some have more than one internal channel; some of the tracts are straight and others are tortuous. All of these characteristics and conditions involving fistulas are easily recognized by a skilful and experienced surgeon. Any additional information garnered directly from the patient will be helpful in the treatments.

C. The Causes are either proximate or predisposing. The former are the same as those described for deep ulcers, because fistulas derive from those ulcers. However, in addition there are causes which are special for fistulas just as there are for certain ulcers. An ulcer filled with granulation tissues can easily be reopened, and be refilled. That is the time when it can become a fistula, as Avicenna wrote (Bk. I, Fol. 4, Doct. 4, Ch. 29).[38]. The same holds for an ulcer from which you pull out a piece of bone before Nature has completely sequestrated it. The remaining diseased bone becomes a source of a fistula. We cite Avicenna to prove that (ibid.) in the chapter which describes bits of bone lying in sordid ulcers.

D. The Symptoms (ie signs) of fistulas: three matters:

1. The General Signs common to most fistulas are given in our descriptions of the various fistulas. In addition there is another certain way to identify a fistula, and that is with the eye of experienced surgeon who seldom errs.

2. When fistulas involve nervous regions or bones we consider two features: a. When we see a fistula go deep and pass near them we assume that they involve them, even more when we know the regional anatomy of the nerves and the bones. b. If

---

[38]   The external surface of the ulcer is closed by scar while beneath the granulations the ulcer continues to erode and become a fistula. This misconception of a fistula working its way from outside-in rather than the reverse was held from ancient times until recent (LDR).

the fistula is longstanding, it will have burrowed deeply and have reached those structures.[39]

3.   Special Symptoms of fistulas which distinguish those in fleshy parts from those in nervous regions and bones. The four Special Symptoms of fistulas in fleshy tissues: a. A recent fistula does not penetrate into the flesh b. It causes little pain. c. The discharge is viscous and thick, muddy and lumpy. d. When you probe with a non-metallic instrument to sound its depth you cause little discomfort.

The four special symptoms of a fistula in nerve-bearing regions: a. The fistula is not very old, perhaps a half-year, and is in a somewhat fleshy site and we assume that it has burrowed to reach the nerves. b. It is very painful. c. The drainage is thin and watery, fetid and dark, d. When you probe the tract you cause more pain.

The four special signs of fistulas that reach bones.[40] a. If it is older than one year we may assume that it has reached bone, sooner or later as determined by how fleshy are the tissues it must traverse and how thick or thin is the body of the patient. b. When you probe the fistula and you touch bottom you cause no pain because bone, like hair, is insensitive. However, the tract itself is sensitive and you cause pain when you touch the sides with the probe. c. The pus is thin, yellow and not very malodorous. d. Your probe easily reaches the bone and it feels it to be uneven and eroded. The probe may get caught (ie in the pitted surface) and not slide unimpeded along the surface as it would if the bone was healthy.

---

[39]   An ulcer in the skin has eroded the subcutaneous tissue but has not gone much deeper, thus it is no longer a deep ulcer but it is a 'recent' fistula (LDR).

[40]   The medieval surgeon believed that a fistula began as an ulcer which, in turn, was an unhealed wound or the residuum of an abscess. He did not conceive that the bony fistula represented a chronic draining osteomyelitic abscess which worked its way from the bone to the surface. See footnotes above (LDR).

II. The Treatments: Three Parts

A. The General Measures

Let the reader first review what we wrote about the General Precepts (ie Rules) for treating ulcers and apply what he finds there as applicable here. I shall not repeat it. In addition I shall describe nineteen precepts for fistulas per se.

1. Before treating an old (more than one year) fistula, prescribe what a physician has recommended against the specific humor at fault.

2. A deep fistula is more difficult to cure, because it is harder to introduce your medications and to extract detritus. Also, a tortuous tract in a delicate patient can be a trying situation for the surgeon.

3. A very old fistula creates difficulties because in time the tract is indurated and the surrounding region is more inflamed.

4. All such old fistulas cause problems because you first must core out the callous lining with a knife or cautery or with caustic medications. Until that has been accomplished, the surrounding tissues can not engender granulation tissues to fill the tract and allow closure by scar.

5. You will soon discover that it is hard to cure fistulas that involve or are near vital organs and sensitive structures or that enter joints or that involve nerves and bones or appear in parts of the body already suffering a bad complexion (ie another dyscrasia) or that drain pus profusely or have more than one tract or have tiny external openings.

6. Also difficult to treat are deep ulcers that cause little pain, those in bone and in ligaments and in other insensitive regions.

7. From the first to the very last treatment you should apply defensives on the surface surrounding the external opening of the fistula. Be sure to apply more of it on the side from which more of the humors seem to arrive.

8. Our standard protocol dictates that we apply cool medicines over all the surrounding region to relieve pain and give support to the patient and to impede the attraction of more humors.

This is more important whenever we instill pain-provoking medicaments into the fistula or carry out painful operations.

9. After a successful cauterization or application of a caustic in the fistula, apply a cool medicine inside and around the opening until the burning pain relents. Then use suppuratives until the eschar separates and is discharged.

10. Do not forcibly pull off the eschar that is formed by your cautery or caustics. Let it sequestrate and deliver itself. Then you can avoid pulling it away from adherent veins and cause bleeding.

11. As long as severe pain continues after a surgical procedure or other treatment, use mitigatives and defensives inside and around the lesion, and apply no other medications.

12. After the eschar has sloughed out and the pain has subsided and nothing else stands in your way, abstain from the further use of the mitigatives, defensives and suppuratives. Then, every time you dress the fistula with suitable medicines, take a mondificative containing honey et al. spread on a thin linen cloth and lay it over the fistula. Use it until the tract is filled with good granulations, then discontinue the use of all detergents.

13. The techniques for bandaging fistulas are exactly the same as those we described in the chapter on treating deep ulcer.

14. When a fistula directly involves or is near a noble or enfeebled or delicate region, or when it is in nerves or in a nervous region, it will not tolerate harsh or corrosive medicines. If you must use them at all, dilute them.

15. Fistulas that burrow into the ear [41] or into the chest or abdomen or urinary system will not accept harsh corrosives.

16. Although sometimes one does well to remove small bits or flakes of bone, the surgeon should not attempt to forcibly pull out large chunks, as they occur in the thigh or leg, or pieces of mandible still adherent in a wound.

17. The Diet for the patient, the six non-natural things [42] and the

---

[41]   Chronic otitis media or mastoiditis. Another example of the misconceived reversed direction (LDR).

[42]   'Non-natural things'. To remind the reader: The term refers to matters affecting the patient which are not part of him or that derive from him. See the Contingents in Notable XIV which introduce Treatise II (LDR).

potions all should be very desiccative and consumptive of the
superfluous humors already engendered and prevent
reaccumulations.

18. After General and Regional Purgations, and after the local
mondification of the fistula, you may prescribe suitable
desiccative potions.

19. The Potions: These remedies are known for their intrinsic
properties and from experience with their use; they are
recommended by the authorities as suitable for treating true
fistulas.[43] Avicenna used roots of scolopendrum (Bk. IV, Fol.
4, Doct. 3, Ch. "Treating Fistulas"). None of the potions has
much effect, in spite of what the authorities say. At least they
do not cause burning pain weakly active as they may be. Indeed,
they are useful in deep ulcers, which the old-fashioned writers
and practitioners mistakenly called fistulas.

B. Procedures in Special Cases

As stated earlier in this chapter, a surgeon who is familiar with the
Generalities can infer nearly all that is needed in treating particular
cases. Here we will describe the three parts of those special treatments.

1. The evacuations, both the general and the particular, and
2. the dietary regimens are those used for treating deep ulcers.
3. The many useful topicals, both the Simples and the Compounds
include defensives, repercussives, sedatives and the dilators that
are either corrosive or non-corrosive, eruptors for opening and
cauterizing, refrigerants, mondificatives, desiccatives,
regeneratives and consolidatives. In all those categories we have
some that are more potent than others. All may be applied as
ointments, plasters, powders, lotions, etc. We have discussed
much of this material in Chapter 10 of Doctrine I, and in Chapter
1 of this Doctrine II, and in the Antidotary. Nevertheless, we
cannot possibly describe the precise amounts to use in every
case and all the ways to apply them and to bandage the regions.

---

[43] As distinguished from deep ulcers. Mondeville derides the surgeons who
cannot make the distinction (LDR).

Galen (*Megatechni,* Bk. III, Ch. 3) said, "The surgeon who is skilful and experienced will give the correct orders here and there, guided by the traditions, his own experience and by the particular circumstances."

How to apply the topicals and how to operate: seven items:

a. When the external opening is small, enlarge it by stuffing it with a drain covered with a paste of elder-tree pulp and gentian roots, or carefully insert strips of sponge. If that does not do enough you will need to use a knife or a cautery or an eruptive caustic medication. I have dealt with the use of the knife and cautery in the chapter on deep ulcers. When I use the caustics I put in enough to serve the purpose. We have described all the corrosives that you will need in the Antidotary.

b. After you have enlarged the small opening you can remove the tough callus that lines the full length of the fistula. Use one of the three methods for enlarging the external opening. When you use a cautery or a caustic medication, follow it immediately with cool medications to mitigate the burning pain. After relief is obtained, use suppuratives to loosen the eschar.

c. After the eschar has come away, use mondificatives containing honey et al. as plasters and ointments and lotions. Use any or all just as I instructed for treating deep ulcers.

d. After deterging, use desiccatives.

e. Then use incarnatives, followed by

f. Cicatrizers.

g. In every stage, use bandages as we did for deep ulcers.

C. Potions

We use the evaporated juice of bryony made into a powder and mixed with honey. A dose the size of a walnut is given with wine once a week. Or we may use the juice of or a decoction of gentian, or a compound made with a pinch of each of agrimony, scolopendrum, olives, all minced and macerated in white wine. The patient drinks a cupful daily at dawn. The potions should be taken until the cure is complete. Of all the above, I favor the last one.

Palliative Treatments For Fistulas

Palliation is indicated in three types of cases.

1. When a fistula really is a priori incurable; such as when it drains urine, or it enters the marrow cavity of a large bone (humerus or femur), or when it involves the upper jaw and burrows deeply toward the brain.
2. Although the fistula may be curable, the cure itself produces an even worse condition, as when you treat a fistula-in-ano that enters the rectum above the level of the muscles which control the passage of feces; you cause permanent incontinence of feces. The same holds for fistulas that complicate chronic hemorrhoids. The cure can make a person quite mad and dropsical, as Hippocrates stated (*Aphorisms*, Part 6).
3. When the patient does not want to be cured because he is unwilling to withstand the painful treatments.

The palliative treatments include four elements:

1. Purgations as we used them in treating deep ulcers.
2. Proper Diets in the scope of the six non-natural things.
3. Potions. Those we used for cure will also serve for palliation.
4. Topicals of three types. a. Mondificative plasters with honey, wheaten flour and the juice of wild celery. Apply it frequently. You may add myrrh, sarcocolla and aloes. b. Ablutions as we used for virulent ulcers, perhaps with added myrrh. c.The defensives we use for virulent ulcers.

III Twelve Explanations

A. A fistula resembles a cavernous ulcer in having a small opening leading to a larger internal chamber. It differs from the ulcer in having a horny lining like a quill from a bird's feather or a reed, whereas the ulcer's surface is soft. The surgeons, including Roger and other Salernitans who did not recognize the differences used the wrong treatments. When they were operating on a cavernous ulcer they thought that they were treating a fistula. They used excessively potent caustics and caused the ulcers to corrode[44] and sometimes they ate through the bottom layer. In the same state of ignorance they would use weak medicines with little effect on the true fistulas, medicines suitable for ulcers.

B. The common folk and the uneducated bucolic surgeons call ulcers, wounds and abscesses, all which require prolonged treatments are St. Eloi's [45] Disease. They compare one with the other, a patient who has recovered after a visit to the saint's shrine with the other who languishes uncured. They blame the latter for his lack of piety in not making the pilgrimage, or that the sickness is not one that fits a diagnosis honored by the Saint's name. So great is the belief among the people that they are driven to make pilgrimages not only for ulcers and fistulas but also for seeking cures for wounds and for abscesses which probably would not need a surgical drainage or would come to a spontaneous eruption of pus. Furthermore, the pilgrims are not only humans; they bring their sick sheep, cows and other four-legged beasts. They believe that the Saint will cure one as well as another.

C. The common folk believe that their ailments did not exist before the sanctification of Eloi. What the medieval writers call a fistula did not exist before his birth. If that is true it would have been better for a lot of afflicted people if the Saint had never existed, especially so since their ailments appeared only after he became a saint.

D. The fistula first got its saintly namesake at the time of his beatification, when many people who gave homage at his tomb recovered from many kinds of illnesses. It so happens that the sicknesses in question derive from cool, raw and undigested

---

[44] The caustics ate into the adjacent tissues (LDR).

[45] St. Eloi is St Eligius. I assume that Mondeville used the French name, as we do here (LDR).

humors, and they are consumed [46] during the pilgrimages, and the sufferers who recovered outnumbered those who did not. It is the former who gave the Saint's name to the ailments, not that he had more curative powers than another saint.

E. The ignorant bucolic physicians who have no excuses for their failures, and who see the extreme faith of the people in the Saint, have come to claim that the wounds and other ailments that they cannot cure are in his province. The naïve peasant folk believe the bucolic therapists. They happily swallow it whole and forego cures without blame or regret. Indeed, they eschew surgical undertakings lest they affront St. Eloi. When they believe they can be cured by the Saint, they will not even allow an initial surgical treatment; rather they may choose not to be treated, claiming that the Saint who gave the disease has the power to cure it if he so wishes. So, under the shadow of St. Eloi, countless thousands of limbs have suffered putrefaction and gangrene, that could have been cured by able surgeons. The wayward surgeons have created St. Eloi's Disease as a refuge for themselves, and the physicians who cannot admit to ignorance will say that it derives from something special; and the theologians who cannot offer an explanation will say that it is an effect of the Divine Virtue; and the Logicians (ie philosophers) who will never admit to not knowing a solution to a problem will say that it is a false consequence (ie of a faulty premise).[47]

F. Some believe that a small sack of medicines which they had collected while reciting certain prayers and hung from the neck, or that simply reciting the prayers when they put the medical necklace in place, will cure their fistulas. The

---

[46] Not eaten! The body's Natural Heat consumes humors or transforms them. The concept is not unlike what the modern physician calls the 'tincture of time' (LDR).

[47] A curious way of saying that the premises of a preceding syllogism, being false, had led them to a false conclusion (consequence) which now is offered as a new premise (LDR).

medicaments are agrimony, hypericon and others. Although I cannot believe that any true fistula can be cured without some surgical intervention, perhaps an occasional ulcer treated with these medicines alone may heal on its own. However, simply to be tactful you may use some of these medicines along with your own. When the cure is obtained the surgeon is praised for a marvelous success. When he fails he cannot be blamed for leaving out the necessary adjunct (ie the necklace) and he will escape criticism for overstepping the limits of his art.

G. When the corrosives cause unbearable pain, obtain relief by removing the medication and by warming the limb near a hearth or bathing it with warm water.

H. The removal of necrotic bone from a wound is described in Chapter 11 of Doctrine I of this Treatise II.

I. When Hippocrates (*Aphorisms*, Part 6) wrote about wounds that remain open for a year, he meant only the deep ones near bone, otherwise it would be a contradiction of what he wrote about wounds in other regions.

J. In adults, whenever a piece of bone has been extruded a depression will remain in the scar of a healed ulcer, which will be the size of the extruded piece. Galen said (*Megatechni*, Bk. III, Ch.1) that in children the defect is filled with new bone, as we described in Notable 6 of Chapter 1 of Doctrine I of this Treatise II.

K. Never pull out the piece of sequestrating bone until Nature herself has worked at it for a long time, perhaps assisted by some of your weaker attractives when the loose bone is exposed. Then, if does not expel itself in due time and when it is loose, you may lift it out. But, if Nature has failed to sequestrate it and has not exposed it, you must not tear out the fragment. Simply abandon it. As Avicenna wrote (Bk. IV., Fol. 4, Doct. 4, Ch. on nerve-wounds), a violent maneuver itself attracts more humors and adds to the ills of the fistula. Another reason to wait for Nature to remove the fragments of dead bone is they may be buried in granulation tissues, and the surgeon cannot assess the situation. A better policy is to defer, lest, as it often happened to the old-style surgeons, you remove something best left alone, and you fail to remove what you have to go after it later on. Nature really does it better in distinguishing what to save from what to expel. Once she makes it clear, the removal is easy.[48]

L.   We described how to extract arrowheads in Chapter 11 of Doctrine I of this Treatise. Avicenna also gave instructions (Bk.I, Fol. 4, Ch. 28, and ibid. Doct. 4). He gave an admirable description of diseased bone and all the diseases that affect bone, their symptoms and their treatments, as well as how to extract corrupted fragments.

Chapter 4

The Treatment Of Cancerous Ulcers

I. The Diagnosis of Cancerous Ulcers: Six Items

A.   The two words of the term, 'Cancerous Ulcer' have different meanings. To know what we deal with here we shall clearly indicate either a non-ulcerating cancer or a cancerous ulcer, to avoid ambiguity.

B.   The Definition of a non-ulcerating cancer: [49] It is an abnormal mass or an inflammation formed from corrupted or putrefied melancholic humors. It contains no exterior openings or wounds. We shall not concern ourselves with them here; the inflamed (ie abscessed) cancers will be an item in the chapter on aposthems.

C.   The cancerous ulcer, however, has an opening, an obvious, rounded, fetid ulcer with rolled raised edges and an excavated interior. The substance is hard and uneven, livid or black in color.

---

[48]   Mondeville's distinctions between deep ulcers and fistulas are not clear. He persists in calling a chronic draining osteomyelitic sinus a fistula (LDR).

[49]   The word 'cancer' was applied to many tumors, ranging from true neoplasms to other lesions, both the ulcerated and those with intact overlying skin. The 14thC French term was 'chancre' (EN). In this chapter Mondeville

D.   The Varieties of Cancerous Ulcers: One type has an internal cause; a burned[50] and putrefied melancholic (ie black bile) humor collects somewhere in the body. Another type has external causes, such as a badly treated wound or ulcer. Occasionally a cancerous ulcer is the effect of both internal and external causes, such as a contusion that becomes a cancer, or when we incise to drain a cancerous aposthem.

Some cancerous ulcers are curable, if they are small, are entirely in soft tissues and are of recent origin. Others are incurable, such as one that arises in nervous regions, or one that grows deep during a long existence, or grows in or near principal or noble organs.

E.   Causes: The curable one derives from a burnt and putrefied melancholy which comes from the combustion of a natural melancholy. the incurable one derives from a burnt and putrefied melancholy which is a product of other burned humors, which by their combustion became a non-natural melancholy. After they have undergone a second combustion, they putrefy and become more malignant.

---

takes great pains to define neoplasms in various grades of malignancy, invasiveness, rates of growth, etc. He is equally 'painful' in assigning to each variety its own causative humors, and is disdainful in his condemnation of the erroneous identification of cancers with other ulcers. However, his own terms confuse cancers, gangrene, etc. as was the common confusion in ancient and medieval times. Bear it in mind that all these lesions are visible, palpable and smellable on or just beneath the surface of the body. The medieval surgeon did not deal with neoplasms of the viscera. Furthermore, we must view Nicaise's translation as the work of man who was newborn in the era when the cellular nature of cancers was first elucidated. Few pathologists in the last half of the 19th C had a solid grasp of the concepts introduced by Virchow. Nicaise's circumlocutions reflect his own attitudes about what he read in Mondeville's pages (LDR).

50   'Burned' does not mean incinerated. One's natural heat may be over active, not necessarily a fever. The natural heat may 'burn' the bile etc. Unnatural heat is defined later in the chapter (LDR).

One kind of cancerous ulcer involves the upper parts of the body. Most often it comes from burnt and putrefied (ie non-natural) melancholy, from burnt bile. It is a very corrosive humor because it comes from a thin fluid that has been heated twice.

The other kind involves the lower parts of the body. It derives from a burnt and putrefied melancholy which was produced by the combustion of true natural black bile. This cancer is more curable and benign and has less corrosive features than the other kind.

One type involves soft flesh, another involves nerves, a third is in bone.

One kind grows deep, another is on the surface. One involves only external structures, another the internal.

Their ages vary from recent to old. Some appear in robust and strong people, others affect the debilitated and the delicate. Not all patients are willing to be cured and they refuse to give themselves to the surgeons and undergo operations.

A drainage incision or a spontaneous eruption of a putative non-ulcerating cancer that really is an abscess is a wound that becomes an ulcer because the treatment was wrong; that ulcer, if treated badly, may become a cancerous ulcer when it attracts more putrefied melancholic humors. At times, the attracted humors have not yet putrefied and do not undergo putrefaction until they arrive at the corrupted site.

Other causes can be inferred from those described above.

F. The Symptoms (ie signs) are described in the section on Definitions. We shall add here the three things that distinguish cancerous ulcers from the sordid ulcers which they resemble. 1. When you irrigate the cancer with lye it becomes even more unsightly than before; its color fades to that of ashes. After the wash has dried, a viscous, sticky stuff remains as an adherent film or membrane. But, when you wash the sordid ulcer with lye, you clean it, and its color improves (ie pink).

2. The stench of a cancer is horrible beyond description. Anyone exposed to it will know the diagnosis before he sees the patient. The odor of the ulcer is bad enough, but not intolerable, and it is very different from that of the cancer, which has the stench of a cadaver.

3. After you apply corrosives on the cancerous ulcer you can see

that you have made it worse. After the eschar separates, the malignancy has increased in size. The reaction of the ulcer is different.

II The Treatments: Three Kinds

A Preventive Measures:

1. If a cancerous ulcer is the result of the faulty treatment of a wound or ulcer, intervene with the measures that we use for the prevention, the cure and the palliation of wounds and ulcers. If you do it properly, you will never see a cancer develop. 2. To prevent a non-ulcerating cancer from becoming a cancerous ulcer, observe the caveat of the preceding paragraph. I shall add more detail in Doctrine II of Treatise III in the chapter on non-ulcerating cancers.

B. Curative Treatments: General Measures:

In addition to all the items in our Precepts (ie rules) for treating ulcers and fistulas which an operating surgeon must know, I shall add fifteen Precepts for cancers.

1. In all cases of both kinds of cancers, for prevention and cure and palliation, always use the general purges as well as those designed for local evacuations, which eliminate the faulty humors.
2. After the evacuations, take pains to identify the type of the cancer that may be curable.
3. If you cannot prevent the accretion of cancerous matter, at least try to fortify the affected region to resist the influx of humors and to reject some of it.
4. For cancers of internal origin, cancers in nervous regions, cancers in the upper part of the body, for old and established cancers that penetrate deeply[51] , for those that derive from the

---

[51] 'Deeply': the term meant the extension beyond the subcutaneous tissues and rarely into a viscus. The cancerous breast was an exception (LDR).

twice-burned melancholic humors, for those that arise in feeble and delicate patients and for those in patients who may refuse treatments, and when a cancer obviously is incurable, in all and any similarly daunting situations, you should know that the treatment for cure, if at all feasible, will be difficult.

5. Both kinds of cancers may be curable: if they have external causes, if they are small, recent and superficial, if they lie in fleshy regions and are not near nerves or important structures, if they appear alone without complicating factors and if they occur in robust persons who are eager to be cured.

6. Both kinds of cancers are incurable: if they are long-established and complicated, when they grow deeply or are situated in nervous or bony regions, if they are near important or vital organs, if they derive from twice-burned melancholic humors and if the victim is debilitated and feeble.

7. When you treat any kind of cancer, the patient must agree to abide by the dietary regimen that is the same for all persons: it contains the six non-natural things. However, the topicals will differ from case to case.

8. You can cure a cancer only by radically extirpating all of it, leaving none of it behind. The malignancy is greater in the roots, as both Serapion (Tr. V, Ch. 25) and Rhazes (*The Divisions*) claimed.

9. It follows from what I have said that you must not undertake a radical operation to cure a cancer unless it is situated in a region where nothing stands in the way of a total extirpation. As Serapion aptly remarked (ibid.), the treatment itself is violent and, if it fails, it will destroy the patient by exacerbating the cancer.

10. It follows, also, that you cannot assume that you can cure a hidden internal cancer [52] when it cannot be encompassed by your surgical operation. Again I cite Serapion (ibid.) and Hippocrates (*Aphorisms*, 38, Part 6) who add that those who are

---

[52] 'Internal' denotes the inward extension of a cancer on the surface, not a visceral neoplasm (LDR).

treated with the knife, cautery or with caustic medicines die sooner than the untreated. The untreated live longer because they have not been burned or incised or injured by painfully caustic medicines; rather, they have received proper palliative treatment.

11. We conclude from the preceding that we should not try to destroy a cancer piecemeal with corrosives, as is commonly done. In fact, while you can burn away one part, the roots become even more malignant. You must eliminate it all at one time (ie for cure), and not in parts.

12. When you try to cure a cancer by radical measures, apply defensives on the surrounding area.

13. The best way to accomplish a cure is to cut it all out with its roots. Then wipe out the blood and cauterize the surfaces (Galen, *Megatechni.*, Bk. XIV, Ch. 4).

14. After cauterizing or applying corrosives, follow the practices stated in Precept 9 in Chapter 3 on fistulas.

15. Deal with the eschar after using the cautery or corrosives as in Precept 10 in that chapter, and in 8, 9, and 10, above.

Operative Techniques

In the preceding sections we have emphasized the incurable types of cancer, in which the surgeon should avoid attempts at radical cures, because palliative measures alone will help the patient. Furthermore, you should understand that curative efforts are very difficult, whereas at times a palliative treatment may cure, if the surgeon prudently conducts it with great care. The surgeon should not undertake a cure until he has been entreated to go ahead with it, and has been assured a larger-than-usual fee.

The three elements of a curative treatment are:

1. The Diet: The diet for both types of cancers should continue without change from the beginning to the conclusion of the course of treatments, until the patient recovers or dies. As Galen wrote (*Megatechni,* Bk. XIV, ibid.), the diet should consist

only of cool foods: well leavened bread; fresh meats of lambs, kids, calves, capons, hens, partridges, small birds of the meadow with narrow beaks; soups made with borage and spinach and broths made with the listed meats to which you may add eggs, which also may be eaten raw. The beverages should include barley water, aged white or red wines which are clarified and diluted with water. The patient should abstain from other beverages and from the flesh of goats, cows and bulls, hares, of aquatic birds and from all garden vegetables except for chick-peas, and from all salted, acidic, roasted and broiled foods, as well as from cheeses and cabbages.

2.   The Evacuations: a. The General Purges evacuate and distract the humors from the cancers. The evacuation is by phlebotomy and laxatives. The distraction is by emetics for cancers below the umbilicus.[53] For lesions above that level use laxatives to empty the bowels and phlebotomies from saphenous veins, and massage of and applications of cords around the legs. b. For evacuating a particular region, the head for example, when the cancer is in the face, after the General Evacuations, use massage, scarifications, cups, and leeches applied near the site. Those maneuvers consume the residue of humors that have infiltrated the surrounding tissues.

3.   Operative Techniques: Two Parts:

A. The Artificial[54]

Having followed the proper sequence outlined above, you incise and excise the cancer with all its roots. Then you get rid of the contaminated blood that floods the defect. Then you repeatedly apply a hot cautery to all the surfaces.

If, in the second maneuver the patient cannot tolerate the hot iron, apply a defensive around the defect

---

[53]   The opposite part of the body is used for the evacuation, that is, for distracting the bad humors from a distance (LDR).

[54]   'Artificial': The surgeon's artifice (LDR).

and lay on a large amount of corrosives on all the surfaces within the wound. Do that once only (ie unlike the repeated applications of the hot cautery). An eschar remains after the terrible suffering has abated, much like that after the actual cautery,

The treatment continues: 1. Apply defensives. 2. Use coolants on the surface of the eschar. 3. Then apply suppuratives to cause the eschar to come away. 4. Then use detergents and, 5. desiccatives, and 6. regeneratives, and 7. cicatrizers.

The medications themselves, how to use them and the special virtues of each are described in Chapter 10 of Doctrine I of Treatise II. The techniques for making incisions and for using defensives and mondificatives are described in Chapter 1 of this Doctrine, on Ulcers. The way to get rid of the blood is not very 'surgical' (ie a mop-up). The techniques for the use of the cautery are in the chapter on cauteries. The repercussives and mitigatives used to relieve the pains after burning, and the suppuratives and the corrosives are to be found in the Antidotary.

B. The Empirical Techniques[55] : Two Methods

1. Medications that act only when carried by the patient (in a sack): the small liver-wort; piloselle, which is gathered in the meadows while reciting the Pater Noster; St. John's herb and agrimony. Use all together or separately. Ceterach works well and some say that it prevents cancers.

2. Medications that act when applied to the outer surface of a cancer: powdered Herb Robert; powdered myrtle-leaves; crushed lesser consolida, alone or with the fat of a castrated male animal; scabious; incinerated dill;

---

[55] 'Empirical': In common use and not Artificial, but not necessarily in disfavor by the professionals (LDR).

crushed clover in honey; juice of honeysuckle or leaves
of sorrel and potentilla. The latter is like tormentilla but
it lacks nodes on its roots.

Palliative Measures

These are similar to what we described for fistulas in the
preceding chapter. Here, too, are three categories.

A. The Diet is the same as when we are treating for cure.

B. The purgatives are the same, with the following restrictions:
Here we use purges only during the spring and the autumn, and only
when recommended by a wise physician.[56] They are repeated about
twice a week, at dawn, during those seasons. Use a small amount of
goat's milk in which you infuse 5 dr. of a good quality epithyme.

C. The topicals: 1. The defensives are those we use for all wounds.
2. Ointments: Equal parts of ceruse and tuthie—well washed to remove
irritants—and oil of roses or the like in which is a quarter-part of the
juice of morel or of a similar plant in an amount deemed suitable at
the time. Grind all with a lead pestle in a lead mortar bowl which you
rub vigorously to obtain some lead in the mixture. This ointment
inhibits the influx of humors, it restricts the infiltrative corrosion of
the cancer and it reduces its malignancy. It allows the sufferer to have
comfortable sleep at night.

You can compound many similar useful ointments, some cool.
some warm; each kind will work better in the cases for which it is
specially designed. They are listed by many authors and we need not
repeat them here.

D. The Humectants are made with decoctions and juices of morel,
purslane, etc. You may use them at every dressing instead of the
ointments, or, if you see any benefit, you may use them together by
placing a cloth or some oakum saturated with one of the humectants
on top of the wound in which you have put the ointment. The warmer
the topicals the more comforting they will be all. Renew them
frequently.

---

[56]   The Physician gave the orders and the surgeon carried them out: the clysters,
cups, phlebotomies, etc.(LDR).

You may temper all the topicals to suit the case. When the disease eats into the nearby tissues and causes more suffering, more pricking and burning pain, and if the patient is a feeble sort, increase the coolness by adding ceruse and other cool substances. You can increase the warming actions by adding tuthie and oil of roses and some white wax.

Never increase the coolness of the topicals unless constrained to do so by the need to relieve the pain. In so doing you will make them thick and fatty, and make the pain worse later on. Similarly, do not add to the warming effects unless you must, lest you increase the invasive tendency of cancers in warmed patients. You should be most circumspect and use only a few well-tested topicals, and you should completely abandon those with which you have had little experience. An error on your part will cause irremediable and permanent damage. The surgeon must use his own common sense in treating these patients; he should not rely on hearsay or on what some author has recommended.

The better to understand this subject, I suggest that you review what I wrote in the General Precepts, and in the chapter on the treatment of ulcers and fistulas in this Doctrine. You should know all of it, and I shall not repeat it here. Besides, an experienced operating surgeon should know where to find what he needs in those pages.

Seventen Explanations.

I. All the famous authorities have written about the treatment of cancerous ulcers. Many of them combine that with the treatment of non-ulcerating cancers. Hippocrates (*Aphorisms*, 38, Part 6), Avicenna (Bk. IV, Fol. 3, Doct. 3),Galen (*Megatechni*, Bk. XIV, Ch. 4) and Rhazes (*Divisions*, at the end; and *Almansor*, Bk.VII) and Serapion (*Practica*, Doct. 5, Ch. 15) all deal at length with this subject, as did Theodoric (*Grand Chirurgia*, Bk. IV, Ch.6) and many others. Yet, none of them has given a complete definition of an ulcerating cancer.

II. A cancerous ulcer is an ulcer. Here the generic term is 'ulcer', and it excludes all other lesions. Other definitions also are exclusive: the word 'apparent' excludes deep ulcers and fistulas; 'round' excludes oval ulcers; 'fetid' is limited to the horrible stench of cancers. When we say, "which have thick, heaped up and folded and raised margins which are separated from the base (ie undermined), or are excavated (ie cavernous), that is, they have cavities beneath the smaller opening atop, or that they are hard and nodular, livid or black", we exclude all the other apparent ulcers.

III. We have four reasons for calling these lesions 'cancers': a. The external ulcer often is round, an unusual shape for other ulcers. Here the shape is like the carapace of the marine animal we call cancer, which in France is called crab. b. It clings to where it lies. c. It is surrounded by many long and tortuous veins that suggest the appendages of a crab. d. It eats in any or all directions, a movement that suggest the sidling of a crab that can move forward or backwards or to a side.

IV. Avicenna (Bk. IV., Fol. 3. Doct. 3, Ch. on Treatment of Non-ulcerating Cancers) claims that it derives from a melancholic inflammation. He said that he encountered this astonishing anecdote in the works of another famous authority. "A woman's cancerous breast was totally extirpated and she seemed to be cured. Almost immediately, the other breast was attacked by cancer." Avicenna reasoned that the second breast had been attacked by the cancer before the first breast was removed, and the humors which had been deflected from it had flowed to the second side. I think that is a good explanation.

V. Avicenna (ibid.) and nearly all the authorities agree that the best treatment for an ulcerated cancer is to keep continuously in place a linen cloth saturated with the juice of morel and never to let it go dry and become warm; re-wet it as often as needed. Hippocrates seems to have had a contrary opinion (*Aphorisms*, Part 1). Avicenna later gave a another opinion (Bk. I, Fol. 4, Ch. 29, and Bk. IV Fol;.3). I believe that those two authorities meant that cold is not good when you are treating the ulcer itself, but it is good when you are trying to relieve pain.

When the pain is really severe, you should use narcotics.

VI. Here I give evidence for my claims that there are two kinds of melancholic humors. One derives from natural[57] black bile; it is called sclerotic, which means hard. We should not call it cancerous. The second derives from non-natural black bile. It is a melancholy which is no longer in its original state. It has been corrupted and burned. That is cancer, and the term itself connotes two types. One derives from the burnt black bile that has not undergone putrefaction during a long existence. That cancer remains unchanged; it does not ulcerate on its own, and it causes little harm. If by chance the overlying skin erodes, and it ulcerates, it does not enlarge because its substance does not putrefy.[58]

The second type derives from a more intensely burned and putrefied melancholy, which causes a cancer that usually does not ulcerate. In the course of time it may ulcerate, indicating different kinds of matter and different locations in the body and different life-styles of the patients. This cancer exists in two forms. One derives from a natural melancholy that is the residue, the lees of the other humors.[59] It becomes the non-natural melancholy when it is burned. The other cancer derives from a non-natural melancholy which is made by burning other natural humors or their ashes, not their lees.

Now let us return to the first category, those formed from melancholy, that is, from putrefied and burnt lees. There are two types. One is formed from a more solid matter, the other from the more liquid (ie subtle). The first, coming from a feebler and unusual substance, is less malignant. It causes less damage and erodes less. The second type is more malignant because it derives from more invasive (ie biting) matter.

A cancer that derives from melancholic humors themselves (ie not the lees) which have been burned and then undergo a second combustion, and therefore are twice-burnt—they acquire evil properties for some reason and they become corrupt and putrefy— the cancer of such an origin is more damaging, more erosive and it does more harm to the patient. It is very hard to cure.

The medical authors describe all the ways that combustion and

---

[57] 'Natural': Black bile from its first beginnings, not the product of the changed or burned humors which were not black bile at their beginnings (LDR).

putrefaction happen, in the parts of their books where they discuss putrid and quartan fevers.

VII. Galen (*Techni*, Ch. 5) and Haly (ibid.) explain that there are two kinds of melancholic humors. One is the lees of other humors, the other is result of the combustion or the ashes of others, of salt-phlegm, blood, yellow bile. When they are burned they become melancholy. You can see it in a child who has a hot liver and large veins; when he becomes an adolescent, he burns his bile to form melancholy.[60]

VIII. Non-natural melancholy has two forms, putrefied and not putrefied.

The putrefied type has three varieties. a. One fills the great veins near the principal organs and accounts for the continuous quartan fever. b. Another is found in varying amounts in medium-size or small-caliber veins anywhere in the body. It provokes a quartan or a double fever.[61] c. The third is found in capillary veins or at their tips. They cause a cancerous aposthem.

There are two types of non-putrefied melancholy. a. One occurs only in soft tissues and causes leprosy. b. The other appears only in the skin. This variety has two classes. 1. One appears in all the body's skin and is called black jaundice. 2. The other is local. It causes black morphia, other dark spots and sometimes cancers, pustules, et al.

IX. Excoriations, inflammations (ie supercalefactions), small ulcers on the gums and on the penis,[62] all of which are shallow and recent, are not true cancers, in spite of what many authors, practitioners and common folk say. Instead, we call them corrosions (ie erosions). They do not fit any of our definitions of cancers, and they should not be treated as cancers in ways we have described in this chapter. Gentle treatments will suffice, as you will find them in the chapters devoted to the them.

X. The persons who believe that cancers simply are ulcers risk serious consequences First, when they cannot extirpate a cancer, they apply strong corrosives, such as used for ulcers, and they exacerbate

---

[58]   Mondeville struggles to explain benign neoplasia (LDR).

[59]   See Tr.III, Doct. I, Ch 3 on Phlebotomy, on the examination of the layered blood-clot (LDR).

the cancer. Second, when they really have an opportunity to extirpate a cancer, they use weak corrosives, too feeble to get into the cancer, and they just stimulate the growth of its roots. On the other hand, they who wrongly believe that an ulcer is a cancer make three grave errors. First they apply corrosives that are stronger than necessary, in too large amounts, as if they were treating cancers, and they cause needless suffering. Second, they destroy more tissue than is necessary. Third, they attract humors which make the ulcer itself erosive and cancerous.

XI. I am amazed at Avicenna's statement (ibid.) that a cancerous ulcer is no longer an ulcer after it has been cauterized with the hot iron, because that makes its surfaces thicker and tougher. That statement is contrary to our General Precepts that a cancer cannot be cured except by a radical extirpation. You should not make it worse with the use of the cautery. Therefore, when a cancerous ulcer has not been totally removed it continues to grow and the ulcer persists. Furthermore, I added this Precept: you cannot cure a cancer piecemeal. Rhazes (*Almansor*, Bk. VII) wrote, "When someone is unsuccessful in an attempt to cure a non-ulcerating cancer, he does nothing more than to convert what previously was curable into an incurable cancerous ulcer. He hastens the patient's death. And when he cannot completely remove a cancerous ulcer he leaves an ulcer which is not just a simple ulcer." My view is that Avicenna did not mean that a cancerous ulcer could stop being ulcerated, a statement contrary to his own teaching and that of the authorities who have

---

[60] I can find no suitable interpretation of this statement. Perhaps Mondeville was explaining the behavior of the rebellious teenager in his own century (LDR).

[61] This curious apposition of terms, 'quartan' and 'double', may confuse the modern reader. John Halle, a surgeon of the Elizabethan era (fl 1550), explained it nicely in his edition of Lanfranchi's *Chirurgia*. "Quarta febris is used either for an intermittent or, for an exquisite continuous fever called quartan continua.....It is rarely seen, and sometimes is known as daily fever." (abstracted from pp 92-93 of his Interpretive Table (ie Glossary), see our bibliography for Treatise V, The Antidotary (LDR).

[62] Possibly herpes of the lips and the genitals (LDR).

dealt with cancer. Rather, he meant that after a cancer has been treated, a non-ulcerated cancer can grow at the same site, because the humors that formerly produced the ulcerated cancer continue to work.

XII. Avicenna (Bk. IV, Fol. 3, Ch. "on Leprosy") said that a cancer in a single place can be dealt with as if it was leprosy simply because it also is beyond cure. When you treat lepers with potent medications you do so to sustain the patient's spirits but you do not affect the lesions. That is not the case for cancers.[63]

XIII. Rhazes (*Almansor*, Bk. VII) wrote, "When you treat a cancer early in its course knowing that it is a cancer that has not ulcerated, perhaps it grows very little at first, and then over time it increases and ulcerates, that is a bad affair. If it impinges the respiratory passages it will lead to a day of anguish and terror and death." etc. And he added that such is a well known sign of incurability. And, further, when a cancer has grown larger, you should avoid heating it and causing it to spread inward as well as on the surface. In his book (*On Divisions*) he advises against the patient lying on his cancer when he is in bed.

XIV. All the authorities and all the books and all the practitioners agree that the sublimate of arsenic is the corrosive of choice for treating a curable cancer; it is very potent and can be precisely placed to corrode more at one time than other corrosives do in two. If need be, it can be attenuated for use in a less aggressive cancer, or for use in a nerve-bearing or sensitive region, or near noble organs, or in debilitated or delicate patients, in hot seasons and in hot countries. Grind it and mix it with guimauve, juice of plantain, et al. (ie to attenuate it).[64]

XV. More about the established methods (the Artificial): Consider here certain empirical facts, as we often do. Many items are thought to be empirical by some people simply because they have no ready explanations for them, and the same items are said to be part of a surgeon's (artificial) armamentarium by those professionals who can offer rationales. The *Aphorisms* of Urso[65] give rational explanations

---

[63]   Where you direct your attention at the lesion (LDR).

[64]   This seems to contradict Mondeville's statements that one can cure a cancer only by a radical extirpation (LDR).

[65]   Urso of Calabria, fl. @1150, A physician and philosopher at Salerno (LDR).

for things that commonly are accepted as unproved, such as why a magnet attracts iron and why a red shield placed over a deeply submerged cadaver in a stream will resist the flow and remain motionless, and reveal the location of the body which could not be found in other ways. In his *Aphorisms* we can find explanations for many things that seem even more astonishing, where the common folk have beliefs and fears, and the savants can identify specific causes.

XVI. In our chapter on empirics we mentioned the medications in the small sacks hung about the necks, etc. We questioned the usefulness of some of them, as testified by Constantine in his book (*Incantations, Magic, Sorcerers, Evil-doers, Remedies Suspended From Necks and Other Parts of the Body*). He offers proofs from authorities, citing Aristotle, and Medical authors. Avicenna (Bk. II, Ch. on Coral), said that Galen hung coral over the outlet of the stomach (ie epigastrium) to treat severe pain, and the pain subsided at once, only to return when he removed the coral, and to ease again when the coral was replaced. He tells of many marvelous experiences with coral and he describes other empirics used successfully in certain desperate cases in which all the standard prescriptions had failed. Therefore, Constantine supported the proposition held by the ancient philosophers, that the vital force (ie virtue) of the spirit controls (modifies) the body. The statements of The Savants prove it. Plato wrote, "When the human spirit believes that something will avail it, a thing that itself offers no benefits, it becomes useful; simply by its effects on the imagination it will help the body. In the same way, something felt to be harmful will do harm, because the complexion of the body befits the virtue of the spirit. That is why, when a physician stirs the spirit of his patient with incantations and the like, and he offers great assurances while he treats the patient's body with suitable medicines, he hastens the cure." Do the incantations etc. have any direct therapeutic effects? We seek answers from Constantine (ibid.) and Ovid (*Ars Amatopia*). The latter wrote, "Whoever asks for help from our Art, extends his faith to include magical beverages and incantations."

The fact that stimulating the spirit also stirs the body is illustrated in the following. Anyone who tries to walk on a narrow rail will fall if he thinks that it will happen, but when that is not on his mind, he will succeed. Similarly, after a bad fall, some will escape while others will die, even without obvious injuries, as I once saw happen at Paris. A man and his friends met someone on the Marmoset Street, and he said to his friends, "Do you see at that fellow? He thinks that I detest him enough to kill him, although I really don't dislike him at all, but I am going to cast some fear into him without touching him." He drew his sword and made a powerful thrust at the fellow, but did not touch him, and the man immediately fell dead.

Here ends Doctrine II of Treatise II

TREATISE III

The Treatment of Maladies That Are Not Wounds, Nor Ulcers, Nor Diseases of Bone, Yet Are Relegated To Surgeons For Treatment.

This Third Treatise, as I stated in my General Introduction, will contain the treatment of all those disorders that are neither wounds, nor ulcers, nor lesions of bones: maladies that affect all parts of the body and which usually fall to the surgeons to treat, by common consent, whether or not they are outside the usual field of surgery.

The first Doctrine deals with certain surgical evacuations that are part of our treatments for diseases and for maintaining good health: incision for drainage, cauterization, etc. Also we include other pertinent topics in our art: the embalming of dead bodies, the amputations of gangrenous limbs and the decorations (ie cosmetic procedures). Besides local lesions in various parts of the body we include red spots on the face (rosacea).

The second Doctrine considers the general treatment of abscesses and the detailed treatment of specific abscesses,in the head, limbs, feet, etc.

The third Doctrine will demonstrate the treatments of some disorders of the special regions, which usually are not found elsewhere, such as tinea of the scalp, blindness, and blockage of the breast ducts with clotting of the milk.[66]

---

[66] Nicaise followed here with the Rubrics for the entire Treatise III. Our arrangement is different. I have placed a long and detailed General Table of

SPECIAL PREFACE

I, Henri de Mondeville of the School of Paris, author of this *Chirurgie* and surgeon to the King of France, was in the service of our King and his court, with others of my contemporaries, responding when the circumstances and the need called us, in default of which we were held to account. My situation worsened after some troublesome regulations were laid upon us at the Royal Palace due to the bashfulness of Her Royal Highness. I had completed the first and second Treatises of this *Chirurgie*, helped only by the grace of God from whom all knowledge flows and all good, and I read them soon afterward in Paris in public in the year of Our Lord, 1312. I read them in the schools without being paid, before a large and very distinguished audience of medical scholars and others.[67]

Then, to my despair, I lost a lot of time in Arras, in England,[68] and in many other parts of the kingdom, following the armies on orders from the King. I hoped that I would be paid what I deserved. However, on another royal order, the debt was cancelled. I returned to Paris and I appeared at Court only at intervals. I really tried to take up this work again bit-by-bit, which I had let go, recalling the folk-proverb, "he who has time and waits for more usually loses what he has."

Contents at the beginning of both Volumes; they include the Rubrics for Doctrines I and II of this Treatise. The Rubrics and the sad Introduction for Doctrine III of Treatise III represent the only parts that have come to us, and they are set after Doctrine II as part of the main text (LDR).

[67] See Frontispiece in Vol;.I. The Master 'reads' in Latin but he comments in French to the students who understand little or no Latin (LDR).

[68] England: Mondeville probably meant Flanders (EN).

Kept busy in Paris, thanks to the broadcast reputation which I enjoyed among the students, citizens, courtiers and travelers, I barely found time to write even one line a day, not counting the need to go to the Schools and to spend entire days running back and forth to earn my keep. Thank the Lord, who was not stingy I resumed my writing; my surgical practice provided what I and my small household needed.[69]

I had three reasons to take up this work again. First was the great interest shown by my contemporaries and by the younger generation. Just as did the generations of our forefathers among the philosophers and savants who offered their own work for the well-being of those who were yet to arrive-on-scene, I, too, directed my labors and my studies (meaning that what I had already learned was not enough for the task and that I had more to gather) to be carried forth for the benefit of generations to come.

The second reason came to me from the Holy Scriptures. "He who sees his fellow man dying of hunger and does not feed him even when he can, he is to blame for his death." So being neither envious nor ambitious nor avaricious and not wanting to take the whole world for myself, and being content with having enough for my needs, I saw that our surgery (ie the new method) was needed by our human family and that it was not being properly disseminated, and I felt myself capable of setting things right. I was beholden to no one (ie having left the royal court), as you may judge in the following. I was not married and therefore I did not have to shoulder the burdens of a wife's expensive tastes or of a palatial household. Thus I returned to the writing of this "Practica"[70] and to turning out other similarly useful works. I feared Divine Judgment, if, for all the above reasons, I was held to account for the ignorance of my colleagues. So it was that I decided again to take up this work.

---

[69] His family included servants, the house itself, etc. Mondeville had no wife (EN).

[70] 'Practica. Here Mondeville mocks himself. The Practicae were simple handbooks or manuals of limited scope and small value as compared to his own major text. See Vol. I, p. 48 ff (LDR).

The third reason was that I had begun a work and I had not completed it, although there was time enough for it. I felt that one or another blame could be cast, such as: 1. If from the very beginning I was unable to continue the work, I should have known the limits of my ability in advance of the undertaking. 2. If I had begun the work and was unable to or did not know how to complete it, I should have had some idea of what would be needed to get it done. 3. If after I began it I lost the will to complete it. Then, before taking it up, I should have found the inner drive to work steadily. It is self-deception that leads one to undertake an enterprise which he sees he cannot carry on to its end, lacking perseverance. Perseverance will find its reward.

The Lord knows the real reason, other than those already stated, why for so long a time I put off completing or editing this *Chirurgie*. It was in order to do a better job, to be able to experiment with the arrangement of topics, to think about it and to see it before putting it out as a well-ordered book. But since I thought my death could prevent it, and since nothing is more predictable than death and nothing is as unpredictable as the exact hour of death, I looked to find a surgeon among my contemporaries who was up to the job, someone who could write well. Was there someone who was not entirely given over to personal gain, who would sacrifice even five sous of his usual earnings just to undertake writing a book that would be useful to everybody? As for me, as I have said, I was unmarried, uncommitted to any person or any job, receiving no subsidy (ie as a courtier) for my expenses, I would not turn away from the task which I set for myself.[71]

I was impelled by the fear of dying, fearing that my death would

---

[71] That passage is important for the biography of Mondeville; he shows us that he was not permanently attached to the king's service, in spite of his title as royal surgeon, and it supports what he stated at the beginning of this Introduction (EN).

[72] H. de M. Tells us in this Introduction that he had completed the first two Treatises of his Chirurgie in 1312 and that he had read them at the schools of Paris. Then he explains why he waited such a long time before completing

leave my *Chirurgie* forever incomplete. If it pleased the Lord. I would undertake a draft of what remained to be written.[72] First I invoked Christ's assistance, that he clarify and expand my beclouded spirit, then in want and ill-disposed; that he lighten the burdens of the work, and that, enlightened and empowered by Him, the work would reach a happy completion, that it would be free of reprehensible errors and be as perfect as possible, and that it would enjoy the praises and glory of all who are in heaven, in the interest of all the people of our time and during the centuries to come.

Now let me explain the cumbersome title of this Treatise.(ie The Treatment Of Maladies That Are Not Wounds, Nor Ulcers, Nor Diseases Of Bones, Yet are Relegated To Surgeons For Treatment).

Frequently it is necessary to erect fences between the inherited properties of brothers or of other members of the same family, to avoid strife. In spite of those boundary-markers, occasionally someone who is misled by cupidity and avarice lays claim to the crops of another. And sometimes he is taken with hatred and a wish for the death of his enemy. Similarly the same (ie 'turf wars') happen between surgeons and medical doctors, and sometimes among the latter alone. The medical writers anticipated the surge of envy, and avarice in their profession and tried to ward off the coming dangers, They knew as did the Philosopher (ie Aristotle) who had written in his *Ethics* about the potter who is a good friend of another because they share a craft, that he will sometimes hate the other, even when blood or friendship had bound them, when it became a matter of one taking the other's riches.

To keep peace, the Physicians on their own, claiming God's Will and Justice and Common Sense, fixed the limits for all categories of practitioners by defining which diseases each may treat. We describe

---

and editing his work Finally he decided "to draft what remained to be redrafted". But we are at a date far enough from 1312, certainly after 1316, established by the death of Louis the Hutin whose embalming he speaks of later on (EN). Therefore he set down a detailed table of contents for his planned Treatise III (LDR).

that in the second Notable preceding Doctrine I of Treatise II. They alloted to Medicine the first two modalities of therapy and to Surgery only the third, and it was agreed to by the majority of their authorities. Accordingly, Medical Doctors must prescribe drugs and order general regimens (ie mostly diets), while the Surgeons may only do manual operations. Therefore, illnesses that lend themselves to potions and diets alone must be treated solely by Medical Doctors. Illnesses requiring operative interventions are to be treated only by surgeons. All illnesses that require both modalities are to be treated by a company of both.

Yet, neither the medical doctors nor the surgeons were happy with those limitations. On the contrary, some medical doctors selfishly wanted to monopolize all methods of treatment without distinction, while the surgeons tried to take away cases from the physicians. That conflict had this result in the West (ie France and the Low Countries), although not in other countries. The Western people declared this to be right r and they decided what has come to be the accepted practice, to wit: All diseases that appear on the external surface of the body as a whole or on a part, such as wounds, ulcers, abscesses, scabies, diseases of the breasts, hemorrhoids, impetigo and the like, surface lesions on the head, arms, thighs and legs, and where the cause is not apparent but the manifestations are external, such as joint pains, dim vision, deafness, painful hands, etc.—all those ailments are to be treated by surgeons. So it has fallen to surgeons from then until today. On the other hand, illnesses within the cranium without external show, or within the body (excepting stones), hydrops and such-like were determined by the people to belong to the medical doctors alone, to whom the sick have sole recourse. That decision pleases all of us surgeons. May it endure through the centuries and remain inviolable. Let no physician try to infringe the pact or contravene it, through his own rash audacity. Let the perpetrator know that by his action and by fiat of the people he will be under sentence of excommunication, from which he will be delivered only by begging grace from the surgeons in the event that he breaks his own thigh.

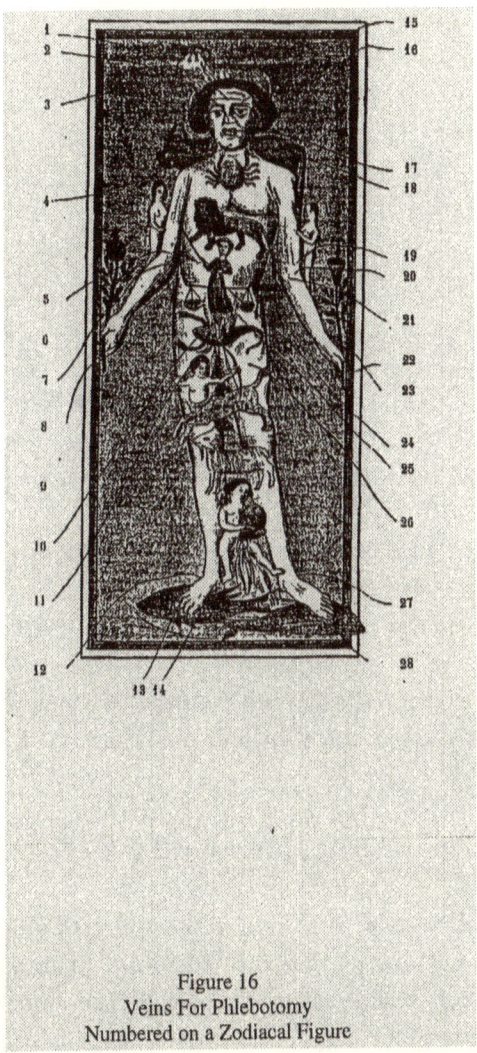

Figure 16
Veins For Phlebotomy
Numbered on a Zodiacal Figure

Figure 16
A ZODIACAL MAN

A Medieval Miniature which demonstrates the Twelve Regions of
the Body and the corresponding Zodiacal Signs
From Fol. 1 of the 15th c Latin Ms 6910A
Bibliotheque Nationale de Paris
see p. 662

TREATISE III
DOCTRINE I

CHAPTER 1.

THE DOCTRINE ON INCISIONS THAT ARE NECESSARY AND USEFUL: HOW TO MAKE THEM ACCORDING TO THE SCIENCE OF MEDICINE AND SURGERY

We will study six subjects in this chapter.

I. The Lesions for which we make incisions:

We use incisions in certain wounds, in some abscesses, in treating ulcers, stings and venomous bites and for the removal of excrescences.

II. The General and The Special Uses For Incisions:

Three General Uses: 1. To separate connected things, such as cutting a thick scar-contracture to loosen it or improve the appearance. 2. To enable one to bring together abnormally separated things, such as restoring parts of a nose which have been pulled apart by making counter-incisions alongside (ie 'relaxing' incisions). 3. To remove superfluous tissue which is growing, for an example, an excrescence.

We have several *Special Uses* for incisions in the treatment of non-ulcerating wounds. We enlarge the openings to find and to remove foreign bodies such as missiles and bits of stone and bone-chips, as in head-wounds with cranial fractures, or to remove some bone in compound fractures in the leg, or to reduce extruded bone in such fractures, and to drain pus when the opening of the wound is too small, etc.

We use incisions into inflamed swellings for several reasons: To drain pus when there is no external vent, before its corrosive action causes damage in the deeper tissues. We drain other such superfluidities as an hematoma or broken down tissues after contusions.

We use incisions to treat ulcers and to drain pus or virus or venom. Also we remove proud flesh and cut away indurated margins from ulcers. We remove solid masses. We sometimes make total excisions to remove excrescences or we make scarifications and introduce corrosives and augment their destructive actions.

III. When To Incise For Phlebotomy

The Physicians say that there are two occasions when to use phlebotomy: one that is elective and the other when it is absolutely necessary. For the first, the Physician waits for temperate weather, when the Moon is not in a bad zodiacal sign, and after a laxative has been prescribed and when all other circumstances are favorable. It should not be when the Moon isin a ruminant sign (? Taurus). But when necessity forces the decision for immediate bleeding, there are no special taboos. When a patient's life is at stake, even when it is a time when you would prescribe a clyster in an apoplectic patient, or the patient is very young or very inebriated, or at an inconvenient hour at night, bleed him. The same sort of caveats hold for the surgeons: there aretwo occasions for making venesections: a time of election and an urgent time of need.

You may decide on an occasion when the disorder is in remission. Then there will be three considerations: 1. The Constellation. 2. The Patient. 3. The Surgeon.

As to the Constellation, the surgeon should be aware that the complexions of the season, of the day and of the hour should be temperate or almost so so far as is possible for the epoch, and that the constellation not be evil, as when the Moon is eclipsed or is in conjunction and is in the same zodiacal sign as a bad planets[73] or has a bad aspect, etc.; when it is late in Libra or early in Scorpio by eleven degrees; when it is in conjunction or opposition with the sun; when it is in a Sign that corresponds to the part of the body that you plan to incise.

The Key to Figure. 16 The Veins for Phlebotomy [74]

1. A forehead vein, used for migraine and diseases of the eyes
2. Bridge-of-the-nose vein for treating excessive tearing.
3. A vein behind the ears for treating chronic headache and depressed spirits.
4. A vein under the chin for treating painful eyes, acne of the face and nose, and painful jaws.

---

[73] See Fig. 16. Nicaise explained: The Ancients had observed that the sun, the moon and the planets were invariant in their motions in the zone they called the Zodiac. It was divided into twelve equal segments called Signs, each carrying the name of a Constellation which they saw there, and in which today we follow the precessions of the equinox.

Humans were considered to be a microcosm in which all the parts correspond to the parts of the macrocosm. Thus the body of Man was divided into twelve parts, each governed by a Sign of the Zodiac, that is, by the Constellation.

The first Sign, Aries, the Ram, governed the head; Taurus, the Bull, governed the neck; Gemini, the Twins, governed the shoulders and upper extremities; Cancer, the Crab, governed the Chest; Leo, the Lion, governed the stomach; Virgo, the Virgin, governed the abdomen; Libra, the Scales, governed the pelvis and the hips (buttocks); Scorpio, the Scorpion, governed the private parts and the perineum; Sagittarius, the Archer, governed the thighs and pelvis; Capricorn, the Goat, governed the knees; Aquarius, the Water-Bearer, governed the lower legs; Pisces, the Fish, governed the feet.

The Drawing in Fig 16, enumerates the veins usually used for phlebotomies, designated by Rhazes as proper for treating various disorders; the right or left side was to be selected to suit the side of the body where the lesion was located (EN).

[74] See Ch. 3 of this Doctrine for details concerning phlebotomy

5. The cephalic vein at the elbow for treating painful heads, eyes, ears and throat.

6. The medial-antecubital vein (the common vein) for treating headache and painful ribs.

7. The basilic vein at the elbow for treating headache, painful shoulders, the spleen, and to stem nosebleeds.

8. The vein between the fourth and third fingers on the dorsum of the hand, called the Salvatelle, for treating headache and quartan fevers.

9. The vein between the thumb and index finger of the right hand for the eyes and head.

10. A suprapubic vein above the penis for treating tympanites.

11. A vein under the penis for treating pain and swelling of the testicles.

12. The sciatic vein behind the lateral malleolus of the ankle for pain from hip to toe. and 14. Veins on the dorsum of the foot for treating pain in the kidneys, ophthalmia, testicular ailments, pain in the hips, flanks and buttocks.

15. A vein in the temple for treating disease of the eyes.

16. A vein alongside the bridge of the nose for vision, migraine and diseases of eyes.

17. Two veins on the neck for treating acne and scabies.

18. Two veins beneath the tongue for treating quinsy, and diseases of the throat and esophagus.

19. The Cephalic vein at the elbow for the head and the eyes.

20. The Cardiac (common) vein at the elbow for the heart and stomach.

21. The Basilic vein at the elbow for the head and the spleen.

22. The vein between the third and fourth fingers on the dorsum of the hand for diseases of the spleen.

23. The vein in the web between the thumb and index finger for the head and eyes.

24. A vein on the flank (side) for diseases of the groin.

25. A vein over the spleen (the tickling vein) for ailing lungs, chest and diaphragm.

26. Veins on the thighs to be bled after meals, to attract humors from the upper body.

27. The Saphenous vein at the ankle for painful hips, kidneys and amenorrhea.

28. A vein over the lateral maleolus at the ankle for sciatica, arthritis, ailments of the testicles, morphia, urinary retention.

That is the reason why Ptolemy [75] in his *Centilegium*, Prop. 20, said that cutting with the knife when the Moon rises in the patient's own Sign is a terrible act.

Selecting the right occasion insofar as the patient himself is concerned consists of never making a large or dangerous incision without letting him or his friends know, for fear that the operation will not be successful. Do not operate on a feeble patient unless there is a great need for it.

The Surgeon's role in deciding when to operate involves not doing it until he has readied all the gear that he may need including the means to arrest hemorrhage and mitigating the pain (Avicenna, Bk.I, Fol. 4, Ch.26). The surgeon must be sober and have a few assistants with him. If the patient or his company request a delay until they their own favored time and if their reasons are good, the surgeon should agree graciously.

When the time to operate is determined by urgent need, the same three elements are to be considered: The astrologic constellations, the patient and the surgeon.

As to the first: If the most favorable astrological time is not due for several days, you may have to overlook that constraint and operate at once. The decision may depend on the patient. Take the case when an arrow or another missile is lodged in a wound, in or near a vital region, and it cannot be removed without an incision, especially when leaving the object in situ could be fatal. And take the case of an corrosive abscess in or near a vital organ, such as a quinsy, when the expanding collection threatens to suffocate the patient. In such cases pay no heed to any inhibiting particulars; operate at once. In urgent cases the surgeon's decision regarding the proper time to intervene may be made in view of the patient's

---

[75]   An Alexandrian physician, a follower of Erasistratus, 3rd C (LDR)

or his friends' demand that he operate at once. He should concede, because if he refuses they may believe that he was afraid of complications because he did not know how to use the knife. Furthermore, the patient would not consent to an operation at a later time. Whatever may be the result of the operation-on-demand, he will have a ready excuse, whereas, if he refuses and operates at a later time, and the outcome is bad, he will have to retire from the case. And if he refuses to do the patient's bidding, another surgeon may be called on to operate and earn the fee and carry off the honors

IV. The Technique: Two Items

A. Two General Principles: 1. In some parts of the body, you can incise in the long axis to drain pus, virus etc. In that way you can avoid injury to important veins, arteries and nerves, which usually are aligned in that direction. A transverse incision may injure them and cause permanent damage and dysfunction. There are three exceptions: a. On the forehead where the long axis extends transversely from ear to ear, perpendicular to the rest of the body. When you must make a deep and long opening anywhere between the temples and you wish to keep it open (ie for drainage), incise vertically, that is, from the hairline toward the face or the reverse. If you make a horizontal incision on the forehead you may divide the muscles that raise the eyebrow, and it may droop forever. That will be of more serious consequence than an injury to some vessels and nerves with a vertical incision. However, when you need only a small and superficial incision, to but not through the muscle, such as a stab wound with your lancet when a suture will not be needed, you may make a horizontal incision without fear for injury to the muscle that raises the eyebrow. b. The second exception involves muscles that lie parallel to the spine. When you need to make a deep incision you should cut transversely, from the back toward the front of the chest or abdomen. In that way you will avoid injury to the important nerves that arise from the spinal cord and pass around the body, which, as Galen said (*de Interioribus*, Bk. I, Ch.7) are beneath the skin and muscles which protect them from injury from outside. However, you may make superficial cuts without risking the nerves. c. The third exception concerns certain emunctories. In the groins and axillae follow the general practice and make curved incisions, in part vertical and in part transverse. According to Avicenna (Bk. IV, Fol. 3, Doct. 1, Ch. 'Swollen Glands') we should attract (ie with topicals) humors more strongly away from noble organs toward the less noble. Failing that, we should use cups to prevent the malignant fumes from returning to the important organs. And only when that fails, do we use the arcuate drainage incisions. The curved ones gape and remain open indefinitely, whereas the straight ones seal quickly. However, when dealing with the emunctories of the brain, that is in the glandular tissues beneath the ears, make your incisions in the vertical axis of the neck, parallel with the great vessels, and lessen the grave risks of cutting them and causing terrible hemorrhages.

2. The second General Principle: Whenever you incise to get rid of solid pieces of such useless or harmful things as darts or bone-chips, or to remove granulation tissues, or to trim away scarred margins of ulcers, to remove excrescences such as warts, nevi etc., or when you think incisions are necessary because there is no other way, you may not always need to make a straight cut. Indeed, when you always do it that way, you may not get the results you aim for. Often, to suit the case, you should cut across or in an arc around the lesion or on a slant or transversely, or in several ways for some cases. Do not be constrained by certain details or fixed rules. Do it the way that will be successful.

B. Special Methods

These are designed to suit the many different conditions. Let it suffice that you know about wounds and aposthems: the incisions will be discussed in the chapters devoted to each of the various disorders. Then apply the General Principles and deduce what you should do. And you should apply the lessons and the precautions gleaned from what follows.

As to the incisions you make to excise the indurated margins of ulcers and to drain superfluous pus, simply do what we taught in the special chapters on ulcers.

As to incisions to remove solid foreign materials that happen to fall into ulcers—fragments of bone, fish-bones, a bit of metal from a nail, a piece of a shell, a sliver of glass, etc.—when you cannot retrieve them except through an incision, there are four methods for you to learn. Sometimes you cut only the margin of the ulcer; sometimes you incise into the base, sometimes you cut along a side-wall and sometimes you incise all the way from the orifice down through the bed at the bottom.

When operating at the orifice, follow our rule and make the incision just large enough to reach and to remove the foreign body. If your initial incision is insufficient and you need to enlarge it, first pass a large needle with a strong thread through opposite margins of the opening; traction on the threads will steady the edges while you cut into the ulcer to allow you to remove what you seek. You will not find a better way, as Avicenna wrote (Bk. I, Fol.4, Ch. 28). But, if this method of traction and dilation is not adequate, you may have to extend the incision in length or width, as you decide. When you incise the base or the side walls of an ulcer, use the methods we teach in the chapters on wounds, ulcers and abscesses, and abide by the precepts given there.

I believe it is never necessary to incise the full length and depth of an ulcer. If the bed of the ulcer is not very much below the orifice, a large incision is not warranted, because a simple opening at the orifice will allow you to introduce to the bottom all the material you may need: detergents, corrosives and surgical instruments (ie tenacula etc.). With properly applied bandages, compresses and other things that the modern surgeon knows about, you can evacuate unwanted liquids that accumulate in ulcers and then you easily can inspect the entire ulcer. The same is true for solid matter that Nature will extrude with some help from your attracting medicines introduced through the enlarged orifice.

However, if the ulcer is very deep, be aware that it will be difficult to make a full-depth incision and not injure nerves etc. Furthermore, the ulcer tract may be tortuous. That is why I think we should seek and use alternate measures before we declare the need for an incision down to the bed of the ulcer to (ie to find and remove foreign bodies), as we stated above. [76]

The operations for removing excrescences vary to suit the lesions: different sizes, cystic or solid, pedunculated or flat, slim-pedicled or broad, involving noble structures, such as the face, or less noble such as the thigh. For small flat lesions with broad pedicles, as are some warts, we usually need only a small superficial incision, such as we use for scarification. We then can introduce corrosives as we have described. For excrescences, large or small, with slim pedicles, when there is small concern for bleeding, as there would be for a cancerous lesion, you should cut the pedicle at skin-level. For those that are flat or concave, or are cystic as are some glands, or are nodes, scrofulas, turtles and those that are large enough to stretch and bulge the skin and subcutaneous tissues we use several kinds of incisions. One that I disapprove of cruciate, and I have five reasons: a. The horizontal incision cuts nerves, muscles and membranes and weakens a limb. b. It cuts arteries and veins and deprives the affected region of nutrition and vitality. c. Often there is hemorrhage. d. The patient is uncomfortable with the incision and e. When masses are removed through transverse incisions that cut through skin and deeper tissues, ugly scars ensue. The mass stretches the overlyng skin which encloses and restricts its growth. After the mass is eliminated, the scar is wrinkled. Others use an angled incision but do not raise a flap, and I object to that, too. Others use a single incision in the long axis of the part but have trouble removing a large mass, and if then they do not remove any of the overlying skin, the scar is wrinkled, and the patient suffers a hundred-fold when you cut the skin again. Others amputate the mass and all the overlying skin, and the scar is ugly.

We modern surgeons, finding defects with all the above, are led to use the following proper method. First we examine and measure the lesion and the amount of skin and flesh over it, and we estimate

---

[76]   In this discourse we infer that Mondeville deals with deep ulcers, such as the tract of an arrow, and the retention of a piece of the arrowhead. The 'ulcer' is not yet a 'fistula' by definition, because the lining has not yet reached the state of induration which characterizes the latter (LDR).

the degree of stretching caused by the underlying mass. Indeed, the excess skin and flesh will remain and make a heaped and wrinkled scar, resembling a purse, as if some of the excrescence remains within.

To avoid that, you must excise the excess skin with the lesion or cut it off with a scissors after removing the mass. I disapprove of that last move for three reasons: a. You double the pain: an incision with the knife and another with the scissors, which is worse than the first. b. The scissors is not as accurate as the knife. c. If you leave the mass attached to the superfluous tissue, you can grasp the mass firmly and make the total removal easier (ie and more accurate).

Therefore, after carefully measuring the lesion etc. I mark the skin with ink to outline the extent of the excision with a semicircular line on both sides of a vertical straight line drawn over the center of the mass. The two arcs meet at the ends of the straight line. That double arc marks the margins of the skin to be excised, leaving behind the right amount to produce a nice scar. Then I proceed to operate so: I grasp the mass and its overlay with my left fingers and pull it up while I cut through the skin and subcutaneous flesh only along one arc. Then I change hands and cut the other side. I take pains not to cut into the lesion, and I avoid entering a cyst. The rest of my instructions will appear in the Special Chapters; here we deal only with the incisions.

V. Precautions to Observe

Eight General Rules

A. "Do not incise if other measures will do", wrote Avicenna (*de Auferenda*, Bk. I, Fol. 4, Ch. 27), and "do not hesitate to incise when the other measures fail." In past epochs that was the case and it was common until not so long ago. But we moderns have all but given up that rule, and we find gentler ways, as Constantine (*The Eyes*) taught in dealing with phlegmons of the eye.

B. When you use the knife, do it as gently as possible for sake of the patient, knowing that pain saps his Vital Forces.(Galen, *Prognostics*, Section 2).

Make your incisions as small as possible, just large enough for your purposes. The reason is self-evident.

D.   Have ready, before you cut, everything that you may need to staunch bleeding, to revive a syncope, to relieve pain and to deal with any complications that may happen after any large incision (Avicenna, Idem. Ch 26).

E.   Before you operate bid leave all the assistants who have no experience with major procedures. 1. A crowd of useless bystanders will get in your way if you encounter trouble. 2. One of them may faint and cause the observant patient to faint from fear. 3. Even if they do not faint, their horrified expressions will alarm the patient who will think that things are not going well.

F.   Do not perform a dangerous operation until you request the attendance of some of the patient's friends as witnesses. 1. That will reassure and comfort the patient. 2. Whether or not they comply with your request to be present, they will be satisfied with it. 3. Whatever may happen, they will not accuse you for operating badly. 4. They will be satisfied that they themselves have seen to it that their friend has been treated properly.

G.   When the dangers are obvious, the surgeon should not operate, unless the patient and his friends plead him to do it, and unless his fee is guaranteed in advance, and whatever the outcome, he will be absolved of responsibility for any complications. Then he may operate with confidence and without constraints.

H.   When the risks are moderate, the surgeon should operate only after the patient has made his confession and has completed the duties required by the Church of any person who faces possible death.

Nine Special Precautions in Certain Situations

A.   When a lot of pus is to be drained, take measure of the patient's capacity, and do not evacuate it all at once. Galen (*Prognostics*)

said, "a complete massive evacuation of pus will deplete the
patient's vitality."

B.   When you want to drain only a part of the pus, make a medium-
     size incision, not large enough to empty the whole abscess,
     and not too small, because the opening will seal too soon.
     When the drainage diminishes and the patient's strength is
     restored you may enlarge the incision if that is necessary.

C.   Whenever you have to make a second incision for drainage at
     a distance from the sealed primary one, you should let some
     pus accumulate in the abscess. The pocket of pus will identify
     the extent for the surgeon and will prevent him from cutting
     too deeply and from injuring the nervous structures beyond
     the pus.

D.   Whenever you make a secondary incision for drainage, insert a
     flat blade (ie a taste) covered with a piece of linen cloth into the
     first incision, to lie beneath the place where you will make the
     second incision. You can elevate the tissues and guide the knife.

E.   The taste should have a perforation at one end similar to the
     eye of a needle. If need be, you can thread a strip of cloth and
     pull it back through the primary incision (ie as a drain).

F.   Before you make an incision for drainage, palpate the region
     around the abscess with firm pokes, to cause pain. That will
     cause more pain than the incision to follow and perhaps the
     patient will not feel the knife.

G.   When you must make a large incision in a region where the
     abscess has distorted the normal anatomy and you cannot be
     sure where the nerves etc. lie, first examine the normal limb
     to orient yourself and make your incision in the diseased limb
     to correspond with the place you would have cut in the healthy
     limb. Usually it is the case that the structures are symmetrically
     placed in both sides.

H.   When you need to remove some skin and subcutaneous tissues
     with an amputated excrescence, don't excise it because you
     think it may be redundant; leave some of it, more than you may
     have estimated at first. Nature usually tends to retract the
     margins of the small opening in the normal skin that you create

when you amputate the lesion; it will enlarge considerably, to be healed with a larger and more deforming scar. The lost tissue will not be replaced.

I. When you have to drain a small abscess, grasp the mass with the fingers of your left hand and pinch it away from the body. That will isolate the pus and empty it quickly when the incision is made.

VI Seven Explanatory Comments

A. A thorough knowledge of the theory and practice of incisions is necessary for every operating surgeon. Indeed, it is more important than any other Doctrine of our Art. Ignorance of even the smallest part of it makes one unworthy of being called a Surgeon. Furthermore, one can understand that it is better and more difficult than any other doctrine of surgery, and that one needs the knowledge before and during his most important treatments. With the precise knowledge of an operation and its theory he can accomplish the most lucrative and the most beautiful cures. Ignorance is the basis of the most serious risks for both and the surgeon and the patient who may suffer death and bring shame. In this claim, my major premise is testified to by the fact that an incision can separate as well as help to rejoin parts; it can extirpate what is redundant in wounds, abscesses, ulcers, bites and venomous stings. With it one can remove foreign matter and excrescences. In a word, whoever knows how to operate according to our principles, in all kinds of surgical disorders, knows how to treat every surgical malady because any of them may be suitable for operative treatments.

B. The two reasons why the Doctrine of Incisions, a subject that is applicable to all of surgery, is placed here in my Text and explained in this one chapter, rather than here and there in bits, are: 1. The material is easy to find and refer to in one place. 2. In one place it is more complete and emphatic. Indeed, all incisions, no matter their differences and the variety of circumstances that merit them, all of them have much in common. The doctrine for one helps explain that for another. It should not be delivered piece-meal in many chapters.

C. Avicenna discussed the art of making incisions (Bk. I, Fol. 4, Ch. 26). In Part II of his *Cantiques* he listed the ailments that are treated by incision, but he did not decribe how to do it. But he does mention it in several places in his *Chirurgia*.

D. Whenever you drain pus or virus, make a small incision in the long axis and drain the matter bit-by-bit. That will do it better than a transverse incision. All things being equal, your control is better and you can squeeze out pus from all parts of the cavity. It is different for solid matter which can be lodged in any direction and you should make your incision in the direction that you deem best for exposure and removal, using longitudinal incisions when feasible.

E. A clever surgeon can change the scheduled time for an operation, moving it up or delaying it by making excuses. He may seek a delay by saying he awaits a more favorable opportunity, or he may insist on doing it a day ahead of schedule when he is concerned that they will find another surgeon somewhere whom they would not call on for care after the operation. The original surgeon may say, "I have anticipated the need for an operation for a long time, but I have not talked about it lest I alarm the patient while waiting for the proper signs. Now, after a long wait there is pus to be drained before it corrupts the entire limb. And we have had to wait until now for the most favorable time set by the constellations. Right now, by Grace of God, is the best time; further delay will cause grave damage. I am ready now, because the operation should be performed at once.

In the other direction, the surgeon may delay the operation byond the appointed time by pretending that he is waiting for a more favorable opportunity, although in reality that could be worse for the patient. He will delay when he has not been paid the expected fee in advance. After the operation the patient may claim that he suffered severe pain or that a fever had intervened and that he expects convalescent care and delays his payment. Or the surgeon may seek a delay because his assistant has been delayed at another case, and he must be awaited. Another reason for him to delay the operation is his awareness of the grave risks, and the surgeon may consult a colleague and they will appoint the day and hour for the operation. Then the first fellow may claim that he has received an important message or a letter that forces him to leave the case temporarily, and that he will return. That is a pretty good excuse, and he may appear to be very upset about having to leave. Meanwhile, the operation is delayed and the surgeon is free to leave. When he returns after the operation has been performed he will know the result. If the outcome is bad he will abandon his colleague and not reappear on the scene.

F. Definitions: We use the word *Incision* in two ways: 1. The physicians call any treatment an incision which is an action of a medicine which eats its way through solid or viscous matter to reach inside a body. It separates and dissociates tissues, hence their term.

2. The second definition is surgical: an *Incision* is a particular surgical maneuver which divides and separates a continuum with a cutting instrument, such as a razor or phlebotome. There are two kinds: One is a chance cut such as a battle-wound. The other is made by surgeons to drain abscesses, etc. That is the only meaning of the term used in this chapter!.

G. Need for Assistance: A surgeon who operates can cause bleeding, a slow ooze or a deluge. Therefore, he should do it in company with a colleague or an assistant who can use a sponge or other absorbent materials to mop away the blood and keep the field dry. Two reasons: 1. To allow the surgeon to see, and 2. To avoid frightening the patient and his associates.

CHAPTER 2

ON CAUTERIES

Their Necessary And Elective Uses Where The Skin Is Intact Including Removal Of Excrescences And Draining Aposthems Anywhere In The Body According To The Arts Of Medicine And Surgery

As indicated in the title of this chapter, we will exclude the applications of cauteries which are not surgical. Our concern is with the metallic instruments and substances such as cantharides (ie vesicants) and eruptors. We shall exclude the treatment of plethoric persons who have not had preliminary purgation and healthy persons who have nothing special to prevent and no diseased matter to evacuate, who simply want a cauterization as a passing fancy. We exclude cauterization as follow-up (ie touch up) measures for healed wounds, ulcers, cancers, fistulas etc in whom some skin had been lost. And, finally, we exclude cases where superficial growths have been extirpated or aposthems have subsided without need for open drainage. We will deal with those topics elsewhere in special sections. Here we shall limit ourselves to the use of metallic cauteries, cantharides and other caustics when they have definite artificial indications.[77]

Five General Items

A. *The Description*: Four Points:

1. Definitions: A Cautery is an instrument for special surgical procedures, which is beneficial for preventing certain maladies in predisposed persons, for palliating and for curing maladies already in effect. We will name them as we come to them. To cauterize, to burn, to cook and to 'filet' as the peasants say, are four terms with the same meaning. To cauterize is to burn or to cook and to filet. The last suggests (ie as a pun) the need for prolonged diligent surgical care, as if he treated his own daughter, 'fille'.

2. The Two Varieties: The Actual Cauteries are metallic, iron or other, instruments that are hot enough to burn tissues. The Potential Cauteries are medicaments: A simple such as cantharides or a Compound such as various eruptors. They are not intrinsically hot nor are they heated for use, but they have an innate burning quality.

3. Similarities and Differences: They are alike in that both are destructive, and both burn, evacuate and consume. There are other minor similarities. They differ in that an actual cautery

---

[77] 'artificial' means 'according to the art of surgery', that is, according to the Rules (EN).

causes less damage; heat is a simple element and its effects are manifest only when its virtue is transmitted directly to a material substance. It works more rapidly and its injurious quality disappears more quickly. Its action is at the place of contact and does not extend beyond the member in which it is used. Because it has no distant effects, it does not attract humors from other parts of the body.[78] It can be applied in all kinds of weather and at any season, especially when the patient has acute pain, and in all sorts of ailments with all kinds of complexions, and the surgeon weighs all the factors and excludes some cases such as certain hot illnesses in which there is nothing to drain.

As I say, it can be used in any weather and at any time of the day in all maladies caused by humors that have not yielded to other measures. The cautery may be applied as a last resource, as stated by Rhazes (*Albucasis*, Bk, I, Ch.1). We apply the iron after all other medicines have failed. Rhazes continued, saying that the 'any time, anywhere' tenet held if other circumstances permitted it. That is, the benefits of the cautery are greatest when the chosen time is correct, especially when you use it to relieve intolerable pain.

The Potential Cautery produces a large lesion because its effects are not immediate; some time is needed for it to do its work. The natural heat of a region has been activated and it waits for the burning potential of the medication to take effect. Thus there is a prolonged conflict between the body's spirit and the burning power of the medicine applied on it. That combat weakens the forces of Nature. As Galen said (*de Ingenio*, Bk. III, Ch. 1, and *Aphorisms*, Part 5) suffering attracts humors from the entire body toward the source. In that way the distant parts suffer by depletion and the cauterized region becomes an inflamed mass (ie an aposthem). Furthermore, the lesion caused by the caustic endures; the painful local inflammation

---

[78] Mondeville contradicts this statement later, when he describes diverting cauterizations (LDR).

cannot just be waved away. All caustics have some venomous qualities, and the evil residues linger in the body.

4.   Comparing the Two Types: three items:

a.   Actual versus Potential Cauteries: From what I have stated above and from what follows you will conclude that the actual cautery is favored as more reliable and easier to use. I offer ten reasons: 1. It is less destructive. 2 The burn heals more rapidly. 3. It is less attractive of humors. 4. It does not harm structures at a distance from the primary lesion. 5. It is useful in more kinds of disorders. 6. It leaves behind no venomous residue. 7. We can terminate the burning simply by removing the cautery which we control with our hands. We are more precise in directing the instrument. 8. Thanks to the foregoing we can also control the intensity of the application. That is not possible with the potential cautery; after it has been applied, it is on its own. 9. When we cauterize to stop bleeding from large veins we must be aware that the eschar will come away sometime later and the hemorrhage may resume. In such a case we can reapply the hot cautery, but we dare not reapply the caustics. 10. According to Rhazes (ibid.) the potential cautery often creates an incurable lesion, such as a lethal cancerous ulcer, whereas that never happens when we use the actual cautery.

b.   Comparing the various kinds of cauteries: We take counsel from the authorities: Rhazes (idem. Ch. 2), Avicenna (Bk. II, Fol. 4, Ch.29) and other illustrious writers all agree that the finest instruments are made of Gold. Then come the silver and then the iron and the steel. All other things being equal, the gold is chosen because it has the most temperate complexion and is more pure than silver. Silver is better than iron for the same reasons. That which is more temperate and more pure causes less damage. However, we cannot assess the

degree of heat in the gold and silver instruments because
their colors do not change when they are heated. On
the other hand, iron changes from black to white. All
cauteries function only when they are hot enough, but
when gold and silver are too hot they bend and they
melt. Furthermore, they cool rapidly when they are
removed from the fire. On the other hand, the iron
cautery retains its heat and retains its shape. For all
these reasons and with concerns for the inconveniences,
I generally prefer the iron cautery. However, if, by
chance an expert with gold and silver can provide the
properly heated cauteries we can avoid some of the
defects and use them. But, all in all, the iron cautery is
the most trustworthy.

c.   Comparing the potential cauteries. In some cases the
eruptors are better than vesicants (cantharides), in other
cases the opposite is true.[79] The eruptors are preferred
when we wish to attract humors from distant regions
and attract bad humors to the surface, with a long-acting
effect. After the eschar comes loose and the abscess is
evacuated, the drainage wound can be kept open for a
long time. Eventually a firm, dry, thick and dark scar
forms.

Cantharides is better when we want only to attract from
beneath the skin, not from deeper tissues. It has a more rapid
action and it does not form a crust; an excoriation remains after
the blister breaks. Those lesions drain suddenly and release watery
matter and they do not have to be kept open. They usually heal
within six days.

---

[79]   Eruptors and cantharides: Some intensely caustic substances eat their way
through the skin and cause the underlying collections of pus to burst through,
hence 'eruptor'. Cantharides, or the Spanish Fly was an irritating vesicant,
blistering the surface of the skin. More detail is given in the Antidotary.
(LDR).

II How To Use Cauteries

The Actual Cautery: Nine General Principles

1. Before using the cautery, especially if the patient is plethoric, evacuate the patient with purges, bleeds, etc., and apply all the proper topicals, ointments, etc.[80]

2. The operator must know the anatomy, the numbers, the location, the complexions and the functions of the site he will cauterize, as well as know the causes, the complications and the duration of the illness of the entire region in which he will cauterize. In some cases he will recognize that the cautery rarely will be of value.

3. In cases where persisting pain has not yielded to treatment with properly used evacuations and topicals, the actual cautery usually will work well after the necessary preliminaries are carried out.

4. Never use a cautery on a patient who is well filled with healthy humors before carrying out a proper phlebotomy, and until he has abstained from wine, and you have not observed improvement. The good humors usually are sanguinous and the phlebotomy and the abstinence are useful. Unless you follow those precepts, your cautery may do harm (see William of Saliceto, Bk. V, Ch.1).[81]

5. In fleshy tissues the cautery easily can penetrate too deeply. In nervous tissues and in places rich in arteries and veins, as at the elbow and in the popliteal region, you should press no deeper than the skin.If a cautery touches a nerve it will cause a contraction, and if it touches a large artery or vein it will provoke a hemorrhage.

6. If some of the old harmful matter persists in the body after an

---

[80]   See foonote 299, Tr. II, Doct. I, Ch. 2 (LDR).

[81]   Chapters 1, 2 and 9 of William's Book V are devoted to Cauteries. Most of that material, in turn, came from Albucasis' book on Surgical Instruments. William. was a teacher of Lanfranchi, who taught Mondeville.(LDR).

evacuation, apply the cautery on the knob in the hollow just below the knee (ie head of the fibula). If the malady persists in only one part of the body, apply the cautery in the hollow below but near the affected organ.

7. Even when properly used, the result of the cauterization may be less than satisfactory. When it is successful you need not repeat an application. However, if the application and the purges and the topicals all had been used correctly and the patient is no better off than he was before the treatments, repeat the cauterization. Hippocrates (*Aphorisms,* Part 2) wrote, "Do everything reasonable". After the second applications always prescribe a rich diet with good wine until the depleted vital forces have recovered. Then repeat the purge once or twice and repeat the cauterization and prescribe a diet containing foods that are contrary to the cause of the ailment. Repeat this routine several times if necessary, until the patient is completely cured. Remember, a persistent drip of water can erode a rock!

    When the initial treatment is partially effective, you can be confident that the malady is curable if you persist and follow the extablished rules. Reapply the cautery in the fontanelle (ie the hollow) where you had the initial success, or in another favorable place.

8. Comparing the actual and the potential cauteries: Never use a caustic if you can find another way to do what is necessary and is tolerable for the patient. If he is not cowardly nor unwilling to accept the hot instrument, and he has a very cool and moist complexion, use it. A caustic, which by its very nature has a poisonous complexion, will inflame and corrupt the complexion of the region where it is applied, especially when it meets up with a potent contrary which is very cool and very moist, and when it is applied during the winter. The place for the application of a caustic should not be near a principal organ in the torso, or near the heart.

9. In consequence if the foregoing, we conclude that the actual cautery is better suited for treating ailments caused by hot dry humors, and we never should use the potential cautery in patients

with those complexions with those ailments. After it has destroyed
the bad substances its actions continue and its lingering effects
are harmful in a body with a warm and dry complexion.

B. The Techniques For Using Actual Cauteries: Two Items

1.   The Instruments in Common Use: Rhazes (*Albucasis*)[82] described
     the many shapes and varieties; William of Saliceto listed six;
     Lanfranchi gave ten. They and others described what each had
     found useful in his day-to-day practices, but they did not describe
     the cauteries for special cases. That is not surprising, because
     every day we encounter new cases that call for novel instruments
     and new ways to use them. The types which they recommend for
     surgeons are those we should have ready-made and at hand
     keeping one or more suitable to the case, adding or eliminating
     some of them, or exchanging all of them for the special kinds we
     will need as we antipate it. You cannot put all of that in a book.
     Really, one can ridicule the idea that everything good already
     has been invented and that there is no room for anything new.
     Damascenus (*Aphorisms*, II, Part 3) said, "Native genius enhances
     our Art, and vice versa."
        The surgeon can and should vary his instruments to suit his
     intentions, and to befit the anatomy of the parts he will treat.
        Seven types of cautery have been in common use from
     ancient times until the present:

     a.   The Nut (ie node), used by the barbers and the
          untrained rustic operators. Usually, they put a perforated
          cold metal plaque on the place they will treat—sometimes
          they will substitute a piece of thick shoe-leather with a
          perforation—and they introduce the hot 'nut;' through
          the hole. That serves three purposes: a. It keeps the hot
          nut from slipping away from the desired spot. b. It avoids

---

[82]   Re Rhazes and Albucasis: see fn 234, Notable XIII, in Tr. II, Doct. I, and the
       Bibliography in the Engl. Translator's Introduction Vol. I (LDR).

burning the nearby tissues. c. A pin or cuff near the tip will be stopped by the plaque and prevent the tip from burning too deeply into the subcutaneous tissues. (see No. 17, p.524).

b.   The Seton Cautery (Nos 19-23). You pinch a fold of skin and subcutaneous tissues with the special perforated tenaculum, with holes set over the place you will cauterize. Push the red hot needle through the holes, and quickly pull it out and insert a cold threaded needle. The thread will remain as a seton after you knot the ends, and it will keep open the cautery tract as long as you wish it.

c.   The Round-Tip makes a slightly deeper burn. In other words, it cauterizes through the skin when you treat a lesion near a nerve. It burns at either end, with a large or a small tip.

d.   A Pointed or Needle Tip makes very small burns, as after the extraction of hairs to prevent regrowth. You bend it to the angle you need.

e.   Some make oblong burns, often large and deep. They are the Cutters, the Olivaries and the Tongue-like instruments, all with at least one cutting edge (Nos. 14, 15, 16).

f.   Some cauteries have a Rod or a flat Blade at the end. You heat it and insert it through a cold tube to avoid burning the surrounding tissues when you insert it to burn only with the tips, as in treating within the nose or mouth. (Nos. 18, 18a, 18b).

g.   One is called the Circular or Dactylary cautery, because it is a flat disc to which are attached several prongs (ie like bent fingers)[83], one in the center and several attached

---

[83]   No. 17 is labeled 'pointed' but corresponds to Mondeville's 'Nut'. with its stop-pin The instrument and its perforated plate is labeled 'dactylary'. What is called 'round' in the text corresponds to Appendix I, Item D in Wm. of Saliceto's Plate 1. The names in Mondeville's text should be used as translated here, whereas the Figures which are taken from Guy de Chauliac's text as written fifty years later see Vol. I. Appendix I(LDR).

to the circumference. The projections are heated and applied through the perforations in a cold metal plate which is made to fit the dactyls (No. 24), allowing the surgeon to apply them easily and accurately. The cold plate protects the surrounding skin. This instrument is most often used in treating pain in the hips and buttocks.

2. How To Cauterize With These Instruments

First gently palpate with the balls of your finger-tips and mark the spot for the application with red ink. If you use a perforated plate set the opening over the red spot. Hold the plate firmly in place and tell the patient to look away while your assistant hands you the red-hot cautery. Insert it for as long as necessary. The cautery may be used without the plate.

3. When, Where And What To Do After The Burn

Immediately apply a mitigative ointment, such as lard or butter or some medication that will ease the pain and comfort the patient. After it has subsided, apply suppuratives such as rancid lard or bitter porridges such as contain liver-wort and the like, until the eschar sloughs. Then put a small green pea in the defect and press it in. Later, after removing the pea, fill the defect with a pill of ivy-wood or similar, a bit larger than the pea. Leave it in until pus appears at the edge of the defect. Then begin a twice-daily change of dressings, gently wiping the area dry with a linen cloth before covering it with a folded (about twelve plies) cloth pad, atop which you lay on some leaves of ivy, oxalis or grape-vine. Place over all a disc of gold or silver or copper metal, or, even better, of sole-leather which you can lace in place with a strip of raw-hide. This will protect the cauterized site from further injury.

The burn wound should drain pus for exactly three months, unless there is a good reason to change the rule. If it drains beyond that time, one of two disturbances, or both, may occur. First, after the essence of the burn has evaporated, the open wound will attract all sorts of humors, good as well as bad, mostly watery liquids. That will revive the pain of the cauterization.[84] Second, there will be a chronic discharge from the wound which the patient cannot bear, and the surgeon cannot stop. He will not be able to discontinue his care of the patient as soon as he may wish, and, even if he is allowed to do so, he would not dare to. As Master Arnold of Villanova said, a very old wound should not be allowed to close lest a serious complication appear. That is, the chronic drainage of humors will no longer find a means of escape unless there is another wound nearby. That is why one should not prolong the procedure beyond the time when the cauterization loses its burning and consumptive quality and its virtue, unless there is a strong reason. Then no longer is it a cauterization, it has become an ulcer.

Rather than useful, it is harmful, except in two instances. First, when the overload of bad humors is too great for the cauterization to eliminate during the usual limited term, and the patient is too cowardly to permit you to make another burn nearby. In such a case, I prefer to keep open the draining ulcer rather than dam back the superfluidities. Second, the draining ulcer may be kept open in a person who had practiced hygienic phlebotomies.[85] In such a case, a chronic ulcer as well as an over-due cauterization may be useful in treating a chronic fistula which had not been cured with other measures.

When the cauterization has been concluded, we must let the wound close by itself, using no surgical measures other than pruning away excessive granulation tissues to allow a scar to form.

Therefore, when it is conducted properly, whether or not it has been partly or entirely successful, and when the hoped for cure has not yet been attained, you may repeat the procedure if there has been some sign of improvement, as we stated in the seventh rule. If the goals of the treatment have been attained, there is no need to repeat. If a good result is not complete, you will do well to repeat the cauterization.

When the initial site was not a good one and the patient had suffered excessively, go to a nearby spot, in the same fontanelle or in another close to it, where you should have burned in the first place.

---

[84]   Edema and cellulitis (LDR).

[85]   Who is familiar with the procedure (LDR).

Sometimes, you may go to an altogether different region when a repeated cauterization is deemed necessary.

Potential Cauterization. Six Points

A.   A Preliminary Statement: The surgeon should refer to the nine General Principles about Actual Cauteries and apply what is pertinent to the following material.

B.   The Medicaments: One eruptor contains 4 parts of quick-lime, 1 part of sticky soot (from a chimney, an oven or a cook-pot). Mix them with French soap (savon) until you have a waxy paste. Another: some cantharides mixed with yeast and vinegar, or sprinkle on yeast and vinegar after you apply the cantharides alone. Another: flammula and crow's-foot applied together or separately. Another: euphorbia (tithymale) and leaves of lupin ground together and applied until the eschar piles up, thick and dense. Another: a peeled clove of garlic with its tips amputated. Others can be found in our Antidotary.

C.   and D. Potential cauteries may be useful for patients with cool and flabby complexions. Indeed, the complexion of the caustic destroys that of the part being treated, unless it encounters an intensely cool and humid contrary.

E.   How to Apply an Eruptor: Form it into a ball of the right size for the case and place it on the site that you have marked, as we described it. Bandage it to secure it for twelve hours. Not much skill is needed for most such applications, except when you use the garlic cloves. For that, you make a small incision and push in the entire clove. The follow-up care is the same as for the actual cautery.

F.   The post-cauterization care is the same for all these cases as it is for the hot iron. You treat alike all cases with thick and dense eschars, except when you produce blisters with cantharides and other vesicants. You open the blisters with a scissors and the watery contents will drain. Then, lay on some leaves of cabbage and dress the area twice a day with fresh leaves until a scar covers the defects.

III. Where, When And For Whom To Use Cauteries: Two Items[86]

A. General Statements: 1. Anent the Region: When treating a recent painful lesion on the trunk of the body, even if the spot is very small, place the cautery directly on the sore place. You do it differently when the lesion is on one of the five appendages to the trunk, ie, the head and the four limbs. On the head you have six indications: a. To evacuate fumes. b. To treat chronic catarrh [87] c. To repercuss harmful humors that flow to the head. d. To evacuate the humors that have arrived there. e. To improve the resistance of the nerve-bearing structures. f. To treat maladies affecting the eyes, ears and other organs of the head.

To treat a. and b. apply the cautery to the anterior fontanelle on top, over the juncture of the coronal and sagittal sutures. As I will explain later, the fumes and gases that engender the catarrhs usually rise to that site. The other four cases are best treated by applying the cautery in a hollow of the neck (ie supraclavicular region) or in a hollow that is closer to the ailing part.

Cauterization of the arms and thighs and the rest of the limbs has three purposes: a. To treat disorders of the entire body as well as of the extremities. b. To treat certain organs, such as cauterizing the arms for lesions of the eyes. c. Cauterizing the legs when treating lesions of the genitalia and of the hips.

Apply the instruments somewhere between the trunk and the affected limb. In a case of a painful joint use a hollow three fingerbreadths proximal to the joint. That is, in a fontanelle or a channel through which the offending humors must pass from the

---

[86]   We note that the recommended sites for applying the hot cauteries often are in hollows or fontanelles, even the most shallow indentations. In the use of the cautery as an attractant, rather than as an incisor or destructive agent, the hollow represented a receptacle for the evacuated matter as well as a position closer to what the cautery intended to attract from below the surface of the overlying skin (LDR).

[87]   The catarrhal drip begins as fumes from the brain and finds its way into the nasopharynxe and descends to the lungs (LDR).

body to the affected part. There are eight such fontanelles for each of the joints, four proximal and four distal, between the ends of the muscular insertions described by the anatomists. You can feel the indentations with your finger-tips.

2. Anent the Advantages (ie The Why). We shall discuss only what is pertinent in this chapter and we will omit the other uses of cauteries. Here we deal with two issues: a. Cauteries alter and restore the complexion of the affected part, which had been disturbed by the bad humors. They resolve and consume the matter, thereby suppressing the bad effects. The affected part is relieved, as we observed in a case when the head and the brain are humid (ie catarrh), and are relieved by a seton-cautery applied at the occiput. b. Cauterization resolves and consumes matter in certain parts, as when a knee is puffed up with cold and gassy matter, we apply one or more cauteries. Indeed, there is no more energetic way to treat superfluidities than the application of the actual cautery. The cautery must be very hot and should hold the heat for a sufficiently long time to accomplish its purpose before its heat is snuffed out. If the heat is weak and the cautery is too small and only a single instrument is used it will not overcome the superfluidity, just as a weak fire will not ignite a pile of green wood.

3. Anent the ailments we treat with actual cauteries:

Let us set aside certain opinions and arguments that could fill our page; I shall limit my discussion to this: I believe that the actual cautery can be useful in all the disorders caused by bad matter when the complexions of the patient or of the affected part are bad. They are especially useful when the causative humors are moist and cool. Those qualities are directly opposed by the cautery's heat and dryness, which enter the patient where the cautery is applied just as all treatments that are contrary to the causes are advantageous. Cauteries are less useful in treating maladies caused by warm and moist matter, such as the sanguinous, and in ailments due to warm and dry matter, as those caused by bile. It has even fewer effects when treating the cool and dry ailments, as those caused by melancholy. The action of intense heat is to resolve and to evacuate the matter that is disturbing the complexion and to treat chronic ailments. However, it is not the only measure that treats by virtue of a contrary complexion; there are medications that evacuate bile and treat three-day fevers and other warm illnesses. After the patients get rid of the causative bad humors, they are cured. Rhazes (*Albucasis*, Ch. 1), said that he had sometimes observed the successful use of the actual cautery in illnesses caused by warm and dry matter where one could question its use. He recommended them as follows, "The human body and its superfluidities may be warm, yet when compared with the heat of the cautery they are cool. Therefore, when one uses it, he treats with contraries."

4. The Most Advantageous Sites for Treatment With Cauteries. In treating chronic headaches, epilepsy, paralysis, tumors, prostration, and all the nervous maladies that arise from the head and its region, all the cool and moist maladies such as migraine, vertigo, obscure brain disorders such as very deep sleep and dull sensibilities, post-nasal drip with harmful effects where the drip touches, such as toothache, gingivitis, sore throat, chest-pains, sick lungs and stomach; for pain and redness of the eyes, eyelids, ears and nostrils; for cough and diarrhea—for all these, first prescribe suitable purges and apply the correct topicals, and then apply the cautery to the anterior fontanelle of the skull. Identify it by placing the open palm of the patient against his forehead, with the wrist line against the base of the nose, and noting where the tip of the long finger touches the center of the scalp. Use a cutting or olivary which will penetrate more easily. If one application is not sufficient to burn through the scalp, repeat it with a hot instrument up to four times until you touch the skull. Keep the burn open with a ball of white wax until the surface of the bone flakes off, allowing the matter within the skull to exhale. If your initial effort fails, repeat the entire cauterization once a month for up to a year. I completely cured a nearly blind man of his blindness in that interval. All the authorities testify to the benefits of this cauterization, especially Jean Mesûe (*Practica*, Bk. V, Sect. 1, Part 2, Ch. 'On Treating Eyes').

In all the above-named maladies, if the head is so filled with bad matter that the cauterization fails, as occurs in lepers, apply two cauteries on the two horns of the head, over the flat sutures between the temple and parietal bones. Use the nut cautery. Some benefit is obtained in all these ailments by cauterizing at the back of the head where the spinal cord leaves the brain. Use a nut or seton cautery in the hollow beneath the occiput at the hairline of the neck to treat the catarrhs, the humoral disorders of the eyes, chronic nervous maladies such as epilepsy and chronic headache.

A round cautery applied beneath the ear or near it is good for treating ailments of the ear as well as the eyes, and toothache. Also, you may place a cautery atop a vein in front of the ear. That sometimes is a marvelous treatment for toothache. A nut cautery placed between the ear and the spine is good for lacrimation and other humoral disorders of the eyes and for paralysis caused by a lesion on the surface of the spinal cord. Cauterizing with a cutting instrument will eliminate superfluous tissue in the eyelids.

Direct cauterization in hair pores will eliminate unsightly hairs and retard their regrowth. A pointed cautery can destroy granulation tissues in the corners of the eyes and can be used to remove nasal polyps. It can be inserted into a lacrimal fistula to burn at the very bottom. It is the treatment of last resort for this disorder after all others have failed.

A round cautery placed beneath the chin is good for infections of the face and for all the ailments of the mouth and its contents. The same cautery placed between the lower cervical and upper dorsal vertebra is a treatment for spasms of repletion that have or will complicate contusions of the head and other nerve-bearing regions. Pressing deeper with the same cautery at that spot is good for treating gibbosities, backaches and flank (renal) pains.

Apply two nut cauteries in the arms three fingerbreadths below the shoulders, one inside and the other on the outer surface between the ends of the two large muscles (ie biceps and triceps) to treat painful shoulders. Those two cauteries or a seton cautery applied three fingerbreadths above the elbow, one inside and the other outside, between the muscle (ie triceps) and the bone (ie olecranon) are also good. The inner surface cautery helps most ailments of the front of the brain, such as dazed states, vertigo etc., as well as humoral disorders of the eyes, cataracts etc. The application in the outer hollow above the elbow serves the posterior part of the brain, such as lethargy and stiff-neck. It also acts vigorously to cleanse the nerves of the neck.

Two or more cauteries applied between the ends of the muscles three fingerbreadths above the wrist joint are good for treating swollen and painful hands (ie chiragra, gout). Four cutting cauteries touched between the fingers after the preceding consumptive treatments (ie relief of pain and swelling) will complete the cure and prevent recurrence.

A point cautery beneath the nipple of a breast will relieve pain in the shoulder and eliminate the slippery matter that is the basis for recurring dislocations of the shoulder. A round or seton cautery at the suprasternal notch treats all kinds of breathing problems and ailments of the lungs. Placing the round cautery on the chest treats ailments of the lung just beneath. Cutting cauteries placed between ribs treat empyema.

A seton cautery over the liver treats chronic illnesses and all sorts of dyscrasias of that organ. Similarly, placement over the spleen treats its ailments. A seton cautery over the stomach does the same for it, as well as treating a chronic weakness of the organ.

Cauterization three fingerbreadths above the navel relieves pain in the region as well as colic and ascites.

Cauterization in the groins with the small round instrument treats pain in the bladder, iliac colic and twisted intestines (ie small-bowel obstruction). If you place small round or a semicircular cauteries, pointed or blunt, directly over the pubis you can treat weakness of the peritoneum and spermatic cord (ie inguinal hernias) if you burn right through to the bone.[88]

A cauterization with a seton cautery into the scrotum that avoids the testis will drain a hydrocoele or a gas bag and will prevent recurrence.[89] Applying that cautery in the hollow below the kidneys relieves pain from renal stones. A seton or nut cautery over the lowest lumbar spine relieves backache and painful hemorrhoids.

---

[88]   The scar may block the inguinal canal or the processus vaginalis (LDR).

[89]   The tragic outcome of draining a 'wind hernia' is not mentioned here. I assume that Mondeville intended to discuss the problem in his planned Chapter 21 of Doctrine III, which was yet to be written at the time of his death (LDR).

Use a dactylic cautery with three 'fingers' anywhere over the buttocks to treat sciatica. A circular cautery is even better.

Treat a dislocated hip joint due to an excess of slippery fluid and relaxed ligaments which have soaked up some of it, by cauterization with a circular cautery in the hollow over the hip.

For painful and swollen knees, one or both, apply four cauteries over each knee: two in the hollows alongside the lower ends of the large muscles (ie the quadriceps) and two more placed three fingerbreadths below the joints. In addition, apply dactylics, with three or more 'fingers' directly over or just alongside the joints. Press them to burn deeply before you release them. Also, apply two nut cauteries behind each knee, three fingerbreadths below the hollow. These last applications also serve well for reinforcing thebody's general health, especially favoring the joints.

A nut cautery on the lateral surface of the leg, three fingerbreradths above the ankle, between the muscles and the bone (ie fibula) serves all ailments distal to the knee.

Cauterize behind and below one or both malleoli to treat ailments of the feet. The medial malleolar application is good for treating the genital structures of both sexes. The lateral malleolar application treats lesions of the kidneys and the hip region. These applications supplement phlebotomies from the saphenous and sciatic veins.

Point cauteries touching the sole of the foot between the first two or between the last two toes treat podagra. You may use the cutting cauteries instead of the points. These treatments may be performed at any time of the day or night. Cauteries placed behind the malleoli and between the toes also treat podagra, in the same way as we use them at the wrist and hand to treat chiragra.

IV. Precautions

A. Use a perforated metal plate to protect the surfaces adjacent to the actual cauterization, at least when you use a seton cautery on a flat exposed surface, as on the head (ie not the scalp), arms, thighs etc.

B.   When you wish to burn in recesses or obscure spots. as in a lacrimal fistula or in the nostrils or mouth, insert the cautery through a cannula.

C.   When cauterizing through the scalp to reach the bone, do not let the instrument remain in contact with the bone.

D.   When cauterizing the region beween the ears and the occiput, take care to avoid burning the arteries and the veins that you will find there.[90]

E.   Use gold cauteries when treating the corners of the eyes and the eyelids.

F.   In those (E) regions, cauterize very lightly and superficially, and a little at a time.

G.   Do not use a cautery if your hands are not steady.

H.   At the occiput a seton cautery is easier to control than the nut.

I.   The actual cautery is better than eruptors when entering a vein near the ear. [91]

J.   The nut cautery is better for treating catarrh by burning at the anterior fontanelle where the coronal and sagittal commissures meet. It penetrates more deeply and reaches the cranium to allow the escape of the fumes from beneath it.

K.   Keep open the defect for a long time.

L.   When you treat for ailments other than catarrh, you may use the cutting or the olivary cauteries.

M.   The cauterization of a lacrimal fistula, after other treatments have failed, includes burning through the bone into the nose.

N.   Applications over the stomach, liver, spleen etc. should be superficial, using the seton (ie the needle) cautery. The same holds for treating near other noble organs and over nerve bearing regions where the subcutaneous tissues are scanty.

O.   After applying dactylic cauteries at the knee, do not insert drains, peas or other solid materials.

P.   The same holds for applications behind the knee, alongside it and below the patella.

---

[90]   From ear to ear, the posterior surface of the head and neck (LDR).

[91]   This makes sense if we substitute 'abscess' for 'vein' (LDR).

Q. When treating hernias, cauterize deeply in the groins to go through the spermatic cord and reach the pubis and keep open the wounds after the eschar is eliminated by inserting a ball of white wax as large as the opening will admit. Delay the contraction of the scar.

V. Twelve Explanations Of Obscure Items

A. Very few physicians become surgeons, and very few surgeons nowadays are well educated. That default has allowed the proliferation of the drunkard-bucolics, stupid and ignorant of all the surgical arts; they abuse cauteries and they apply them before they purge the patients contrary to what is proper.[92] The results are abscesses and ulcers in the cautery-burns, which attract more of the bad humors. And they use cauteries on persons with warm and dry complexions, provoking serious dyscrasias and violent fevers. Those abuses and the subsequent bad results have evoked unjustified criticism, leading to an undeserved abandonment of cauterization by modern surgeons.

B. The great authors and experts agree that there is no better method for consuming the bad matter than the use of the hot cautery, when it is hot enough and large enough. When it is too small and not very hot and when the potential cautery is used in cases that are very cool and wet, the burning quality is snuffed out. It is as if you compare the effects of a brightly shining sun to that of a shaded lamp. The hot sun consumes everything that is exposed to it; not so when its light is dimmed. Furthermore, during the spring and autumn the dimmer sun brings more violent winds.

---

[92] The cautery-burn that was used to attract the humors away from the site of the primary ailment also could attract bad humors from the intestines, and they could drift through the region of the ailment, to its disadvantage. Therefore, the intestines were purged to evacuate their bad substances before the cautery was appliesd (LDR).

C. They say that the cautery is properly used in two ways: 1. As judged by appearances; you can tell that it is used according to our art by observing it in action. However, that is not the whole tale. You judge the result by the relief obtained and by how well you have achieved what you set out to do. 2. On the other hand, the statement is true, because when a cauterization is made in the correct way it cannot fail to do some good. A failure indicates a faulty technique.

D. Renew the heat frequently during the procedure and you will succeed more rapidly and more completely.

E. When you cauterize a swollen arm or leg, remember that the limb suffers from repletion, which means that you should cauterize the opposite side to evacuate the ailing limb by diversion.

F. Take note here as in other chapters that a cauterization, actual or potential, will produce an eschar in two ways: 1. As a concomitant of a burn caused by fire, or by a heated instrument, or by boiling water or oil. That eschar in itself contributes nothing to the health of the patient, be he well or sick. You should get rid of it as soon as you can. 2. In some cases, the eschar is a preventive or curative thing; it is not an accident, and it is produced by design, in two ways: a When there is no preexisting wound, it works in three ways: 1.To block the entry of bad humors into a sick member, as when you apply a cautery at the knee to prevent or treat podagra. 2. When you impede the passage of bad humors from a sick organ to another, as when you cauterize at the neck to limit a malady to one eye. The cautery is more effective when the humors are moving. 3. At the lesion itself, as when you cauterize the buttock to treat a painful hip. In all three cases, do not cauterize when the ailment is on the wane and or before the purgations and other measures are seen to fail, as we have repeatedly emphasized. b. When you create an eschar in a wound to solidify (ie by charring) some residual or adherent matter that you want to get rid of. The techniques and the various suitable caustics are described in our Antidotary.

G. Cauterization to clear out old accumulations of humors is best made near a large vein that drains the affected region, as you would apply it near a cephalic vein in an arm to clear the head.[93]

H. Abnormal accumulations of old harmful matter that cannot be evacuated through natural channels can be consumed by cauterizing directly over the affected site.[94]

I. By keeping open the burn-wound beyond the specified correct time, you continue to attract the more liquid humors, both the good and the bad kinds, and that is harmful in two ways: you attract the bad humors to the site of the cauterization and you deplete the body of good humors.

J. The cautery succeeds by its virtue, the burning. All the medical writers confirm that fact, because they never order an incision (ie including the cautery) when treating medical disorders or as a hygienic measure unless there is a wound or an excrescence or an abscess.

K. When the patient obviously is weakened by the cauterization or for any other reason, terminate the application.

L. Sometimes a surgeon will apply leaves of ivy or similar immediately after the burn, or will use a many-folded piece of linen cloth. Others place a metal plate on the burn. I think that the ivy leaves are not necessary in cases where you have used a perforated plate, nor are they any better than many other kinds of leaves, unless they are very stiff. The leaves are used for two reasons: they are soothing and they soak up the pus like the cloth pads, and that pleases the patient and his assistants. But the leaves dam the fumes and vapors and they condense them. Thus thickened, the fumes become pus in increasing amounts. That is why a cauterization as observed by

---

[93] The medieval anatomist conceived a centrifugal flow from the superior vena cava via the jugular veins into the cephalic veins of the arms. That is route of attraction of humors from the head to the arm (LDR).

[94] An example is the treatment of recurrent dislocations of the shoulder by cauterizing nearby to attract the slippery superfluidity from the joint (LDR)

the patient and by his attendants, seems to attract and expel pus. That is wrong, of course. The leaves neither attract nor expel; they retain and condense.

The leaves most commonly used are ivy, cabbage. oxalis and grape; all are cool. Yet, according to the authorities, the truth is that cool substances do not attract, rather they repel and condense.

I use the cloth pad on the fresh burn and place the leaves over the pad where they do not block drainage into the pads. When the dressings are changed twice each day, the soaked pads have wet the overlying leaves, and that is interpreted with pleasure by the patient as the result of the added attractive powers of the leaves.

CHAPTER 3.

PHLEBOTOMY: THE THERAPEUTIC DIMINUTION OF BLOOD, USEFUL AND NECESSARY FOR THE BODY, AS IT SHOULD BE PRACTICED IN INDIVIDUAL CASES, ACCORDING TO THE ART OF MEDICINE AND SURGERY[95]

Phlebotomy is a medical measure that uses a surgical operation. A physician determines the need and the surgeon carries it out; both of them have the same goal, the health of the patient. Long ago the physicians ceased doing it themselves as beneath their dignity, so they say, and ceded it to the surgeons. More recently, the surgeons have ceded the operation to barbers, with two excuses. 1. It does not pay well. 2. It requires little skill. For these two reasons and others: 3. Today, few people other than those with surgical ailments seek a surgeon's recommendations about when to bleed. Most of those are well-to-do, or part of the nobility or are prelates in the care of physicians. The common folk put themselves in the hands of the barbers. Since all the medical and surgical hand-books deal with phlebotomy in great detail, I simply will briefly review the subject in this chapter, and I will include a summary of the indications given by

---

[95]   See Ch.1 of this Doct., Part I On Incisions, and Fig. 16 (LDR).

other authors. If you seek more complete information go to the authors whom I will cite below and to the chapters noted there.

I will consider thirteen issues here: 1. The definition. 2. The uses and its effects. 3. For whom it is suitable, and for whom it is not. 4. Certain General Principles. 5. Which veins to use, and for what ailments. 6. The qualifications required of a good phlebotomist. 7. The proper timing. 8. Preparing the patient. 9. How to phlebotomize. 10. The care after the bleed. 11. Examining the shed blood. 12. Precautions. 13. Explanations of obscure points.

I. Definitions

A bleed is the therapeutic removal of a small amount of blood from veins and arteries. It really serves the purpose of a General Evacuation to rid the patient of a multitude of humors. The definitions, all and parts, are well known and need no explanation and amplification beyond what is found in the hand-books.

II. The Uses And The Benefits

We bleed for three purposes: 1.To preserve health. 2. To prevent illness. 3. To treat maladies already in effect. The first two have elective bleeds, the third is a necessity.

The time for the elective as well as the necessary, and other related matters are taken up in Chapter 1 of this Doctrine I. As stated there we have six uses for phlebotomies: 1.To evacuate blood and humors that are offensive in quantity and quality or both. 2. To induce distraction or a separation of humors such as when you treat an abscess you bleed from the opposite side to divert the bad humors from the aposthem by attracting them to he site of the bleed. 3. Simply to attract the bad matter, as when you bleed from a saphenous vein to attract blood to the womb and induce menstruation. 4. To modify some humors, as when you bleed because we sense that the blood in the veins is too hot. 5. To prevent certain disorders that are suspect or threatening. 6. To lighten Nature's burden and improve its curative powers.

III. Indications and Contraindications; Two Parts

A. When Bleeding Is Not Indicated, 6 causes: 1. Factors relevant to the Causative Humors: Too dense, too thin, too viscous, too adherent, too cool, too lumpy, too watery. For example, if the matter is too dense, the bleed will attract only the liquid parts and will leave behind an even thicker residue. The other features will reveal the objections and you will draw your own conclusions. 2. When the patient's vital force is feeble, a bleed will deplete it even more. 3. Certain aspects of the disease itself: here are three negative reasons: a. When the causative humor is poisonous or violent, as in anthrax. b. When the pus already is too abundant, as in quinsy. c. When the lesion is very serious, as an injured or inflamed eye. 4. When complications take effect, and the bad humors suddenly are disturbed. When Nature already is fully occupied in defeating the illness and has enough to do to digest the pus and evacuate it. A phlebotomy at that time will distract Nature and the illness may prevail. 5. Certain elements in the Complexion: the air, the lifestyle the geographic region, the weather, the customary use and disposition of the body (ie vigor, robustness, etc). All the foregoing deter you from bleeding lest you further enfeeble the vital force. For example, a melancholic complexion is a sign of anemia, and the same holds for phlegmatic humors. One easily can review the other features and find more information in the works of other authors. 6. Certain conditions affecting the stomach and other internal organs. If the patient has overeaten or is anoretic, or if he is constipated or has diarrhea, you need no help to understand why you should not bleed, or why it is not necessary (ie to bleed an already 'evacuated' patient).

Children younger than nine years should not be bled excepting in some urgent circumstances such as when they are suffocating due to an excess of blood as in quinsy. Even then, the physician must not be rigid; he must leave the final decision to the parents and friends. Old folks who are slipping into the decline of their dotage, after age seventy, especially if they are not robust;

convalescents who have passed the critical phases of their illnesses; pregnant women, especially during the first and thirds trimesters; young but washed-out people with small and hard-to-see veins and scanty beards; those with clumped humors and scanty blood. All those persons should not be bled; let them treasure their limited supply.

The same policy will exclude persons who have been poisoned— internally or externally; those who have early cataracts and dropsy, excepting those with amenorrhea or hemorrhoids. Exclude women during their menses and persons who are obese (ie difficult veins) and who have recovered from colic or from vomiting, lest you attract bile into their stomachs. Exclude those who have recovered after another evacuation, or after long spells of insomnia and after a spell of hard labor. Exclude those with delicate stomachs and livers who often fall prey to cool maladies. There are many more conditions that you will encounter.

B. When Phlebotomies Are Indicated:.We bleed patients who suffer from recurring abscesses and boils, from anthrax, fevers and overindulgence with meats and cheeses and wine and sweets, and those who are idle and foppish, and those who eat too much broiled food and then generate too much blood and those who drink too much heavy wine, and those who have too much melancholic blood, that is, the red humors which accompaniy blood as it flows through the body, and those in whom you suspect there may be an internal overheating (ie combustion) of humors and the like, or who may have an inflammation on the surface of the body accompanied by a fever, and in those who may have had a prolonged course of treatment of a malady during the warm seasons, and those with strong vital spirits and a sanguinous complexion and who grow bushy hair. We include those who suffer gout and other kinds of inflamed joints, and those with paroxysms that occur with sanguinous disorders, who have continuous fevers and large inflamed internal or external masses or pleurisy or buboes, and those who suffer spasms of repletion and various similar maladies.

IV. Precepts Taken From Older Treatises

A.   We bleed for one of three reasons: to preserve health, to forestall an illness and to treat it.

B.   All therapeutic (ie 'artificial') bleeds aim to reduce an excess of sanguinous humors or remove it if is defective, or both.[96]

C.   When you bleed during the springtime months, take special care to avoid collapse of the patient.

D.   Victims of torture, flagellation, rackings with body blows should be bled immediately, using standard methods, to prevent aposthems from appearing (ie hematomas becoming abcesses).

E.   You will have more success if you bleed a person to prevent a suspected illness than if you wait for it to develop fully. That is, bleed at the very onset when the humors are in flux toward the site of the malady and are well mixed with healthy blood, before the malady is fully under way. Later, when the signs indicate that the sick humors are already being digested, after the stage of evolution has passed, you may bleed even then, using our standard techniques.

F.   When the sick humors already are agitated (ie but not yet digested) and on the move, withhold the bleed lest you make matters worse by impeding the Natural process of digestion of the bad humors.

G.   When the malady is past the stable state, discontinue your phlebotomies. However, when you think it is worthwhile, you may use small bleeds which do not deplete the patient. For the same reason, you should not bleed during the cool months of fall and winter.

H.   Diversive bleeds, that is from the opposite side, will most often succeed in arresting the flow of the humors to the site of the disease.

I.   Avoid bleeding when it is very cold or very hot, and soon after coitus.

J.   Repeated small bleeds are better than a single large one. When the arm you have used for those bleeds is inflamed, use the

---

[96] Normal venous blood consists mostly of sanguinous humor, carrying with it various amounts of each of the other three (LDR).

other arm and apply naturally cool topicals, which you have warmed, to the inflamed arm.

K.   When you bleed from a cephalic vein do not repeat too often at the same site lest you cause an inflammation.

L.   Bleeding from the index finger-thumb web is good for illnesses of the liver and the diaphragm, Galen advises it.

M.   When you bleed for ailments in which sanguinous humors predominate do it as you would for other humoral maladies. Bleeding evacuates all humors, and if the cause is only sanguinous, bleeding alone is a sufficient treatment. When other humors also are involved, bleed as appropriate for the affected part as well as to evacuate the body as a whole.

N.   Scrawny persons are more liable to faint during a phlebotomy because they have less blood.

O.   You should not bleed when there are signs of repletion (ie spasms) or when raw humors are in excess (ie an undrained abscess affecting nerves).

P.   When you have to bleed in those cases, they should bathe beforehand or take exercise or take some relaxant.

Q.   In weak patients, use several small bleeds rather than a single large one.

R.   A bleed through a small incision removes more of the thin part of the blood and leaves behind the thicker. That is better for the summer months. A larger incision is favored for the winter when it is serves better to sap the vitality. A major bleed from a large opening is suitable only when the vitality is energetic, when the blood is thicker as during the cold months. Smaller incisions are suited when the opposite conditions prevail.[97]

S.   Even when it may seem unnecessary, you may bleed to treat certain fevers, to enhance the patient's vital forces in their action to get rid of the residues of the bad humors and cure the ailment.

T.   Always obey the 'two diameter rule' as Avicenna insisted (Bk. I, Fol. 4, Ch. 3.)[98]

---

[97]   The temperature and other elements are more fully discussed in Section VII of this chapter (LDR).

[98]   Bleed from an opposite side or the opposite half (upper or lower) but do not cross both diameters (LDR).

U. Bleed only after you know the correct diagnosis.

V. Bleed when the veins are distended with blood, before the blood becomes corrupted, against established medical principles as stated above.

W. You may bleed at any time in an acute situation, even when you may impair the patient's own natural forces which are digesting the bad matter.

X. Do not bleed when the patient is constipated or when the air is windless and heavy (humid).

Y. Use the cephalic vein for conditions of repletion in the head and neck. Use the basilic vein for repletions below the neck. For ordinary repletions use the median vein. The exceptions are cases when another vein is selected by the referring physician.

Z. It is unwise or harmful to bleed for cool and dry ailments and further deplete the patient's blood.

AA. Galen said (*Megatechni*, Bk. XIV, Ch. 3) that after the bleed have the patient sleep for two hours.

BB. If the veins are full, the blood may putrefy if you do not evacuate it.

CC. Bleed when an overabundance of humors causes putrefaction.

DD. Bleed a febrile patient on the first day, at the latest on the fifth day. If you wait until the seventh you will create a crisis.

EE. Bleed a patient who harbors an aposthem. Nature then can combust and digest and get rid of the residue after the bleed.

FF. A disorder with several humors at fault should be treated by bleeding. When only one humor is at fault, purges alone will be sufficient.

GG. A phlebotomy is better than a purge when treating pleurisy, and it cannot do harm.

HH. Phlebotomy is the best treatrment for ailments caused by blood-born humors.

JJ. Even when the humors are in the large veins, a phlebotomy (ie which uses small veins) can get rid of some of them.

V. The Selection of Veins and Arteries for Bleeding

Before proceeding, let us review these facts. Veins of all sizes derive from the kylous (the many-branched) vein as the trunk of a tree which comes up from the roots gives off many branches, as we described it in our Anatomy. Among the many veins are the few that we commonly use for phlebotomies (see Chapters 6 and 12 and Fig. 5 in Treatise I).

In each arm there are three veins used for most phlebotomies: 1. The Cephalic Vein is opened in two places: at the elbow and in the web between the thumb and the index finger, near the base. Bleeds from that vein treat warm affections of the head and the neck. Bleeds from the web are preferred and are more effective (by weakening the patient) for disorders of the head, yet they are not dangerous or subject to error (ie missed by the phlebotome). 2. The Basilic Vein (hepatic) is opened on the inner side of the elbow-crease and between the small finger and the auricular finger [99] on the dorsum of the hand, and in between, in the forearm. The bleeds are for treating all cool maladies of the body below the clavicles. It is useful in preventing disorders of the head. Use the right hand for disorders of the liver and the left for the spleen. 3. The Median Vein derives from branches from the other two. It also is called the Cardiac, the Purple, the Brown and the Black Vein. It lies centered in the elbow fold. One bleeds here to treat maladies anywhere in the body, but especially those of the heart and the chest.

Bleed from a vein on the forehead for illnesses of the rear of the head to divert the bad humors and as curative measures for illnesses in the front of the head. Sometimes we use it to treat delirium, and for chronic disorders of the head it is better than the other three evacuative measures.[100] A bleed from this vein near the vertex of the scalp is good for treating ulcers of the scalp, saffati and tinea. Use the blood to rub on the tineas while it still is warm.

---

[99] This conflicts with the classical terminology: the little finger is auricularis (for cleaning the ear !); the fourth is the ring finger, annularis; the middle finger, impudicus, is the signal of contempt; the first finger is index; the thumb is pollex. (LDR).

[100] Laxatives, clysters, cauterizations (LDR).

Bleed from the arteries as well as the veins on the temples to treat migraine and ocular disorders. It adds support to other treatments in preventing recurrences. Sometimes, after a bleed you may apply a hot cautery or a corrosive topical to keep open the wound.

The veins beneath the ears (ie external jugulars) are useful against pustules on the scalp and for migraine. Veins just between or beneath the nostrils are useful against delirium and other maladies of the head. The sublingual veins up front near the teeth are used for treating quinsy, tonsillitis, throat disorders, acute and hot inflammations of the eyes, itching, pustules of the nose, scotomas and vertigo. In such cases first bleed from a cephalic vein.

Bleed from veins on the inner surface of the lips to treat small ulcers and pustules in the mouth (ie aphthous), gingivitis and the defects after dental extractions. Use a vein between the chin and the lower lip to treat halitosis.

The guide veins (ie external jugulars) are useful against leprosy and suffication due to congestion of neck veins.

Bleed from the large vein on the inner surface of the thigh about one foot above the knee, which feeds the entire leg and fills the varices and causes stasis ulcers (ie mal-mort). It will treat all those maladies. To stop bleeding from that site insert a small pill of sublimated arsenic or similar into the phlebotomy. It will corrode the vein and form a kind of flesh in the opening, but will not shut off the vein itself.

Bleed from a vein in the popliteal hollow to treat disorders of the uterus or to induce menstruation.

The Saphenous Vein between the ankle and the heel is bled for uterine problems in women and for ailments of the testicles, groins and penis in men. Corresponding to this vein on the lateral surface of the foot is the Sciatic Vein which is very good for treating pain in the upper thighs and sciatica which extends from the hip to the toenails. Also, it is useful for internal disorders of the kidneys and nearby organs within the body and on the surface. Bleeds from both the saphenous and the sciatic veins have been used for delirium.

Bleed from a vein between the fourth and fifth toes to treat salt phlegm, stasis ulcer (mal-mort), cancer, esthiomene and all kinds of melancholic disorders of the legs.

I know a Parisian physician who is unusual only with his experiences using certain difficult treatments with which he has claimed miraculous cures. He bleeds from the same single vein in patients whom he personally knows, and always had bled (ie hygienic) when they were in good health. He believes that, whenever they are sick, Nature will send the faulty humors to the same spot where the customary bleeds had been made, and nowhere else.

VI. The Necessary Qualifications For A Phlebotomist[101]

He should be of middle-age, have sturdy limbs and steady hands. His eyesight should be good so he can identify the veins commonly used for phlebotomy and recognize arteries and the nerves that lie beneath them so he can avoid them. He must not bleed a sick person unless that has been requested by a physician or a credited surgeon. Nor should he bleed a child or a domestic or other kind of servant, nor the mistress of the house without her husband's permission, especially if he is a rich or famous or noble person. He should own a good supply of sharp, clean and polished lancets with narrow blades set at angles of various degrees. He must be able to use all of them skillfully, as suited for each case. He should profess himself to be as skillful in the procedure as any other, or at least state that one cannot find another phlebotomist as capable as himself because he then will be trusted above others. He should make that claim before performing the phlebotomy and other simple and commonly performed operations. In that way the patient's imagination and confidence will help, and will assure success without doing any harm.

VII. The Timing

A bleed may be scheduled (ie elective) or be required on short notice (ie necessary). I explained those terms in Chapter 1 On Incisions.

The elective bleed aims to preserve and conserve health. It is performed at the wish of the patient during the spring and autums because the weather is more temperated in those seasons. That is what Hippocrates prescribed (*Aphorisms*, Part 6.) The precise time is determined by several factors. For example, when you are concerned that the humors will putrefy more than showing a few bubbles [102],

---

[101] Here it is clear that Mondeville is instructing lay practitioners and barbers (LDR).

[102] The appearance of the blood spilled into the collecting dish (LDR).

then bleed early in the spring. When the opposite is the case do it later in the season. Many other things are involved in that decision, subjects for the surgeon to judge. If the temperature is rising, wait for a north wind, preferably during the morning. But if the day is cool, bleed while there is a southerly breeze, toward noontime. At what epoch or age or during what phase of the moon? The four phases of lunation are known to all: The first quarter belongs to springtime and adolescence. The second phase is warm and dry and is associated with summertime. The third is cool and dry and associates with autumn and age. The fourth is cool and moist and is tied with winter and senility. Therefore, since Nature rules and the physician is only the servant, so it is that young girls want normal menstruation during the moon's first quarter, young women want it during the second, mature women during the third and older women during the fourth, just so do we decide in prescribing phlebotomies, taking into account the patient's age and the quarter of the moon. That is what the Poet (ie Ovid) meant when he said, "The old moon needs old people, the new moon needs youths, and the middle-aged moon wants middle-aged people. That is how women purge themselves."

As to the time of day, Avicenna (Bk.I, Fol.4, Ch. 'On Phlebotomy') said that when we bleed to evacuate blood (ie sanguinous humors) do so from early in the third hour (9 A.M.) to the end of the fifth. Others have said to wait until a meal has been digested and the bowels have been emptied and the weather is most temperate. Therefore, we should bleed when we are near the time when the blood is moving, that is, between sunrise until the third hour, because a humor in motion is more easily evacuated.

If we wish to evacuate blood that is bilious, we should bleed between the first and the ninth hours. For melancholic blood, do it between the ninth hour and evening, and the same for phlegm.

As to the day itself, some recommend that we follow the Egyptian calendar (see Guy de Chauliac, p 566), where some days are excluded.[103] I have five good reasons why that choice is unwise. a. The plagues and the persecutions that befall us today come at different times (ie than in ancient Egypt) and unlike the influences of astrology

---

[103] The days of the Ten Plagues that preceded the Exodus, as described in the Pentateuch (LDR).

and of Nature on earth, theirs were caused by miracles. b. The effects then were limited to the Jews and the Pharaoh. c. They affected only the Jews in Egypt, not others d. The calendaric effects did not exist before or after other such(ie historical) plagues. e. The Egyptian calendaric pattern has no effects on any other medical or surgical procedures that are just as serious as phlebotomy. Therefore, there should be no prohibited days for bleeding. Whenever the patient or somebody else says something about suspected days before you begin your procedure, defer the act on the advice of the referring physician. If you go ahead with it and something bad ensues, even when no fault can be found in any part of the treatment and the complication was fortuitous, people will blame only the physician for negligence in knowingly permitting a surgical phlebotomy on a banned day.

Some persons claim to have documents or writings that determine when a bleed on such and such a day and month will lead to a specific person's death, and when the event can happen without a complicating illness or death. Don't believe any of that drivel, because even the Heavens (ie astrologic) lack such general influences, so fixed and determined. Indeed, the influence changes continuously as the Earth moves. So it is that on the first day of May or of September its influence will change from year to year, and the same is true for every day and every month. The reasons are clear.

In summary, when the Moon is embarrassed in any way, as when it is in opposition or conjunction with the Sun, or is in the sign of Gemini, and when any of all the other ways that we must know will prevail, then we should abstain from phlebotomy. We discussed them in greater detail in the chapter on Incisions. You will find there the protocol for necessary (ie emergency) phlebotomies.

VIII. The General Regimen For Patients Who Want Phlebotomy

When a patient needs a phlebotomy to relieve a thickening of humors, he should bathe during the preceding evening to cause his blood to be thin and to flow more readily. If his stomach is weak, let him eat a mouthful of bread soaked in the juice of oxalis leaves or similar until the belly tightens a bit and is comforted and repels bile. For those who fear phlebotomy or who have had no experience with it and may be cowardly, offer bread soaked in wine or syrup or sugar of roses (rosat), etc. Let those who can do so chose the day when they can be free of such emotions and bodily disturbances. They should fast during the preceding evening, including dinner, If they insist on eating something, let them simply nibble. They should not be constipated on the day of the bleed, but if they had a considerable (ie debilitating) evacuation during the night before or in the morning, such as a major sweat or other, you may delay or cancel the operation, then and there.

IX. How To Perform A Phlebotomy

When he is at work, the phlebotomist should exhibit a smiling and agreeable face and he should please the patient. Preparations are made near a source of water and should include a hand-towel, a strip if cotton and a wood rod (ie to twist the torniquet), some oakum pads and some hemostatic topicals. He should eschew many assistants for five reasons. 1 They will get in the way. 2.The patient may be alarmed by a hovering crowd. 3. One of them may faint and present an unwanted added patient. 4. They are not accustomed to the fearful reactions of the patient, which are not uncommon. 5. Some people believe that if more than one incision fails to strike a bleeding vein, a serious or lethal outcome will occur in their presence.

When everything is ready, search for and find the site for the incision and insert a blade in the usual way, observing all the precautions already described and others to follow. When the patient is a poor man, let the patient himself hold the twisting rod in his other hand. A physician should stand by and instruct the operator, perhaps to tell him to make a larger incision during the winter, and to inform him when the humors are cold and thick, and to tell him how much to bleed, because they are more readily expelled through such an incision. In warm weather and when the humors are thin and are flowing briskly and the patient's vitality is already feeble, the smaller incision will allow less vital spirit and heat escape. When the patient is weak, let him lie supine during the bleed, supported by his back, the strongest part of the body. As Avicenna (*Canon*, Bk. I) and others have asserted, the body is built up from the spine, as a ship is built from its keel. To pursue the analogy we call the vein that courses along the spine 'the keel vein' (ie inf. Vena cava).

X. The Postoperative Care

The patient should rest in place where the air is fresh, mild and dry, in a room that also is dry and not rheumatic, where it has been warmed by a fire if it is too cool. If it is too warm let it be cooled with cold water and clumps of green plants. Keep all the doors and windows closed to keep out the heat.

For two or three days he should eat small amounts of easily digested foods that are known to generate good blood, such as hen's eggs, small chickens, roasted or cooked without water. In spite of the admonitions of Isaac[104] urging abstinence from wine and offering a profusion of food, I say that you should diminish both when the patient's vitality is low (ie due to the bleed), but restrict the food rather than the wine.

The patient should rest in bed on his back or on the side opposite the bleed for an hour or longer. Then he may begin to eat a little, but he must remain in-house for three days, abstaining from other evacuations (ie purgation, sweats, etc.) and from tub baths. All of this is for two purposes: 1. To prevent the humors that were attracted by the bleed from returning to the diseased site. 2. Lest the heated and the thinned humors which are caused by bathing will break open the sealed vein.

To prevent complications caused by disturbed emotions, the patient should keep himself happy and satisfied by enjoyng the company of his companions, playing at dice and bones to gamble for food and wine.[105] He should not be irritated or be bored. He may hire a musician who can play several instruments, but who is of his own sex to avoid suspicions and nasty gossip.

As to sleep, etc.: Galen said (*de Ingernio*, Bk. IX, Ch.2) that he should sleep for two hours after the bleed, giving two reasons. 1. The patient should be nursed during the weakness caused by the bleed. 2. Sleep will

---

[104] Isaac Judaeus, the Elder. Tunis. fl. 10thC (LDR).

[105] Here we see that games with dice so common in today's bars and clubs are indeed very ancient (LDR).

come easily for those who frequently nap during the day. 3. For a healthy and robust person with no organs susceptible to disease, whose nerves are not too cold, the post-bleed nap may be unnecessary. However, Avicenna seems to have had a different view (*Canon*, Bk. I, Fol. 4, Ch. On Phlebotomy), claiming that sleep is not at all necessary. But I agree with Galen. Avicenna gave three reasons for his opinion: a. Sleep immediately after a bleed will induce contractions of the nerves in persons who have cold and weak nerves; it produces deep within them—the common folk say, "in the very marrow of their bones"—a painful heaviness not unlike that at the onset of a quartan fever. That heaviness causes the heat and the natural spirit and humors to withdraw deep within the body. Thus, the already debilitated nerves experience a new coldness, and from that comes the painful contractions. b. Anyone with a debilitated member is at risk for aposthems. The humors which are withdrawn from the body by the bleed or are driven inward by the sleep, become more liquid and are attracted to the ailing member and engender an aposthem. c. All persons who have been bled fear that in their sleep they will roll over on the incised arm (ie and reopen the incised vein).

When the illness requires a copious bleed, as in a case of quinsy or a continuous fever or some other very intense affair, or when it is caused by an excess of blood, such as in inflammations of the shoulder, take measure of the vital force at the time and if the vitality is strong, at one session remove as much blood as is necessary. But if the patient is feeble, use repeated smaller bleeds at fairly long intervals (days). The latter method is used with more care and is called secondation. The phlebotomist can use his fingernail to reopen the vein in the depths and release a thin liquid blood (ie the serum) and leave behind the thicker elements (ie the coagulum).If he uses a knife he will release the corrupted clot. A second sharp incision should always be used if the patient can tolerate it rather than the fingernail method.[106] When the bleeding makes bilious humors flow to the patient's tongue and cause thirst, slake it with barley water.

---

[106] The tradition was carried until recent times. Even in this century, many surgeons kept one of their index fingernails long to assist in their dissections, notwithstanding their use of rubber gloves (LDR).

XI. Examining The Shed Blood

This examination is necessary because it provides real data with which you can assess the vital spirits and the material body; previous conjectures become facts. We learn the condition of the spirit because, as Aristotle said (*de Animatibus*, Bk XII) a thin blood with a temperate quality (ie warmth) is a sign of common sense and good intelligence. It also indicates the condition of the body, that it has a good digestion in the liver and the heart.[107] Therefore, examine the blood three times during a bleed.

A. At the very beginning, four assessments. 1. Observe how the blood first escapes from the vein. When the arm-band is tight enough and the venesection is large enough and yet the blood simply trickles forth, that means that a feeble, aged or very young person lacks blood. When the patient is pale and phlegmatic the trickle means that the blood is too viscous to run freely. When the patient is melancholic, dark-visaged, etc. the blood may be too thick. When the patient is obese, even if he has large vein, they are compressed by the tissues which bulge into the incision and cause the flow to be sluggish.

If, to the contrary, the blood gushes forth, it is a sign that it is acidic and too penetrating, too watery, too liquid. You judge all these qualities according to the complexion of the patient, etc. and by his color.

The moderate rate of flow is what you want.

2. Examine the way the blood collects in the receptacle. If it comes in bursts and makes waves in the dish, matching the pulse, and the patient feels as if the arm is flooding and the surface of the blood in the dish appears like the wind-blown surface of a lake, and the blood surges like surf against the sides of the basin, that means, when other things are consistent, that you have opened an artery, and you must stopper it at once.

---

[107] The Reader is reminded that the Liver performed a second digestion on chyle to produce the venous blood of nutrition, and some of that went to the heart, which made it's own second digestion to convert nutritious blood into the arterial blood containing the vital spirit (LDR).

3. The color is important. When it is pale and contains solid bits,[108] you should arrest the flow because the escape of the watery matter will make what remains thick, and even more crude, having lost its vitality. Similarly, if the color is a delicate pink and glistens and is clear and it flows briskly, take that to mean that it comes from an artery. When the color is dark red, assume that it is melancholic or cool. Adjudge other colors as you do these. Therefore, when you have withdrawn almost enough blood and the color of the emission improves toward normal, that is when to stop the flow.

4. Examine the blood for material qualities to see how it moves in the basin. I have seen blood stand for an entire day and not coagulate, meaning that it is too liquid and watery. Sometimes it is thick at the very onset and becomes thin at the end of the bleed, or the opposite may occur.

B. Examine the blood in the basin before it coagulates, tilting the basin this way and that and observing the play of colors, just as you see them in the neck feathers of a pigeon when it twists, or as you see in the precious materials imported from Asia (Tartary) which we Frenchmen call velvet. That is a bad sign when we see it in blood. Rapid coagulation is a good sign; a slow process means that the blood is thin, a bad sign.

C. Examine the clot: 1. Its color: Pink blood with a red clot at the edges of the dish indicates a predominance of sanguinous humor and a sanguinous complexion. A pale grayish clot indicates a phlegmonous complexion and a deficiency of blood and an inadequate digestion. The phlegmatic pallor is either uniform or has the appearance of egg-white. When it is uniform, it is a sign of a humid phlegmatic complexion; when the pallor is mixed with a metallic sheen it indicates a mixture with a cool humid melancholy. A very red or clear purplish clot is a sign of a strong bilious complexion. A grayish clot, ashen or cadaveric, indicates that it has undergone combustion. When the clot has a color of ashes mixed with lard, no

---

[108]   Perhaps Mondeville describes a 'secondary' bleed. The vein may contain flecks of coagulum as well as thin serum which escapes first, followed by larger clots, which Mondeville calls 'crude matter' (LDR).

matter if the patient's color is good or if he is yellowish or whitish or mushroom-colored, it is a very bad sign of corruption and rotted humors, of an incurable indigestion and a bad life-style and diet, a febrile pestilence, a doddering invalidism, a general corruption, a gourmandizing and lazy indolence. You must change the life-style and diet and purge the patient several times, doing that very carefully. When the clot is milky and when you place a bit on your finger-nail and it neither adheres nor gels, that is a sign of failure of digestion of the crude humors.

Consider in detail all cases in which the clots have some white elements, and determine if it indicates failed digestion or combustion, or if the patient is exhausted, or if there is a fever, or if the present condition was preceded by great anxiety or by exhausting labor, or if the body is yellowish and if the urine is clear and watery. In all such cases, a pale clot may indicate a combustion and a patient who is subject to melancholic ailments.

When the colors that we have described are not accompanied by the symptoms listed, then the irritation and the coldness are causes, and they predict that a putrid fever, aposthems and hydrops will ensue. A dark, tawny or purplish color or a vague and indefinite hue are signs of a very cold complexion. A uniformly dark clot indicates a combusted humor.

2. You may learn something from the intact clot after its watery part, its urine-like fluid, has exuded. A foamy deposit on the surface that was not caused by blood spattering into the basin is a bad sign: either of gas in the blood or of its having been overheated or of being condensed or of viscosity. The surface of the clot should be smooth and have a uniform sheen as if it had been polished. That shows that it has been uniformly digested and is clean and not shadowed, that is, it is slightly translucent as well as mirror-like. You infer the opposite if it lacks those good qualities. When the covering membrane resists a dull-edged instrument or can withstand for a long time the weight of a solid object, that is a sign of density and viscosity. A clot that can be cut into easily or is easily squashed and is watery, is a sign of rawness, or of an incomplete digestion and a depleted natural heat. It is a sign of bad blood, because when good blood is digested it thickens. Thick

but not viscous or adherent clot is a sign of good nutrition and digestion.[109] A clear warm clot is better than others. A rapid coagulation, all at once, is moderately tempered and is of nearly pure blood, lacking other humors.

After examining the intact clot, split it with a dull hard edged instrument and examine it in four ways. Observe all of its parts as you move it about in the basin. Refer to materials dealt with in Chapter 1 of Doctrine II which follows (The Generation and The Separation of the Humors). The authorities have voiced three opinions about the locations in the clot of the humors removed by the bleed, as they are found in the basin. We are concerned here only with the four principal humors.

The first humor is the white froth on top. Avicenna and Haly claimed it was bilious, because bile is lighter than phlegm, and will float above it. A second opinion is voiced by some practitioners who believe that the foam is phlegmatic, claiming that the lowermost layer, the residue of all the other humors, is melancholic. The red membrane beneath the foam is bilious, and the mass of clot beneath the membrane is sanguinous. They conclude that the phlegm is the foam because there are only four humors and foam and phlegm are similar in substance and color; the same for bile, etc. etc.

A third opinion seems more rational, holds that phlegm is neither froth nor membrane,but is found in the whitish firm bits that are mixed in with the blood. Like bits of the sclera of the eye.[110] The rest of the phlegm is completely mixed with the blood. My reasoning is based on direct observation and cannot be questioned, and it confirms the third opinion and is consistent for all regions. Thus, phlegm has no single specific place in the body, as is the case for the other three humors. Rather it is distributed everywhere as a substitute nutrient when there is a deficit of blood. Therefore, outside the body it has no special place in the clot, as do the other humors. It is mixed with the blood; it neither forms froth by itself nor does it form a membrane; it

---

[109] After the serum has exuded from the clot one considers it to be different from the 'milky' fresh clot described in the preceding paragraph. The 'dry' clot lacks 'stickiness'(LDR.

[110] Probably these are flecks of clotted leukocytes (LDR)

participates in both. So it is, when the bile boils it forms the froth and to the degree that it bubbles it forms more or less foam, which finally appears when the blood is agitated by splashing into the basin. Frequently there is no foam at all, which refutes Avicenna and those who believe that phlegm forms the froth. I apologize for having to say that Avicenna was in error; even in the absence of froth we know that there always is some phlegm in the blood. A phlebotomy is a universal evacuation, and even when the evacuation is known to be more phlegmatic, there may be little or no froth. Therefore, since phlegm may be in a lower or middle or upper level of the blood in the basin, it always is mixed with other humors, and not in fixed or equal amounts.

After the above examination of the clot as a whole, we now open it and judge its various characteristics. First we will pour off the urine-like watery serum that may obstruct a clear view of the rest of the clot, and we judge whether the blood had been too watery. Then we will use a wood blade, about as thick as three pennies, not sharp edged nor too dull, to divide the clot carefully and slowly from top to bottom. If the clot yields too easily to light pressure that is a bad sign of watery and undigested blood. If it resists, and the membrane remains intact when you lift the clot from beneath, that means that the blood is too viscous and dense, or it is a sign of dryness due to excessive combustion of the humors. A moderate resistance to the blade's pressure is a good sign. After cutting through a clot that is too watery, as described above, decant the fluid and examine the clot, part by part in order, very carefully.

People discuss the foam and the membrane but they fail to mention its flavor, which should be slightly bitter, nor its thickness which should be that of two or three small pennies if the volume of the bleed has been enough, and the humors are in correct proportion. But if that is not the case, and the bleed has been too meager or too excessive, the film will be too thin or too thick, and you may conclude that all is not well. Those indicators tell you what is better and what is worse. You can see for yourself how thick and how dense is the membrane, and how much or little is the volume of the blood, beneath the film. Then you go to the blood itself.

Judge the humoral mass in four ways: 1. Its Substance should not be too liquid nor too dense. Place a bit of clot on your palm and rub it with a finger. If it feels slightly oily or greasy that portends leprosy or excessive obesity. To differentiate between them, use two tests: a. If the rubbed clot contains solid bits, the size of millet seeds or grains of sand or tiny pebbles, that is a certain sign of impending leprosy. If there is no granularity, it means that obesity is on the way. b. After the blood has fully clotted, wrap a bit in a piece of linen cloth and run water through the cloth until the clot is nearly washed away. Untie the cloth and shake it to remove clot, retie the knot and wash it again in running water until you cannot see any blood. Untie the cloth and rub it beween your fingers. If it squeaks that is a sign of leprosy. If not, it indicates obesity.

A good blood clot is known by its Color: it should be a moderately dark red, uniform, purplish and it should have that color of natural blood in the basin when the other humors have settled out. If it is too red, that indicates an admixture with bile. If it is too pale or has traces of white, it is mixed with phlegm. If it is very dark, we recognized the melancholy in it.

We can also tell a good blood clot by its Flavor; its sweetness will predominate. Bitterness is a measure of its bilious content; if it is insipid it is mixed with natural phlegm; if salty it has salt-phlegm; if astringent it has natural melancholy; if sour it contains acidic (ie burnt) melancholy.

What we stated about flavor can be applied to Odor. A good blood clot has a pleasant scent. An unusual or fetid or foul (corrupt) odor indicates some bad humor or some corrupt and malodorous foreign matter, or perhaps the blood has sat too long in the basin after the bleed.

The third part of the clot is the humors at the bottom of the basin, beneath the other two layers,[111] is the melancholy. It is the heaviest and densest layer of humoral residues. It has four signs to show why it so dense and fatty, since all residues are denser than the original substance. If the residue has the liquid quality of blood it is a bad sign; Nature has failed to separate the melancholy from the rest of the

---

[111]   The other two layers: the membrane and the coagulum (LDR).

humors. If that liquidity involves the entire clot right to the top, all the blood must be considered to be melancholic.

The natural melancholic humor is recognized by its intensely dark and impenetrable color. If it is non-natural melancholy, after a combustion, the dark color is even more intense and glistening, like the wings of a raven. The flavor of natural melancholy is astringent with added bitterness like the taste of an animal's spleen. A combusted melancholy has an equally astringent flavor with added sourness. I cannot comment about identifiable odors.

In addition to all the rest, sometimes we can learn something from an examination of the serum, the urine-like seepage on the surface of the clot. A moderate volume is a good sign; too much of it bodes ill, indicating a humid body; too little means too dry. That will tell you to treat with contraries.

Twenty Three Precautions

A. When you bleed from a neck vein or artery, use a hood with a draw-string at the neck.

B. The same holds for bleeds from hot cauterizations or spot applications of caustics over a temporal artery.

C. When treating toothache or painful jaws, bleed from a vein in the hollow near the ear. Use that same vein for treating surface lesions on the cheek.

D. When he bleeds from a jugular vein the phlebotomist or the surgeons who assist him should provide in advance, all the hemostatic topicals and other materials he may need to arrest the bleeding. He should follow all the instructions and rules given in Chapter 7 of Doct. II of this text.

E. When you make secondary renewals of the bleed you must enlarge the incision if you plan to use the finger-nail method.

F. In that case, lubricate the inserted finger.

G. When you use a lancet for the secondary bleed, the wound can be small.

H. The knife is preferred because you release the clot as well as fresh blood, whereas the finger-nail releases only blood.

I. A secondary bleed may be performed on the same day in a robust and brave patient, and it may be more copious than the first. A secondary bleed from a feeble or fearful patient must be less than the initial amount.

J. When you have doubts about the vital resources of a patient or you sense some imminent complication or other danger, it will be safer to use your finger-nail for a secondary bleed, lest your motives be falsely accused (ie as venal) especially in cases when you should bleed two or three times in the same day. You may do that, or even more often, if you use your finger-nail. You will not be blamed for that.[112]

K. A diversive bleed is more effective if performed at a distance but on the same diameter.[113]

L. When performing a diversionary bleed, interrupt the flow several times by stoppering the opening with a finger. Meanwhile have the patient cough and spit. By prolonging the bleed and stimulating the movement of the blood the humors are more readily attracted from the distant site of the lesion.

M. The more often you interrupt the outflow, the better you will sustain the vital forces.

N. If you think a syncope or other loss of vitality will occur, have the patient lie supine during the bleed.

O. Always open the vein in its long axis when you do not wish to obstruct the flow of blood, except, perhaps, when the vein is tortuous, as is often the case in the legs.

P. Whenever you bleed from a vein in the antecubital fold, the middle of the upper arm should be bound moderately tightly during the entire procedure.

Q. When you bleed from a cephalic vein, a mass will appear (ie a hematoma) unless you make a large enough incision in the skin (ie for drainage).

---

[112] The surgeon's knife was feared, and the fees for operations with the knife were criticized (LDR).

[113] 'Diameters': Transverse: divides the upper half of the body from the lower; Vertical divides the left from the right halves (LDR).

R. Be careful when you open the cephalic vein not to prick the tendon that lies beneath it.

S. Likewise, take care when you use the cardiac vein to avoid the two cords alongside it.

T. When you bleed from the basilic vein, avoid the artery which lies beneath it. U. When you use veins in the hands or feet, immerse them in warm water for one hour before the procedure, and continue all through it.

V. Bleeds from the hands and feet are more successful when performed immediately after a meal rather than later on.

W. When you bleed to relieve a chill caused by an overabundance of thick humors, do not warm the patient near a fire. Rather have him exercize moderately before the procedure.

XIII. Eight Explanatory Comments

I.   Galen (*de Sanguine*) and Avicenna (*Canon*, Bk. II, Fol.4, Ch. 20) both discuss and deal with the subject of phlebotomy. Master Jean de Saint-Amand (*Treatise on the Antidotary and Revocativium memorieae*, at the letter F) and all the authorities and writers of the hand-books of medicine and surgery cover the subject in detail.

II. The serum which is the urine-like water should have the same color as the urine passed by the patient before he is bled, if he has followed a good regimen (ie diet, etc). The reason is this: if the serum had not been emitted with the bleed, it would have had to be discharged by the kidneys and bladder as urine. When compared side by side (ie in urinoscopist's jars), they seem alike and a physician can diagnose with more certainty from the blood than from the urine. I have explained how to separate the serum from the clot.

III. The most able and experienced physicians and surgeons generally disagree on two items relevant to judging the blood. First is the Color. One says it is red, the other says that it is reddish, including all shades of red. Second, I say that even when they agree on the colors they differ in assessing the

good and the bad qualities of a specimen of blood. One says
that pale blood has been burned, the other says it has been
digested, that it is phlegmatic and uncombusted.(ie crude).
They also disagree about diagnosis from urine, and the
physician advises examination of the blood as soon as it is
collected, saying that a delay would make it worthless, for fear
that a delay would give time for another physician to pass
judgment and opine differently.[114]

IV.   Blood undergoes putrefaction more easily and more rapidly
than other humors, because it has the two qualities that favor
putrefaction, heat and moisture. The dryness of bile resists
putrefaction, coldness protects phlegm and melancholy
resists better than all because it has both of the resistant
qualities.

V.   Avicenna (Bk. V, Fol. 3, Ch. 1) made this distinction which no
other author had recognized. "When you anticipate ailments
of the head before they have appeared, bleed from a basilic
vein as a preventive. When the ailment is starting but is not yet
fully recognized, bleed from the median vein which will
evacuate the matter before it reaches the basilic vein, because
it receives it via its communication with the cephalic vein.
After the malady is in flower, bleed from the cephalic vein,
which evacuates only the matter that has flowed directly from
the head. When we wish to divert the humors with a potent
bleed at the very beginning of an sickness, use the saphenous
vein and veins near the small toes, and apply cups on the legs
and the toenails and scarify the shins." Those distinctions of
Avicenna are applicable in many cases.

VI.   When you use purges and phlebotomies in the same case,
refer to Treatise II. Doctrine I, Chapter 1, Part 5, (Potions and
Evacuations In Injured Patients) for the discussion of the
proper order.

VII. A surgeon often is asked to bleed from one side or the other.
Remember this rhyme:

---

[114] Another comment that irked the Physicians of Paris (LDR)

> In Spring and Summer use the Right,
> In Autumn and Winter use the left.

The reason is this: During the warm seasons the warm humors are more abundant and it is correct to use the side in which the humors are more abundant, which is the right side. The opposite is true in the cooler seasons.

If the patient suffers from an ailment of the liver bleed on the right. Use the left side for splenic disorders. All of this is as Avicenna recommended.

VIII. Rhazes (*The Totum Continens*, Chapter on Phlebotomy and Cupping) states that in parts of the body where veins are not easy to see, one should apply a bandage containing a mixture of nuts and a substance from the Lebanon (not identified), the evening before the bleed. In the morning even the small veins will be nicely apparent.

CHAPTER 4.

CUPPING

You may think me foolish to repeat much of what we wrote in the previous chapter on phlebotomy. However, I think it is proper, although there are many points of similarity. Here we will examine seven General Questions.

I. Preliminaries:

I have found no author or book that says cupping is not useful. I define it as follows: It is a commonly known surgical procedure, which preserves health, prevents certain maladies and treats them when they are present; it sometimes is used only for the sake of appearances.[115] Most authors describe two kinds: Some include

---

[115] The meaning is unclear. Perhaps a surgeon applied a cup as a harmless sham, merely to suggest that he was actively treating, (LDR)

incisions over which they place the cup, others do not. The first has a stronger action, attracting the humors from deeper beneath the surface. After the withdrawal of blood, it cools and dries the deeper tissues. Without an incision the cup has stronger suction and it warms and dries the surface where it is applied. Both kinds attract from distant regions and reverse the directions of flow of the blood and the humors, either from below up or the opposite. They should be used only after a suitable purgation and when the condition has plateaued, at least if you do not want to augment inflammations at the emunctories.

II. Fifteen General Rules

A. Never use cups on cloudy days or when the wind blows from the south.

B. Apply cups when the Moon is full, because that is when the humors are most abundant. Do so near the third hour (9:00 AM) when they are most liquid.

C. Cupping cleanses the skin and the nearby tissues better than do phlebotomies.

D. It attracts thin blood better than the thick, and it will be of little value when used against thick humors, and frequently it will be harmful and will weaken the region.

E. Do not use it after a bath unless an hour has elapsed and the patient does not have thick blood. The deeper the preliminary scarification the better the result.

F. Do not apply a cup directly on a breast. It may suck the breast into the cup and be very difficult to remove.

G. When the cupped surface swells excessively it may be difficult to remove the cup. In that event, foment all around the cup for a long time, using a cloth or sponge wet with warm water.

H. Apply the cup immediately after making the incision. The first time wait briefly before removing it. Wait longer after you reapply a second, third time etc., according to how the patient tolerates the procedure.

I.   The patient may eat one hour after the cupping. If he is bilious, let him eat some cereals and drink a little grenadine wine.

J.   Never use a scarification-cup combination until you first use the cup alone at the same site. If you fail to do that, the scarification-cupping will withdraw blood feebly and slowly.

K.   Remember this: cupping over the occiput can cause more damage than elsewhere.

L.   When a bleeding evacuation is called for and you cannot use a phlebotomy because the patient's vital forces are weak, you may use cups, but never on an infant less than two years old.

M.   Do not use cups in maladies dominated by cool and dry humors, because the cups will attract them to the surface where they will be cooled and dried further by the exposure there, and the illness will be worse.

N.   When you use cups without scarification to bring an abscess to a head, purge the patient beforehand.

O.   If you want to deflect and disperse the matter in an aposthem (ie prevent suppuration) apply cups on the opposite side on a line from the site of the aposthem.

III. Ten Uses and Advantages:

Three for cups alone: 1. To warm a cool region. 2. To reduce gassiness, as in a colic. 3. To reduce a dislocated, a twisted or a contracted part, and to align a depressed fracture of a rib.

Five for cups with scarifications: 1. When used on a limb it is a general evacuation of humors and reduction of repletion. 2. It will evacuate the thinner part of blood that has accumulated at the surface. In addition there are three uses along with those of phlebotomy: 3. They both evacuate local collections of humors. 4. Cups do not evacuate as much of the vital spirits. 5. They do not evacuate the principal organs.

There are two advantages in common for both types of cupping: 1. They both can reverse the flow of bad humors, right to left, from below up, etc. 2. When applied over an abscess they both can bring it to a head.

IV. Where and For What Conditions:

Thirteen places for cups without scarification: 1. When they precede a cupping with scarification. 2. Placed over venomous stings and bites. 3. Over an ear to suck out a solid foreign body. 4. To extract pus from a fistula or deep ulcer. 5. Placed beneath the breasts to prevent menstrual blood from returning to the uterus and at the nostrils.[116] 6. Near the umbilicus to pull up a displaced uterus. Place them opposite the side of the displacement; below the navel if the uterus has prolapsed; for a high uterus[117] place the cups just above the pubis or over the vulva. When placed over the navel it favors the reduction of instestines trapped in hernias and it will induce menstruation. 7. Set the cup in the hypochondrium to arrest a nosebleed by diverting the blood. Use the side corresponding to the bleeding nostril; use both sides if blood appears at both nostrils. 8. Put a cup over the liver for bleeding from the right nostril. 9. Place it over the spleen for bleeding from the left nostril. 10. Place them over both organs for bilateral nosebleeds. 11. Place the cup over the belly to treat a colic by attracting the gas. 12. Place the cup over the site of the pain as a stone passes from the kidney to the bladder. Move the cups along to pull the stone into the bladder. 13. Place cups around the anus alongside hemorrhoids and anal fissures.

Cups with scarifications are used to bleed patients who are too feeble to tolerate phlebotomies or who are too old or too young. If they are older than three or four years they can tolerate cupping. In such cases also you must avoid phlebotomy.

There are seventeen places for cupping with scarification: 1. To drain collections of blood under the skin.[118]. 2. To attract pus from ulcers and fistulas. 3. On the vertex of the head to control general agitation (ie mania), ocular diseases, vertigo. Some say that it retards the graying of hair; others say that it hastens the decline of the intelligence. 4. Place cups on the two lateral occipital humps (ie

---

[116]   To support lactation, and to treat nosebleeds (LDR).

[117]   Perhaps a large leiomyoma or other pelvic neoplasm (LDR).

[118]   Hematomas after contusions (LDR).

horns) to treat tinea and ulcers and pustules of the scalp. 5. Place cups up front on the skull at the hairline to treat maladies at the back of the head. In some persons who are weak-brained in the frontal region, cupping there will worsen the mind and impair vision, hearing and reasoning. 6. A cup in the hollow below the occiput (also called the stern of the ship) is useful in treating headaches, infections on the face and ocular disorders. 7.If the occipital brain is weak, a prolonged cupping can destroy the memory. Any one who has a strong occipital brain can tolerate one or two cuppings there when that is necessary. 8. Cupping between the shoulders controls palpitations of the heart, syncope of sanguinous origin and maladies of the head and eyes. Prolonged application will harm the heart, stomach and structures near them. 9. A cup over the point of the elbow treats scabies of the arms and hands. 10. A cup over a kidney treats inflammation of the thighs, including scabies, and deflects blood carrying humors from above, and threby treats hemorrhoids, elephantiasis, gas in the bladder and uterus and itching of the back. 11. Cups on the hips or buttocks relieve pains there, in the thighs and hemorrhoids. It suppresses hernias and podagra. 12. On the buttocks, cups relieve backache, pain in the flanks and itching everywhere, including scabies. 13. Cupping on the anterior surface of the thighs treats abscesses of the testicles, wounds of the legs and thighs, provokes menstruations and diverts humors from the upper body. It helps chronic pains in the kidneys, the uterus and the bladder. In women who are pale and soft and who have clear blood it is better for initiating menstruation than a phlebotomy from the saphenous vein. 14. Place cups on the legs to treat abscesses and pustules of the buttocks. 15. Cupping behind the knees treats renal, uterine, testicular ailments and the organs of nutrition which produce blood. However, a large evacuation of the body may enfeeble the patient as much as a venesection. Cupping there treats pulsations behind the knee of bilious origin.[119], and malignanat ulcers of the legs and feet. 16. Place cups at the ankles to treat amennorhea, sciatica and podagra. 17. Cups placed on the soles of the feet are said to remedy ailments of

---

[119] A popliteal aneurysm (LDR).

the feet. Perhaps that may give brief respite, inasmuch as cupping may affect early stages of maladies anywhere in the body by diverting the humors.

V. How To Place Cup:

A cup is a simple round glass jar with a small opening and a broad base, familiar to everybody. The site for the application should be warm and rubbed clean. The skin should be soft and should accept the cup without discomfort. Place a fluff of finely teased charpie in the cup and ignite it with a straw or a candle. Quickly set down the cup so that it adheres snugly and encloses the tissues it touches. It will not come apart on its own. To be effective place several in the same manner.

Then, if you want to scarify before placing a second cup, make your incision at the site of the first cup and immediately cover it with an ignited cup. In that way you can remove about an egg-shell full of blood. Then you can remove the cup and replace it with another until you suck out more. Continue the repeated process until you have removed as much as you wish. Afterwards, dry the surface and compress it repeatedly with a warm sponge.

VI. Ten Precautions

A. Shave the head or other hairy surfaces where you want to set your cup.

B. Whenever you set a cup let the skin be as loose as possible to favor adherence.

C. For facial lesions and inflamed eyes, apply the cup over the occiput, while you sprinkle cold water in jets over the face and eyes to refresh and enliven the spirits and impel the blood toward the back of the head.

D. During the application in C. the operator should gently press the face to massage the blood to the rear.

E. If the patient needs repeated cupping at the occiput, first use a cup without a scarification. Then repeat the application about half a diameter lower, covering the lower half of the area of the first cup, and touching the skin of the neck below. Then follow the line of the vertebrae and apply a succession of cups downward until you reach the the shoulders, where you make the first scarification-cupping. In that way the first cup attracts from the head, and each one after it attracts from the one above and finally the bad humor is removed. The occiput is free of any traces and the memory is no longer troubled.[120]

F.   When a painful scarification-cupping fails to withdraw enough blood, rub the incision across its axis with the lip of the cup until the blood runs.

G.   When you cup at the breasts lift them and set the cup under them to get at their roots.

H.   When you seek a suitable site for cupping at a distance, arrange one that is downstream from the malady. For example, to extract pus from a fistula with an opening at the knee, which drains a bone lower in the leg, elevate the leg after you place the cup over the orifice.

I.   When feasible, place the cup near a large vein that passes near the site of the lesion, so better to evacuate or divert. An example: when you use cups to evacuate or divert from the right ear place them near the cephalic vein on the right arm.[121]

J.   Cups used in children serve better when they are placed near the heel.

VII. Explanations of Obscure Points:

The physician and the surgeon must exercize nearly as much care in their use of cups and leeches as when they use phlebotomy. That is not because they are as serious or risky operations of themselves, but because they are not so commonly used. When complications

---

[120]   The occiput is seat of the memory. This is an interesting way to obliterate a troubling memory (LDR).

[121]   Follow the Humor: Down the external jugular vein to the subclavian. Then via the cephalic vein over the shoulder and down the arm to the elbow where the cup is placed (LDR).

occur, the inexperienced common folk will cast more blame on you. So, be diligent and assess the state of vigor of the patient; is he feeble or robust. If feeble, draw less blood (by cupping) in face of the increased risks of chills, epilepsy, discoloration, hydrops etc. If, on the other hand, the patient is robust you may withdraw more blood and his vitality will withstand it.

CHAPTER 5.

LEECHES

We shall make twelve General Statements:

I.   Blood-sucking leeches perform evacuations of blood when they bite and hold on. The leech is an aquatic worm, well-known everywhere.

II. A leech withdraws blood from deeper within the body than a cup, but less deep than a phlebotomy.

III. Their uses can be inferred from what we have written about cups and phlebotomy.

IV. Leeches are useful for nearly all the ailments for which we use cups, especially for lesions in the skin, as safati [122], pustules, infections, and things that accompany corrupt humors, such as itching, scabies, dandruff, impetigo, morphia, albarras, malignant ulcers, cancers, cancrena[123], mal-mort (ie stasis ulcers), delirium, mania, depression etc., and for all 'unripe' aposthems,

V.   Place the leech where you would use a cup, especially over aposthems to bring them more quickly to a head. At times I have seen them set on the face and the nose to treat blemishes, and I

---

[122] 'Safati': an Arabic term for a contagious infection of the scalp, causing loss of hair. Perhaps a form of tinea; or perhaps due to lice (LDR).

[123] 'Cancrena': a foul, necrotic, invasive ulcer. Also esthiomene. Also gangrene (LDR).

disapprove of that. When a leech swells, I have seen them as large as a soup ladle, it still may not drop off. Also, I have seen people apply them around the joints of the hand to treat scabies, and around the anus to treat early stages of delirium. And I have seen them set near the heels for the same ailment and on the soles of the foot. When you use them to treat aposthems, place them atop the apex rather than around the base. For ulcers, place them around but not inside the lesion.

VI. Select them small, well-shaped and slim as a rat's tail, with small heads, pink abdomens and with some shiny streaks on their backs, like orpiment in color. They should be taken from a clear stream near rocks, with a good current and a sandy bottom, where there are many frogs. Large leeches with fat heads are venomous. They come from fetid ponds and they emit a foul odor when disturbed, They may be covered with froth and be multicolored

VII. Prepare the leech by not feeding it for a full day and night and then offering it a bit of some animal's blood. Then wash it in fresh water,

VIII. Foment the site for the leeching with warm water, massage it until it reddens and apply a cup without an incision. After removing it, anoint the site with philosopher's clay or some animal's blood and apply the leech.

IX. If the leech does not adhere on its own, hold it in place with a straw or a reed.

X. To remove the leech put salt, aloes and warm vinegar on its head. Or you may lift it with a horse's hair slipped between the patient's skin and the leech's head head.

After removing the leech, apply a cup over the spot to remove the rest of the blood, or simply it wipe it off with a sponge. Stop the bleeding as we taught in Ch.1 of Doct.I of Tr. II, in Section 2, titled 'The Arrest Of Bleeding.' Use a tight bandage to hold a coin over the spot and press it on the opening. With that in place no further bleeding is possible until you remove the coin.

Chapter 6.

Methods For Amputating Extremities And Sawing Diseased Bones

(Nicaise introduced this chapter with the following long note)

During the Middle Ages and continuing until the end of the 17thC amputations were performed only in cases of gangrene with methods not much different from what Mondeville described in this chapter. However, as a disciple of Theodoric and as a military surgeon, he was more intrepid, and he used amputations in certain cases of ulcer and cancers. Guy de Chauliac (ca 1350) amputated supernumerary digits, but, in respect of amputations for gangrene, he wrote. "the physician who considers himself honorable does not amputate, because afterwards he is left with his own remorse and a patient who declares that he would have preferred to die." Mondeville comments on amputation through joints and in the limb between joints and says that amputations that disarticulate are less serious. He disagrees with the opinions that impute that to the damage and to the loss of the bone-marrow. He discusses the influence of phlegm in those cases, but he emphasizes especially that the differences in the results are due to the lack of attachment of the soft tissues to the bones after amputations that are not at the joints. He describes a single rapid circumferential incision using a red hot iron cautery, whereas Guy used a knife when removing supernumerary digits.

After dividing the bone, Mondeville used the cautery and ligated the major vessels. Guy also used the cautery and boiling oil, followed by the red hemostatic powder mixed with egg-white, and direct pressure. Hemorrhage played an important role in all operations because surgeons knew little anatomy and they lacked clear instructions for choosing which measure to use for controlling bleeding. The lack of anatomy complicated matters even more.

Now I shall offer a brief account of hemostasis and narcosis as practiced in the 14thC.

Hemostasis

First, note that the surgeons who followed the ancient methods did not seek to stop the hemorrhage immediately after a wounding; they allowed a certain amount of free bleeding in the belief that they were preventing inflammation. Mondeville insisted to the contrary: arrest hemorrhage at once!

His procedures varied according to the nature of the source and according to the skills of the surgeon. There were five types of measures to be used: compression, hemostatic medications, cauterization, direct ligation of vessels, and closure of wounds with mass sutures.

Compresiom: One used a finger tip placed on the opening in a small vessel, holding it in place for up to an hour. If there were many small bleeders he packed The wound with oakum tents soaked with egg-white and warm wine or vinegar, diluted with water.

Hemostatic Medications: He put them in the open wounds, including those with some loss of tissue. He used Simples and Compounds, different ones to suit the site of the bleeder. They were powders dusted on wine-moistened pads of oakum that had been dipped in egg-white. I shall list only a few: 1. On small vessels he used a powder of 3 parts of quick-lime, 2 parts of frankincense, 1 part of sangdragon. 2. For larger vessels: Galen's powder: 2 parts of white gum of frankincense, 1 part aloes and enough incinerated mullein and egg-white to make it the consistency of honey. 3. Guy de Chauliac used his own red powder: One part each of bol d'armenie and incinerated oak galls; 1/3 part each of sangdragon, frankincense, aloes and mastic. Then he added 1/4 part of chopped mullein. Others are described in the treatises of both men.

Cauterization: Mondeville used it frequently in open wounds and on bleeding surfaces. He applied a red-hot iron or used corrosives like couperose, sublimated arsenic and vitriol. Sometimes he used boiling oil as advocated by Avicenna who claimed it prevented inflammation as well.

Ligation of Vessels: Although there were several methods, few surgeons practiced ligation before the 17thC.

1. *The Ancients passed a threaded needle through the skin a under the vessel on both sides of the bleeding site (ie or both ends of a severed vessel), and knotted the threads on the surface.*
2. *Avicenna isolated the bleeder and caught it with a hook and ligated it.*
3. *Lanfranchi incised the skin and soft tissues to identify the vessel, twisted it and ligate it, or both ends if severed.*

What about the knot? Mondeville objected to burying the knot because it could be difficult to find and to remove it later on. Guy left it alone, to extrude itself. Mondeville tied the knot outside but it included only the vessel which he had dissected free of the nearby nerves, etc. Later, he proposed a second method for knotting within the wound in such a way that he could remove it with a simple maneuver.[124]

Close the Wound With Mass Sutures: This was the most common method when sutures were used. Mondeville advised placing the sutures to approximate the ends of the vessels as accurately as feasible, but gave few details. Guy said, "if we do not sew them separately, we bring them together with the other tissues." Separate (interrupted) sutures were used (ie not deeply inserted) if the blood vessels did not spurt. Pressure pads were laid over the sutured wound. Running sutures with deeply inserted bites were used and pulled tight if the bleeding was brisk and if it spurted.

In all cases, the limb was elevated.

Anaesthesia

I am astonished that Mondeville did not mention the measures used by his Master, Theodoric, to try to narcotize the victim before amputating the limb.[125] *Guy said that he soaked a new sponge in a mixture of opium, juices of morelle, hyoscyamus, mandragore, ivy, hemlock and lettuce. It was dried by sunlight and, when needed, he dipped it in warm water and had had the patient inhale until he fell asleep. He awakened the patient with another sponge wet with vinegar, placed under his nose. Or, better, he applied the juices of rue and fennel in the nostriuls and ear canals.* (See Appendix).

Sometimes he tried to narcotize the patient by dosing him with opium.

(Now we return to Mondeville's text of Chapter 6 On Amputating Limbs and Sawing Bones. LDR)

We shall consider seven points: The signs of gangrene (corruption) that requires amputation; where to amputate; the armamentarium that the surgeon should have at hand; how to prepare the limb; the operative technique; treating the wound afterwards; explanatory comments.

---

[124] Mondeville discusses and describes these issues in full detail in Part IV of Ch. 1 of Doct. I, Tr. II. See p. 395 and fn 282 (LDR).

[125] I cite CG Cumston for Mondeville's method, Buffalo Med. Journal, Vol. 58, 1903, p.655. "Suntque chirurgici dant medicinas obdormativas ut patients non sentient incisionem, velut opium, succus morellae, hyosciami, mandragorae, cicutae, lactucae. Imbibunt ineis spongiam novum, et permittunt

I. The Signs (Symptoms) of gangrene.

The gangrenous region has a leaden hue. When you press firmly on the affected part, you dent only the skin, and the hollow persists after you remove your finger. The part is completely numb, and, in time, it turns black, as if it was charred by a flame.

II. Where to Amputate

When a surgeon encounters gangrene that has resisted all other treatments, he must amputate the limb to save the patient's life as well as to arrest the advance of the gangrene. Thus, if the end of a digit is gangrenous, amputate through the next joint; that is the rule to follow elsewhere. For example, if the gangrene reaches the palm, amputate at the wrist. If it involves the forearm amputate at the elbow. But if it extends into the upper arm the patient cannot survive. What I have said about the arm applies as well to the toes, the foot and the lower leg; if the gangrene reaches the thigh the patient will die. All the Medical authors who have written on the subject and authors of surgical handbooks prior to our own era have followed that doctrine, and that group includes Avicenna (*Canon*, Bk. IV, Fol. 4, Ch. 4). He wrote. "The treatment of a corruption consists of local excision or wide resection. If it has entered the bone marrow one can cure it only by getting rid of the entire bone and its marrow, where it joins the next bone." Rhazes agreed, in his *Albucasis*, Bk.II, Ch.8.

---

eam ad solam exsiccari; et quando erit necesse, mittunt illam spongiam in aqua calida, et dant eam ad odorandum tantum usquequo patients capiunt somnum. Et postea cum alia spongia in aceto infusa, naribus applicata expergefaciunt etevigilant eos."

Cumston used a 14thC Ms of Mondeville,clearly different from that used by Nicaise. I believe that Mondeville knew and used the methods of Theodoric, who in turn knew the same techniques cited in the Bamberg Surgery (Corner, GW, On Early Salernitan Surgery and Especially the Bamberg Surgery, Bull. Instit. Of Hist. Of Med. Vol. I, pp.1-32, 1937) and by Gariopontus,who used Arabic sources (see Bass-Handerson, op.cit, Vol. I, p. 262) (LDR).

However, both of them seem to me to contradict themselves in stating that you must amputate at the elbow for gangrene limited to the forearm near the hand. They suggest that you cannot cure gangrene that occurs between two uninvolved joints, and then they recommend and suggest that we saw the corrupted bone. Now, it is agreed that we should not saw bone at the joints, but we should disarticulate by cutting ligaments and cords. The reason for that policy is the observation that the gangrene extends, and I claim that it is not the fault of the operation but of the ignorance of the surgeons who apply suppurative topicals (ie on the sawed amputation stump). We moderns do it differently. We have seen great numbers of cases in the past and we continue to see them now in our daily practice, persons who recover and form perfect scars, who have been amputated by sawing at the joints as well as in between. However, they heal more rapidly after disarticulations.

III. The Armamentarium

The surgeon should be prepared to use strong hemostatic medicines, and he should have large square (ie in cross-section) needles threaded with tough threads with which to ligate the artry when that is necessary. He will need sponges to mop up the blood, and many hot iron and gold cauteries, and sharp cutting tools in sizes and shapes to suit the case at hand. He will need fresh water and rose-water to revive the patient who faints. He must have a few assistants and several attendant friends who know what to expect in operations of this type.

IV. Preparing the Patient

Encircle the limb with two tightly wrapped cords or towels. Place one just below the joint and the other on the healthy side of the site of the amputation. Two aides must grasp the limb securely, above and below, to enable the surgeon to do his work in a stable field. The tight bindings will reduce the patient's sensibility. The limb should be elevated to lessen the loss of blood.

V. The Technique

The Operation should avoid causing unnecessary suffering. Disarticulation is not so difficult and requires less skill for an experienced surgeon who knows his anatomy.

If you must saw a bone do as follows: Make a circular incision between the two bindings. Cut right down to the bone with a hot iron or gold cautery, as broad and as slim as a knife blade. Then cover both surfaces of the soft tissues with damp cloths to spare them injury by the saw. Use the correct tool and saw through the bone with deft, light and smooth strokes.

VI. Treating The Wound

Stem the hemorrhage as I have described. In any case, use hemostatic topicals. At the second dressing, follow the practices I described in the chapters in treating wounds

VII. Two Explanatory Comments

A. Avicenna (Bk. IV, Fol. 5, Tr. 2), in discussing fractures, said that an opening into the marrow is fatal, as is commonly taught, because during life the marrow is liquid and thin, like fat melted in a fire. When it is exposed it solidifies as does melted wax. The *Totum Continens* (Bk. VI, Part 2) repeats everything Avicenna said. But I think they did not include limiting the time of exposure to air. When the marrow, damaged or not, is exposed it suppurates. Its normal state is warm and loose, especially if you treat the wound in the old-fashioned way.

B. Contrary to the common belief, the wound or saw-cut into the marrow is not the cause for death after amputations through the forearm or elsewhere outside of joints, nor does it delay the recovery, when compared with those who have undergone amputations at the joints. Indeed, if you remove all the marrow that escapes at the time that you remove the bone, the healing is not delayed. You do better to remove all of it than to leave it to suppurate and to infect the stump. It will attract fluid and wet the wound and delay or prevent healing.

The real sources of the increased risk of amputations are the divided tissues: in disarticulations they can find new attachments, whereas, after the bone-end and marrow have been removed, the divided muscles cannot take root. The tissues regenerate very slowly and the wound seems not to heal. Then the blame falls on the incompetent surgeon and the recalcitrant patient. If you say that disarticulated stumps heal slowly because Nature delivers moist phlegm into the joint, you admit that healing will not occur until the region is dry, then you can say that wounds into joints heal more slowly than others because the phlegm accumulates. But that cannot happen when the joint has been eliminated. But, really, it is not a matter of accumulation. The humors cannot accumulate any more there than elsewhere, because the humors circulate in periarticular tissues as it does everywhere. It is the same for amputations through bones in the hand as at the middle of the forearm or at other sites.

CHAPTER 7.

PRESERVING AND EMBALMING CADAVERS

The same surgeon who is asked to correct an infant's congenital imperfections also is called upon to preserve cadavers; it is an honest and lucrative occupation, and it will serve you well to learn it.

There are three categories: one requires that you do very little, as when you need keep the body of a pauper or some more wealthy person until it is interred within three days after death in summer time or four days in winters. Another way is needed to preserve the body of a prosperous person for a longer time; a mid-level nobleman (ie a baron) or an upper-rank military man, for a month or longer. A third case is the body of a king or queen or a ruling pope or a high-ranking prelate that will lay in state with its face exposed for a long time. In this chapter we will omit the description of embalming a pauper; it neither is useful or necessary, and it pays nothing.

I. Short-term (3-4 days)

To preserve a prosperous person's body, with or without exposure of the face will be almost the same for both.

When a surgeon is called upon to insure that no deterioration in the appearance will occur before the funeral, and when the fee has been agreed to, he will do as follows: He will have a large amount of a red powder that contains equal amounts of frankincense, mastic, sandragon and bol d'armenie and fine wheaten mill-dust flour in an amount equal to all the others. He will need about twenty bandages made from a fine strong, closely woven cloth, each about a hand-breadth wide and six yards long. Also he will have some clumps of hempen lint with which to make small cushions, and three small packs one of them about the size of a suppository which will be shaped to a point at one end, and two will be about the size of the small finger. He will need at least ten long, thick and square (ie four edges) needles threaded with a strong thick thread (ie waxed), and about ten yards of a wide waxed linen cloth.

When thus supplied, mix the powder with some egg-white to make it as thick as honey, and saturate the bandages, pads and packs. Insert the large suppository into the anus and pack four layers of pads against it, held securely in place with bandages which will make a belt around the flanks, with four turns, two from the front and two from the back, coming down from the belt and over the pads at the anus. Then close the mouth with sutures and stuff the nostrils with the two small packs. Place pads over them and bandage them in place, carefully and securely. Then begin at the feet and bandage the legs separately up to the hips, then wrap the torso up to the shoulders. Make it tight and fitted such that they will remain in place after they are dry. When you are satisfied, cover the body with two layers of the waxed linen, sewing the seams with waxed thread to seal them. Put the body in the casket and surround it with flowers and herbs and twigs and other sweet-smelling materials, If the face must be exposed, do not suture the mouth or pack the nostrils. Instead, insert about 6 drams of mercury into each side and then stuff in some silk strips to prevent its escape.

II. To Preserve A Rich or Important Person's Body

For as long as a year before interment, or when it must be transported a long distance to his home, with or without an exposed face:

In addition to all the above, the surgeon will need a compound of colocynth and borax boiled with honey and water with which to impregnate the bandages, lint packs etc. He will need 1 lb. each of myrrh and aloes and 1/2lb. each of camphor and salt. Grind the solids and mix them to the consistency of honey with rose-water and vinegar. Add 3 dr. of mercury and a small amount of costus and 10 lbs of wax.

When the mixture is ready. The surgeon begins by inserting the wet suppository into the anus. Then he rolls the body side to side, tilting it head down and feet up. Then he presses the belly until the suppository is expelled, followed by as much fecal matter as he can squeeze out. Then he anoints the entire body with the mixture before he reinserts the soaked suppository and packs. He bandages and encloses the body in the waxed cloth, as before, but here he adds a second layer of the waxed cloth, and sews the seams of each layer on opposite sides of the body. Then he wraps the body in a rawhide sewed with waxed thread. Finally he wraps it in a thin lead sheet, weighing at least two hundred pounds. He secures the folded ends with two metal strips running lengthwise over the body, the ends of which are joined by (pinched) metal rings. Then he uses two more metal strips laid crosswise over the body and the longitudinal strips, and he uses more iron rings to pinch and hold the ends. Then he places the lead-sheathed body in a casket. Enclosed in that way, decayed or not, the body can be preserved and transported across the universe without emitting the slightest stench, until the Day of Judgment, unless the the enclosure is breached in a collision or by another external violation.

III. A Long-Term Preservation with an Exposed Face

Here the task is difficult because of the irrepressible upward diffusion of fumes from the decaying body. Indeed, the body which in life is vigorous and robust, begins to decay and putrefy within eight hours after death, lacking nourishment and vital spirit. Some bodies soon turn black or brown, some soon turn cold and dry and remain firm.

When death occurs in a cold and arid region and during a cold and dry season and when the moon is in its third quarter—when it, too, is cool and dry—the bodies seem to maintain themselves for a longer time and can be preserved from corruption if great care is taken. When some or all of those conditions are opposite, all efforts are opposed. A young person, who in life had a pale or pink color, who was vivacious and relaxed in manner, deteriorates rapidly after death. It is different with old and wizened people with somber gray coloration, etc. They are not so rapidly disfigured as the vibrant youths, and their bodies can be more easily preserved.

As we described it, you know how to bandage the feet and legs, first separately and then together up to the hips; how to evacuate the feces and plug the the anus and nostrils, etc. We now add two items: 1. Anoint the face with the balm to preserve it. Although it is claimed that it is effective, I myself have not tried the methods described by physicians in their handbooks, except for one source, The Practice of Simple Medications, titled *Circa Instans*.[126] I have little faith in the others. Besides, I embalmed the bodies of two French Kings and saw little or no benefit from my use of the balm. Perhaps it was because they were so loose-bodied, with soft and delicate and beautiful faces. Perhaps the balm was stale. 2. When the body must be preserved for longer than four nights, and when granted dispensation by the Church at Rome, you open the abdominal wall with an incision from the chest to the pubis in men and with a double incision from the substernal notch into both flanks in

---

[126] Nicause said this was attributed to Matthew Platearis, a Salernitan physician, fl. C.1130-1150. *The Book of Medical Simples* in the chapter titled Circa instans negotium, often was referred to with Mondeville's name (EN).

women, and then fold back the flap based on the pubis (ie the sexual parts). Through those incisions, remove all of the abdominal viscera down to the anus. Then liberally apply the following powder to all the inner surfaces: Equal parts of myrrh, mummy, aloes, spices and herbs that prevent decay and lessen bad odors, including roses, violets, camphor, sandalwood and musk. Add salt to equal the total of the others. Fill the abdomen with sweet-smelling herbs such as camomille, melilot, pennyroyal, mint, menthastrum, balsamita, etc., enough to fill out the normal contours. Then sew the incisons and complete the embalming process. If you must preserve the viscera, cover them with the powder and place them in a silver or lead jar which you seal with many layers of waxed cloth.[127]

CHAPTER 8.

A SUDDEN GASSINESS THAT COURSES THROUGH THE BODY AND CAUSES CHILLS AND FEVER[128]

Three Parts:

I.   *Definition*: The Title of the Chapter is self-explanatory.
II. *Causes*: The proximate and most immediate cause is an infectious and toxic substance somewhere in the body—perhaps in a foot or a thumb.

---

[127] Nicaise wrote this about the two kings embalmed by Mondeville, from which we may determine the date of his writing this treatise. Philip the Fair's father (Philip III) died at age 40, in 1285, when Mondeville was very young and long before he became a Royal Surgeon. In 1304 he was teaching Anatomy at Montpellier. Philip the Fair (IV) died in 1314 at age 46 and Louis the Quarreler (ie Louis X) died in 1316 at the age of 27, and after him Philip the Tall (V) died in 1322, age 29. What Mondeville wrote about the complexion and manner of the kings could suit all. However, we can deduce that Mondeville wrote this treatise in 1316 or soon after. (EN)

[128] The anaerobic gas-forming infection described here seems to come from an undetected small wound. The author wastes little effort in describing his failures and fatalities (LDR).

III. The *Symptoms* (ie Signs) are a rapid spread through the affected limb, like an arrow in flight. Often there is fever, but not always. Pain and heat usually are limited to the place of origin.

Treatment

I   Diet: The victim should abstain from salty, spicy and dry foods, from roasted meats, legumes, spiced meats, cheeses and foods like them, as they are listed in the medical handbooks.

II.   Medicines: The bad substances at fault are digested by cool and greasy topicals. Use purges that suppress the acuity and the gassiness. Follow the purge with a theriac or an opiate of equal potency.

III. Operative Procedures: When the measures in I and II fail, do this: Wait until the gas collects in the least dangerous place. Then wrap a tight band wet with the theriac above and below the site to trap the gas. Incise deeply between the bands, using a hot cautery, and release the gas. Keep open the wound for a long time. Then treat it in stages as for any wound. If you fail to cure the patient completely within a reasonable time, use the cautery and incise again between two restrictive bandages.

CHAPTER 9.

THE TREATMENT OF BLOWS (IE WHIPPING) AND OTHER CONTUSIONS IN PERSONS WHO HAVE BEEN HUNG, RACKED OR SUBMERGED OR OTHERWISE TORTURED.

I. The Protocol of the old-fashioned treatments:

A. *Diet:* Insistence on light foods in small amounts for three or four days, followed by a gradual return to the patient's customary meals.

B. *Evacuations:* A diverting phlebotomy should be made soon after the injury. If the upper body is damaged, bleed from the foot. If the legs are damaged, bleed from the hands, from the opposite side from the injury. If the injuries are in many places, bleed from sites chosen by the surgeon as he deems best. If the patient remains constipated for three or four days, prescribe a suppository or a clyster.

During the four or five days after the insult, in addition to the above measures, prescribe these potions: For constipation, take 2 dr. of rhubarb and 1 oz. of syrup of roses or violets. Or offer a tea of large comfrey. Or, better, provide a good chew of comfrey roots; that is the remedy commonly used by the peasants. It works.

C. *Topicals:* At the outset, use warm oils of myrtle or roses with added myrtle powder for three or four days or until the contused site is past concern for suppuration. When that is behind you and the bruise no longer is enlarging, thanks to the topicals, and after the phlebotomy, anoint the parts before the morning and evening meals, with this ointment: 3 oz. of wax, 6 oz. of resin, 8 oz. of terebinth, 2 lbs. of oil, 1/2 oz each of frankincense and fenugreek. Combine them all. After the inunction, immerse the victim in a tub bath containing resolutive herbs. On leaving the tub, anoint him again. Repeat the bath every four days until the patient recovers, but use the inunctions daily.

II. The Modern Method

I described this in Treatise II, Doctrine I, Chapter 12 'On Contusions', and in Chapter 4, dealing with contusions of the head without open wounds. I refer the Reader to those chapters.

III. Topicals in Common Use

Cover the entire patient excepting his head in a heap of horse-manure for three days and nights, or wrap and sew him in a warm sheep-skin, or use a recently flayed horse-hide, immediately after he has been released from the flagellation and other tortures. I do not approve of those measures.

IV. Explanations

A. When there are multiple and various injuries and the injured victim hurts badly or cannot move or is numb—one or all of them—or when he is comatose or has collapsed, especially after a severe fall (ie thrown from a wall!) or beating, and when you see dark blood at a wound or a dislocated or fractured bone, the treatment should include all that we use for contusions.

B. When in addition to the bruises there are open wounds or dislocations or fractures, treat them as described in our chapters dealing with those subjects. If there are more than one such lesions, refer to all the pertinent chapters.

CHAPTER 10.

THE RELIEF OF PAIN IN EXTERNAL STRUCTURES THAT ARE NOT DAMAGED AND PAIN IN INTERNAL ORGANS. THE SURGICAL PROCEDURES

I. Four Considerations

A. *Definitions*: Pain is a sensation caused by an abnormal or harmful action. If that is sudden or intense, one feels pain.

B. *Varieties*: There are three types of pain: One causes an illness, such as a syncope, or appears at the onset or just before an illness. The second is a malady of itself, such as a headache which does not precede or linger after another illness, as after a fever that has subsided or a headache which accompanies a fever. The third is a complication of another illness, such as one I just described which accompanies a fever. Avicenna claimed that there are several more types of pain, but Galen said that all of them can be put into these four categories: 1. Ague which is ulcerative or prickly. 2. Extensive. 3. Serious or inflammatory. 4. Inflative (*Aphorisms,* Part 2.)

C. *Causes*: Two General Causes: 1. A sudden change in the complexion. 2. A wound. These differ in that one includes external causes: a blow, a fall, hard labor, coitus; the other has internal effects: heat, cold, dryness, humidity, with or without matter (ie pus) or gas, etc. At times we cannot ascertain which of them is the cause for the pain, the above-named and many others, such as a cramped position in bed, a fall that happens to a drunkard, and other things that the patient may not be aware of. Usually he believes, and the surgeon, too, that the cause is an internal one. The surgeon should be aware that the initiating cause may have been external whereas the immediate precipitating cause cause may be a bad humor. For example, a laborer may release cold humors tucked away somewhere inside himself and he develops a cold, non-inflammatory illness, such as epilepsy or paralysis, but the real cause was the laborious activity and the proximate cause was the phlegm (ie the cold humor). The external causes are uncountable, whereas the internal causes most often are the four humors and gas (ie wind).

D. *Symptoms (signs).* The diagnostic signs are fairly well described in the preceding section. It will be sufficient for the surgeon to add what his hands and his eyes tell him, and what information he obtains from the patient and a few other details. The pain caused by humors is localized and moves little or none at all, whereas pain due to gassiness (ie colic) shifts continuously. However, on occasion a chronic pain of humoral origin may be dislodged and felt in another place, the result of a change in the weather or the good effects of the topicals or other such reasons. The shift occurs gradually and the pain itself is not violent. When it happens you take it as a favorable sign that the ailment is not stubbornly rooted.

The pain of internal causes is described in this mnemonic: The faulty humor penetrates, it stabs and spreads, it increasess and wanders. In other words, sanguinous humors penetrate; bilious humors sting; phlegm spreads; melancholy aggravates and wind travels. The symptoms are distinctive. More on this subject may be found in our chapters on the formation of humors and on aposthems.

Sometimes there is another symptom besides the penetrating pain that we described for ailments due to sanguinous humors, and that is its burning quality, like a burst of fire that spreads across the chest. That tells you that there is a lot of corrupted blood in the chest and if you do not treat it a putrid fever will ensue.

When a patient suffers from severe heavy discomfort, as if his body is overburdened and he cannot bestir his limbs, that indicates that the cause is several bad humors.

II Treatments: General and Particular

A. General Measures: All pain can be relieved with one of three methods:

By evacuations, when the humor is polluted as well as being excessive, being repleted and of poor quality to begin with. By anodynes (ie as topicals) that warm and dry the plethoric patient. By narcotics when you fail to ease the pain or realize beforehand and it is beyond the scope of the other measures, as in desperate cases of colic, painful joints (ie gout) and in violent cramping in the legs. Furthermore, all pains are treated by opposites, by evacuations and by suppressing the causes.

B. Special Measures: For relief of pain due to external causes, refer to the chapters on wounds, contusions etc. and to the two chapters in Doctrine I of Treatise II that deal with Contusions of the Head without external Wounds.

For relief of pain due to internal disorders: When the pain is not intolerable and not inflamed and is penetrating, as is caused by sanguinous humors, use a diverting phlebotomy and moderate abstinence from wine and a dietary regimen that is contrary to the cause. We described it in Doctrine II of this Treatise III, in Chapters 2 and 3 where we consider aposthems that derive from single unmixed simple humors, and we describe preventive treatments. Inflamed pain of sanguinous origin is treated as above, by a strong bleed, abstinance from wine and a very strict diet. In addition, we must consider the vital force, because that kind of pain is unrelenting and it forewarns you of more serious complications, such as a synochic (ie continuous) fever.

In addition to these precepts, I repeat my suggestion that you consult the chapters cited above.

Bilious pain is prickly like stings of nettles or pins. It shifts about from place to place in the skin and is prickly and it itches. If it is not too severe, treat it by avoidance of all acidic, fried and salty foods, by massage and by a general regimen contary to the causes. If it is more than local and the comfort is intense, first scrub the whole body and apply the topicals described above on the clean skin.

For phlegmatic pains use vigorous massage and repeated baths and abstention from meat. If the phlegmatic discomfort is widespread, first bleed the patient and follow that with purges which are effective against phlegm.

For Pain due to melancholic humors use rubbing, phlebotomy, baths with fresh water, suitable purges, and a dietary regimen contrary to the causes.

For pain due to gas that begins suddenly and is extensive and shifts about the body, we will describe treatments later. For belly gas (ie tympanites) refer to our chapters on dropsy (ie ascites) and on gassy aposthems.

Pain of sudden onset with no discernible external cause, that may be due to any of the corrupted humors. If it is violent and unbearable, perhaps accompanied by other symptoms, is treated with diverting phlebotomies as described by Galen (*Aphorisms*, Part 1), whether the cause is a sanguinous humor or others, to the degree that the patient's vitality will allow it. Indeed, severe local pain acts to attract humors, both the good and the bad kinds, and causes an abscess at the site. If the pain or other distressing symptoms is the result of excessive congestion of the veins with humors that are not yet corrupted, treat that with phlebotomy in the popliteal region for quick relief. Then anoint the affected region with rosat oil and other warm anodynes.

Pain that follows the strains of arduous labor is relieved by simple rest and the applications of analgesic ointments, such as we have described. However, if that treatment is inadequate, use phlebotomy, generous diets and the medications which are time-tested for their good results. The surgeon will use his own experience with them and will choose those that befit the case.

Pains of internal origin such as colic, kidney-pains etc., and pains due to both internal and external causes (which causes themselves are considered to be surgical), and headaches, pains in the flanks and back, because they are not entirely surgical problems, will not have special chapters in this book. The same is true for medical illnesses, as epilepsy, both kinds of dropsy, jaundice, continuous fever, tertiary fever, and others like them in great numbers, about which I have collected and filed away my observations (experiences), admirably and accurately and with certainty, some fairly recently, at great monetary expense to myself. I plan to write about them after I finish this treatise, in many chapters, God willing. I defer that now for three reasons. 1. At present I am entirely occupied with surgery. 2. I will write later on when I can take up the topics and arrange them. 3. I fear that the reigning Masters at the Faculty of Medicine will not find it acceptable at this time.[129]

III. Six Explanations Of Arcane Matters

A. Take note: In the definitions we note that a single cause of pain, for example a bad complexion, can affect the patient in three ways. It may have very little effect on the pain, as in malnourished patients. It may have potent effects, as in ophthalmias. Or the effect may be extremely potent, as in apoplexy.

B. In this section we will not repeat what already has been written about pain itself as a malady (ie dyscrasia) that can play a role in other disorders.

C. Pains without detectable causes may be prodromal signs of other illnesses. They deplete the vital forces, they cause inflammation and sometimes syncope. Whenever a severe pain strikes a patient he cannot function as long as he suffers.

D. Three medical measures can be used in times of need, with marvelous effects in easing suffering, in calming the patient and curing him. They are phlebotomy, clysters and the sweat-box (calefactory).

A properly conducted phlebotomy immediately eases some kinds of pain and relieves such illnesses as continuous fevers. A clyster will relieve a colic and the like. A sweat-box unit will ease some strong pains, such as pleurisy. That may be a clue as to what next to do. If it fails to give relief, at least it does no harm, and it tells you to go on to a general purgation (ie a phlebotomy or a clyster etc.).

E. Avicenna (Bk. I, Fol. 4, Chs. 20 and 21) wrote, "During a phlebotomy on a very cool-complexioned patient during an episode of severe pain, you should observe him carefully. His

---

[129] That telling statement may explain why Mondeville's great book languished during the decades and centuries after he died The clerical physicians who dominated the medical establishment were less than pleased by the barbs tossed at them by their acerbic surgical colleague (LDR).

humors may be activated by the pain as well as by the phlebotomy and, when the bleeding ends, they may be attracted to the seat of the pain more than beforehand. But Galen seems to have given a contrary opinion (*Aphorisms*, Bk. I, No. 24), "When there is a spell of severe pain, in acute inflammation, in an attack of feverish chills, there is no medicine that is better than a bleed, etc.) The resolution of that argument is found in Avicenna's words (*Canon*, Bk. I, Fol. 4, Chs. 1 and 3). He said that there is nothing else available other than phlebotomy when pain is unbearable. You must bleed almost to the point of syncope or until all the bad matter has been expelled, to prevent a total collapse of vitality caused by the pain. It is the only available means for relief. But when the suffering is not extreme, a diverting bleed or clyster is not necessary for treatment. The pain caused by the bleeding balances out the relief (ie obtained by it) and the humors agitated by the bleed are strongly attracted to the lesion. The old-time practitioners followed their own dictates. Today, the physicians and the operating and consulting surgeons hold an opposing opinion, even though there may some truth in the old one. Wherever he may be, if a person suffers, you should bleed him at once, of course in consideration of all the Contingencies. Even with that good advice they added this counsel: If you advise phlebotomy in cases of excessive and oppressive pain, and the patient dies, then the patients entourage can blame you for the death. If you recommend phlebotomy for less severe or moderate pain, to be repeated until relief is obtained, the patient may not agree to it because it would be to no avail to treat pain that no longer esists.[130]

F.   All fatty topicals ease pain, especially one that contains the fat of a drake. The same holds for all similar medicaments that dull the pain; they include most of the maturatives.

---

[130] Therefore, bleed before the pain overwhelms the patient and you will avoid the complication, and bleed before it may be refused and avoid the loss of your fee (LDR).

CHAPTER 11.

EMBELLISHING MEN, NOT INCLUDING ITEMS FOR CERTAIN PARTS OF THE BODY AND THOSE THAT ARE USED FOR WOMEN RATHER THAN MEN

Two Parts:

I. Four Preliminary Considerations.

A. A surgeon needs to pay attention to relevant matters in this as well as the ensuing chapters that deal with itching scalps, impetigo of the limbs, morphea (vitiligo), albarras (scaly skin), and all the chapters in which facial maladies are discussed.

B. Here we will deal only with a few matters concerned with the improvement of the appearance of the face as a whole. In Doctrine II of this Treatise we will deal with lips, teeth etc.

C. Inasmuch as pure embellishment of itself is contrary to God and to justice, and usually it has nothing to do with illness, and it serves to hide something fraudulently, I will spend little time with it. Furthermore, I do not like the subject. However, an urban or provincial surgeon who must attend wealthy women or members of the royal court may garner a good reputation for his skills in this field. He will become a favorite of the women, and that is no small matter these days. Indeed, one cannot get ahead in this profession without their favor, especially among their men-folk. In some cases it is more desirable than the good favor of the Pope or of The Lord.

Here are four General Principles (ie for treating men). 1. To correct unsightly blemishes on the face, apply a corrosive under the chin, after suitable purgation. 2. Scarifications work just as well for lesions of the legs. 3. Cups applied in the hollow of the neck (ie the posterior triangle) also work well. 4. Make your applications when the patient is in his tub bath or fumigator or sweat box.

II The Techniques

The facts are that men's faces develop crops of blemishes, especially among the rich, the nobility and the lascivious ones. Those crops are lucrative sources for cultivators of the crops, the impatient peasant therapists. The six lesions are treated as follows.

1. *Excessive Redness* : Treat with general purgation, including laxatives and bleeding. Also use locally diversive evacuations such as leeches applied alongside the nose and near the ears; cups applied between the shoulder blades; spot-applications of corrosiove topicals. After placing a cup beneath the chin apply a plaster to the facial areas, containing ashes of burnt shoe leather and honey. Leave it on during a full day and night; the benefit should last a year. Another plaster: 3 oz. each of powdered iron-rust and burnt beef hooves and spleens, 2 oz. of salt-free fresh lard, juice of plantains and rose-water. Boil it down to the consistency of a plaster. Another plaster: fresh cheese mashed with the juice of henbane leaves.

2. *Pale Areas*: To add color use this: Dried and powdered white bryony (viticella) mixed with rose-water. Dab it on with cotton pledgets. Another: mash a hard-cooked egg-yolk with some mild red wine; filter it and apply the liquid. Another: white mustard (seeds) and red arsenic, or powdered melon rind. Mix with milk and apply it at bedtime. In the morning wash the face with a tea of melons or dried violet bulbs. Repeat daily for a full week. Another: Take sawdust of brazil tree wood[131] with a small amount of alum. Wet it with rose-water or violet-water and dab it on the face.

3. *Sunburn or wind-burn* : Prevention: Apply egg-white alone or a decoction of seeds of vetch (orobe), or the two together.

---

[131] Bresillet: A strange anachronism, almost two centuries before Brazil was 'discovered' and named more properly it was Sapanwood. See Antidotary (LDR).

Another: Equal parts of a wheaten flour paste and egg-white. Treatment: Crushed lily-roots boiled in water and mashed with fresh pork-lard, melted and filtered. Then add a powder of two parts each of mastic, olibanum and white lead (ceruse). Then add one part of camphor, and enough rose-water to make a paste.

4. *Dark and Unsightly Blemishes:* Apply a decoction of orache. Another: Roots of lovage or warm rye-bread and add rose-water. Another: Anoint with a water of white tartar with added tragacanth or flour of vetch mixed with honey. Another: duck-fat with rose water. Another: oil of fenugreek or myrrh beaten with egg-whites until foamy. I knew an experienced practitioner who swears by dry hysop made into a potion to restore a good color.

5. *To Permanently remove Unwanted Hair.* See the next chapter.

6. *Lack of Facial Hair.* If we exclude eunuchs, treat by massaging with abrotonin or a pumice as you would use for preparing a sheet of parchment.

CHAPTER 12.

EMBELLISHMENT IN WOMEN. ARE THERE MORE ISSUES IN THIS SUBJECT? YES. AT LEAST TWO CATEGORIES.

A. The five Measures for the body as a whole.:

1. *The Sweat Box:* Some are public as are found in some rural retreats and villages that are available to all women. Others are private in private homes, and they are equipped like the public ones, but they have been set up to escape public exposure of the woman. Often they are seen in houses for the mistresses of wealthy men. Sometimes they provide for women who are banned from the public places and who have none in their own homes. In some the sweats are induced in a barrel. First, they heat tiles or river-stones in a hot charcoal fire and place them at the bottom of the barrel.

Over them they lay a floor-board with many holes, and they pour water through it onto the hot rocks. The woman, covered with heavy wraps, squats over the perforations. She remains until she is well-sweated.

2.   *The Massages:* One may use hands or, if preferred, a soft sponge which gently attracts the impurities and absorbs them.

3.   *The Baths:* After the massage, the woman enters a fresh-water bath in which is submerged a sac filled with bran and the flours of fava-beans and lupins. If she desires, she may steep sacs containing lovage, lily-roots, guimauve and other herbs in the bath water. Albertus Magnus (*The Spirits*) cited Pliny who wrote that Poppea, Nero's mistress, bathed in asses' milk to whiten her skin. If one does not have enough of that milk, I advise her to bathe in a decoction of a certain fish, the sting-ray or of a sea-urchin.[132].

4.   *Ointments* : The following are added to a waxy base: Oils of roses and violets and chicken fat. Melt them and add white wax and keep it bubbling until it thickens. Then add finely ground white lead and place it all over a gentle fire while you mix in some minced nutmeg and cloves to provide fragrance. Some add camphor, which I disfavor because its pungency thwarts the libidinous effects of the rest. Ovid wrote, "Camphor in the nose makes men chaste." Another ointment is made with white tartar, white tragacanth and some powdered abstersives.

5.   *Lotions (ablutions).* Boil lovage, flours of chick-peas and fava-beans, radish seed, white tragacanth in water, and add some milk. It will soften the skin of the whole body. An erotic man may apply the lotion on his mate, and she may reciprocate, if so desired. When the bath water or the lotion are objected to because they are too thick or too cloudy, you may filter them through strips of cloth.

---

[132]   Is this another of Mondeville's teases? The Reader will detect Mondeville's distaste with the cosmetic treatments and misogyny in his wry asides (LDR).

CHAPTER 13.

PARTICULAR EMBELLISHMENTS FOR SOME PARTS OF WOMEN

A. Preliminary Comments

I. The General Care of the female body as a whole (ie Chapter 12) came before the care of particular regions. A reverse of that order would nullify what came before it. So, too, we must have a suitable order for presenting all the matters of care for special regions. Those will be: 1. The sexual parts. 2. The breasts. 3. The armpits. 4. The hair. 5. The face. 6. The neck. The hands will not be included here.

II. Avicenna (Bk.I, Fol.7, Ch.1) described a depilatory, but neither he nor any other author of medical or surgical works or practitioners have described an eradicator (ie permanent) of hair, although some have given false claims. 1. Because a depilatory causes a skin-irritation that lasts for several days. Then the hair begins to grow and the surface of the skin is rough with the stubble as if it had been shaved by a razor. 2. The misled practitioners mislead their patients. After the depilatory they tell them to apply medicines to plug the hair-pores and to use narcotics to prevent the growth of new hair. All of that is useless because it does not eradicate the roots or prevent their growth, as we do when we prune trees in such a way as to prevent new growth.

III. The depilatory of Avicenna and others is made with three parts of quick-lime, one part of orpiment and one-half part of aloes. Pulverize them and boil them in water until a dipped feather sheds its plumes; then it will be as thick as a porridge. If it is too thick, add some of the patient's own urine. Apply it freshly made while warm when the patient is in her bath or sweat box or after a long steam-bath, so that the skin-pores are open. When it is made as described (ie as an adherent porridge) it burns the skin (ie too hot for too long), so I make it differently. I use boiling water in a basin set atop another (ie double-boiler) and add the powders and apply it at once, before the patient has a chance to scratch or rub herself (ie it cools). I leave it in place long enough to recite the miserere twice through. Then I try to remove the hair. If they can be lifted away, I irrigate the area with warm water while gently massaging the surface with the palm of my hand to remove all the hair. If you try to remove all the hair of a large area at once, you will excoriate the skin. Afterwards, wash the place with warm water followed by bran-water. Then apply a soothing ointment of henna and egg-white to lessen the discomforts of the treatment and to smooth the skin. Then use a detergent lotion.

IV. Some practitioners and some handbooks describe at length a way to treat the whole body. First come the baths and the sweat boxes, then the application of the depilatory to the whole body. That stuff needs the art of a master to fashion, at great expense with long effort, all to little or no avail for the patient. Besides, it is not without its risks. If it is over-cooked or if it is applied when it is too hot or for too long a time, it will burn and excoriate everywhere. The female practitioner will tell the patient's husband or lover that her mistress had taken too hot a bath and had not been careful about the depilatory or had not complained. So it is, if one cooks this one too much or too little, and applies it as ordered or otherwise, or makes it as instructed or not, the untoward results occur. If too little or too brief there will be no effect, etc. etc. After the application,she takes to her bath and uses the ointments and the ablutions in that order and follows the entire tedious protocol. An entire day is not enough time to treat the body and its limbs. However, the prudent 'artiste', by setting a great value on personal beauty and by flaunting his work can earn great profits, because the more laborious the process the better the women deem it to be.

V. In summary: The depilatory described above (Avicenna's) has two unpleasant effects: 1 It is malodorous and unpleasant for the woman and her companion. Even after a careful toilette the odor may persist. 2. The skin is burnt and is excoriated.

You may lessen the odor problem by adding aromatic powders: Roses, Cumin, Camphor, Frankincense, Mastic, Musk, Nutmeg, Cloves, etc.

The second problem is relieved with the ointment of henna, as above. You may moisten the depilated skin with vinegar and rose-water or use an ointment of white wax and rosat plus the aromatic powders. Also, you may moisten the skin with the juices of morel, plantain, crassulas, etc. The therapist should keep ready a supply of all the above.

B. The Cosmetic Care Of Seven Parts
Of The Woman's Body

I. The Sexual structures are dealt with in two sections: internal and external. As to their internal parts. some older prostitutes have over-large vaginas by nature or as the result of too much free and well lubricated sexual activity, and they want themselves to be as they were when young, or at least to lose the obvious feature of their trade. Also, some young unmarried women want to restore intact hymens and appear to be virgins when they marry. This is what they do: They put some ground glass on the vulva just when they begin coitus, or they insert some sang-dragon and then a pad or oakum moistened with rain water, or a lotion of astringent plants (rose-petals or anthers, sumac, plantains etc.), or they apply leeches at the introitus taking care not to let them crawl in. When they are removed they leave scabs which are rubbed off during coitus and the blood stains everything (ie as desired). Another trick: they insert a bit of sponge soaked with blood or a thin bladder taken from some fish which they fill with blood. Afterwards they wash the vulva with the juice of a the large comfrey.[133]

Some women devote much attention to the external parts to make themselves more attractive to men. The have three methods. 1. To prevent the growth of hair when it first appears at puberty. 2. To get rid of the hair that has appeared. 3. To prevent regrowth after depilation.

The first is accomplished by rubbing the fuzz away with the blood of a hairless mouse. That was tried on a young servant girl in my own household. Several witnesses told me that it was successful.

The second, the depilation, is performed in 6 ways. a. They use scissors. b. They use razors. c. They avulse hairs with forceps or with their fingers which they make sticky with naval pitch melted on a nearby stove. d. With depilatories which I have described. e. With a piece of cloth covered with naval pitch.[134] f. With a very well-tested

---

[133] This was a well-known astringent with hemostatic properties. Its application here reinforced the man's impression that the blood was the woman's own! (LDR).

[134] Applied, adhered, avulsed (LDR).

depilatory of recent invention, which gets at the roots of the hairs and causes little or no local damage. It has no unpleasant odor and, as far as I can determine, it has not been described for this use anywhere. The medication has been known for a long time as diachylon. and has been in common use. It leaves a white surface after it is removed and it, or something like it, is the only depilatory that delays the regrowth of hair. Apply it so: Spread the diachylon on a piece of strong linen cloth of a size to cover the part. Warm it by the fire and lay it on for several hours during the day; it will do no harm (ie if longer). Use a scissors to clip the surrounding untreated hairs before you remove the cloth to avoid the painful avulsion of hairs whose roots have not been treated.

To prevent any regrowth after the above, use an ointment of psyllium (plantain), vinegar and lacustrine frog-meat. Another: the blood of a turtle—some say it is really a snail and others say it is a tortoise. Another: an oil in which a turtle or a sea-urchin has been cooked until they disintegrate. Another: a decoction of walnut shells in water. Another: sprinkle some iron filings on a glob of coughed up sputum. Let it set on the part for an hour; wash it off and repeat the application seven or more times. Another: apply psyllium soaked in cool water or in the oil made from hyoscyamus seeds that have been heated between cabbage leaves or in ordinary oil. Another: a decoction of the same seeds. Another: ant-eggs in oil. Another: a decoction of sea-urchins. Another: powdered sublimated arsenic diluted with water in which it stays for three hours. Decant the water and mix it with oil of violets for an inunction.

When only a few hairs are the object, as on the woman's face, hands or feet (exceptions are the eyelashes), you may avulse them one at a time and immediately insert a hot needle (ie a gold cautery) into the pore or apply a bit of a corrosive topical as described in Chapter 15 to follow. Someone once told me that you should perform that treatment only when the moon is in a hairy sign (such as Taurus, etc). The hairless signs are Cancer, etc.

II. *The Breasts*: Sometimes you need to prevent their enlargement which embarrasses women or she may ask you to reduce what she has. For those purposes, use the juices of crushed leaves of hemlock or calamint, or simply foment the breasts with rose-water and vinegar or apply plasters of white lead, bol d'armenia, cimolia and terra sigillata— all together or one at a time. More potent are hysocyamus, white poppies and a mucilage of plantains etc. Another: warm vinegar thickened with the powder of the green coccus [135]. This is moderately astringent. Another: powdered cumin with vinegar or honey. Another: bol or terra sigillata or plain clay, or green oak-galls—all together or separately—tempered with warm vinegar. Apply it on a bandage which is left in place for three days.

Some women have a sense of shame which will prevent them from exposing themselves to a surgeon. They fill sacs of a shape of the breasts but smaller. In the morning, having bound their breasts tightly underneath, they affix the sacs inside their undershirts. The women of Montpellier wear tightly laced corsets over their breast but do not lace them tightly over their sexual parts, fearing to do harm if by chance they should fall.

III *The Armpits*: Two problems: 1. The eradication of the pests called lice by Avicenna (Bk. IV, Fol.6, Doct. 5). We will consider them later, in Ch. 20. 2. The relief of bad odor. Do as follows: First recommend foods that are contraries, and prescribe suitable evacuations. Have her chew wild celery and then chestnuts and drink a small amount of good wine. She should eat the roots of teasel and asparagus which cleanse the body of corrupt humors and eliminate a lot of fetid urine. After the purge she should bathe and wash the armpits with a decoction of iris, alum, litharge. myrrh, rose-hips etc. Then she should deodorize the armpits by rubbing in a lotion containing musk, sandalwood and roses, all powdered and mixed with rose-water.

IV. *The Hair*: The cleaning, coloring and scenting. All other matters anent hair will be dealt with in Doctrine III of this Treatise.

1. Clean the hair with a simple and commonly used lye-soap (lessive), or with warm water containing raw egg-yolks if the woman refuses the lessive.

2. The most popular and desirable color is that of saffron (ie straw-berry blonde), which you can provide as follows: First wash the hair. Then take equal parts of white lily-roots and roots of the bush called meadow-sweet (ie filipendula)(vignette, in France). Add some cumin and 1/4 of a flower of yellow saffron and crush a spiny flower of the large thistle if you can obtain it. Grind them and add some lessive and mix (ie shake) in a sack. Use it to rinse

---

[135] 'Coccus': Unidentified. Perhaps a scribe's error for 'costus' or 'crocus' (LDR).

the hair, occasionally adding powdered pomegranate-rind, although that may make the lessive too thick. You may use walnut flowers or the very bitter outer rind of the walnut fruit, ground together with the teasel. Soften the hair with the juice and fluff it with a towel moistened with the same. The hair will take on the color of the saffron, no matter what color it had before, blonde or dark; the color will persist unless it is washed out or the scalp sweats excessively. Renew the dye occasionally to color the new growth from hair-roots. You can also color the hair with oat straw and licorice, crushed and mixed with lye.

3.   You can scent the hair with musk, clove, nutmeg, cardamon, galingale and others like them.

V. *The Face*: Apply what you learned in Chapter 11 on Embellishment in Men to the care of a woman's face, and what is in the ensuing chapters here and in Doctrine III.[136] In that section we discuss seventeen facial problems other than the four parts taken above. There are thirty-three ocular disorders, twenty-five disorders of the eyelids, ten ailments that involve the eyes and eyelids together, seventeen disorders of the nose, fifty-seven disorders of the five parts of the mouth. All of them concern the face. The surgeon who deals with a woman's face should be able to refer to that information when he needs it. In addition to all the above, there are certain matters for our attention: 1.In respect of all women, no matter the presence or the absence of facial blemishes or any other defects, no woman is so beautiful that she is content with what she has. 2. The corrections of blemishes that are neither flat discolorations or papules. 3. The smoothing-out of wrinkles. 4. The elimination of bad odors in her nose. 5.The correction of bad breath. 6. The restoration of a youthful appearance.

1.   The Face in General: Review our chapter on men about the treatment of a dark complexion, and Chapter 12 containing the

---

[136] We have only the Rubrics for that Doctrine III. However, Mondeville's intentions were clear (LDR).

recipes for waxy ointments. Some women demand more than is given there. They use French soap (savon) in warm water to wash their faces. Then they apply a bran lotion. Then they anoint the dry face with an ointment of oil of tartar, repeated daily for eight days. Then they soften the skin by washing with warm water containing oil of almonds and chicken fat. Some women go beyond that. They think that they can create miracles by using a depilatory medicine which I strongly disapprove of; it is much safer to avoid it. In addition they make use of lotions made from fava-beans in the pod and a decoction of fava-flowers, lilies, convolvulus etc.

2. The blemishes: We use a humectant of water of tartar thickened with tragacanth. The tartar is calcined by soaking it and egg-shells in water for five days in a humid place, as in a cellar. Remove the tartar and work it smooth with the serum of egg-whites. Then add some camphor and distil it in the same way as you make rose-water. Another recipe: 4 oz. of litharge, 1 lb. of good white vinegar; bring to a boil and let it sit. Then add three parts of liquid to one part of oil of tartar. Apply this to eradicate the blemishes and improve the appearance.

3. Wrinkles: One type is natural, the other is acquired. Treat the natural kind as follows: soak eggs in their shells in vinegar until the shells soften. Then add white mustard seeds and clean lily roots, finely ground. Use it on the face or anywhere on the body. For the acquired wrinkles, usually the result of exposure to the sun, mix a powder of crystallized vernis with the lard of beef or deer. Use it on the face.

4. Foul Odors In The Woman's Nose: This is difficult or impossible to completely eliminate. However we may ameliorate it with a compound containing a cinnamon'chew' (masticatory). Take 2 dr. each of marjoram-seeds, cloves, small basil, nutmeg, cinnamon bark, aloes, storax calamint and amber. Add 1/2 dr. of musk and grind them all. Moisten with a small amount of fresh rose-water and make small pills. Swallow two pills every morning and retain two in the mouth. That will correct halitosis and lessen the perception of bad odors.

5.   Bad Breath: That is palliated in the same way as in 4., above. Also, review what I wrote about malodorous armpits. If the halitosis is due to a rotted tooth, extract it. If it comes from an ulcer or a putrid cancer or infected gums, treat them as will be described in Doctrine III. If the halitosis comes from the stomach and the cause is warm, let her eat peaches; if it is from the chest, use diahysop, diairis, diacalamint etc.[137] Furthermore, no matter what the cause may be, treat all cases with a general purge to eliminate offensive matter from the intestine, and prescribe a healthy diet of contraries. Then have her chew nutmeg or bay (laurel) leaves with some musk, and retain a little under her tongue. Or, she may prefer absinthe, mastic or wild thyme. All of them are well-tested remedies.

6.   The Pretense of Youthfulness: Older women, especially the courtesans, who try to make themselves appear younger, have already been treated by physicians with purges, diets and a marvelous array of prescriptions and the use of 'altivoli'—a compound devised by Bernard of Gordon, and described in his littlebook *On Theriac*. The surgical topicals for this problem include a frequent facial application of the juice of bryony-roots mixed with honey. Some women use a fine razor to peel away the outer layer of skin and sometimes they apply a depilatory which eradicates all the facial hair. Others use a corrosive of cantharides and yeast that produces blisters over the entire face. Then they await the new skin.

C. External Ornamentation

We offer no medications or instructions. They seem to be subtle enough on their own as part of a life-style, and all of it is a game. It supercedes all other interests; one fad leads to another. Furtthermore the players are the older courtesans and pimps who are experts at ornamentation. New styles of clothing every day: colored shoes, sashes, silken and batiste hoods, all kinds of pins and clasps, shiny trinkets of

---

[137] The prefix 'dia-' means 'it contains' (LDR.

glass, golden crowns, hats, bonnets, tunics and scarves of all sorts. If she is not red enough she wears red silk or scarlet; when she is too red she wears black, green or white. Women have had this expertise since ancient times. I turn to Ovid who wrote this in his little book *Ars Amatoria.* "We are overwhelmed by ornaments. Everything is covered with gold or precious stones. The woman herself is the smallest part of the display."

In these matters, you need only to know an excellent, simple, tested and necessary way to scent the clothing, the hoods and all the fine cloths. Use many violet flowers, and wash the clothing with lessive from time to time, and rinse them with fresh water containing finely ground iris roots. Then dry them.

CHAPTER 14.

THE ITCH AND SCABIES[138]

Here we discuss three General Subjects

I. The Diagnoses, Causes and Process

Not much need be said because everybody knows enough to say that he itches or has scabies, and that he feels them in a cold or a warm place. Everybody can distinguish them. However, it is worth our efforts to know how they come about and what causes them.

---

[138]   This chapter and the following group describe a large group of cutaneous disorders, most of which were characterized by shedding of scales, crusts and flakes or by a discharge of sebum, or by various discolorations. The names overlapped the diagnostic distinctions, and a modern dermatalogic nosology is a difficult fit. Nicaise translates Mondeville's term 'scabiei' as 'gale' or gratelle and Mondeville admits to a confusion of terms in the third Explanation at the end of this chapter. For the convenience of this translation the term 'gale' will be taken to mean scabies wherever that seems consistent with the text. Although the relation of scabies to the sarcoptic mites was not known to Mondeville, Nicaise stated that they had been discovered by Avenzoar in the 12th C, but he did not establish their relation to scabies (LDR).

The causes are *Materal* (precipitating) and *Efficient* (underlying). The Efficient cause often is a weak digestive system but it always is a weak expulsive force.[139] The Material cause is twofold: Distant: a bad regimen: too many sweets, too much acidic, salty, roasted, fried etc. foods; too much sweet and heavy cloudy wine. Proximate: salty, bitter and burnt humors.

The process (ie pathogenesis): Nature sends the humors to nourish all parts of the body. If some parts are not healthy, the humors are not accepted and they flow into the subcutaneous tissues. The thicker humors remain close to the skin and produce scabies. The thinner and drier humors pass through the fatty subcutaneous layer and enter the thin epidermis and cause itching. Thus the two ailments differ in where the humors lodge; the thicker humors stay in the fatty layer and the thin humors stay in the epidermis. However, in both disorders, the humors are alike in their saltiness, alteration and malodorous character and dirtiness. They differ in that the scabetic humor is thick and fatty and cannot move from the layer in which it is lodged, whereas the excessive thin humor of the itch can spread in the skin. They are alike in that both derive from imperfectly combusted non-natural humors.

The treatments are similar in many ways, and most authors prescribe baths, sweats, fomentations, massages, epithems, ointments etc. The medications differ as better suited to one or the other at the time, as we shall see.

II. Treatments: Two Elements

A. Evacuations: Phlebotomies may be indicated in both cases.[140] The same purges are useful in both cases if they evacuate salty, sour and burnt humors. They are: juices of fumitory clarified with sugar. Drink

---

[139]   These lesions remain in the skin because Nature's expulsive force could not push the bad humors (matter) all the way through and discharge them (LDR).

[140]   The Reader will note how often Mondeville gave the nod to the Physician (The Long-Gowned Academic) for prescribing a phlebotomy. The academic surgeon seldom overstepped the bounds of his own turf to risk offending his medical colleagues, no matter that he lacked respect for their abilities (LDR).

some daily for five days. Another: the juice of scabious (ie the herb Centaurea) or acedula with some goat's milk. During the seasons when those herbs are not available for fresh juices, use a syrup of fumitory or scabious or a decoction of one or all of them (ie previously concocted and stored). Sweeten them with sugar. If the patient is penniless use honey instead of sugar. You may clarify the liquids with egg-white as do the apothecaries. Dose it as above. The surgeon should not turn up his nose at these simple, successful, safe and inexpensive surgical cures.

When they are not sufficient, take 1 lb. of goat's milk, 1 oz. of the ground shells of the yellow myrobalans and macerate the shells in the milk in the open overnight. Then filter it in the morning before adding 2 oz. of sugar of roses and dosing it. A wealthy victim of scabies may first be referred to a physician for purging if that is the local custom and afterward be returned to the surgeon for treatment with his topicals.

B. The prophylactic and curative dietary regimen eliminates and avoids the efficient and material causes noted in the preceding sections, and all else that burns and reduces humors to ashes.

C. Two kinds of Topicals: 1. Those used for both scabies and the itch, including fomentations, sweats and ointments. Foment with decoctions of boiled fumitory, scabious, acedula, mauves, chick-weed, marrubium and inula. If the scabies is wet, add some vinegar. If it is dry apply a decoction of oxalis, wild celery, wheat-bran (in a sack). This decoction also is useful against the itch. Use the same herbs in the properly prepared sweat box, when that treatment is called for. The Ointments should not be repercussive, rather they should attract the matter to the surface. They should be detergents and abstersives, such as; 1part each of old oil of laurel, pork lard, wax, frankincense and mercury, and 4 parts of finely ground sea-salt. Mix with some juice of fumitory or plantains, as much as can be taken up by shaking. If you add a bit of soot moistened with vinegar the ointment then is suitable for all sorts of skin lesions. You must use a low flame when you compose the ointment. The mercury must not be near a flame whenever it is used, and use it only when it has been mortified (ie with saliva, etc.). After the body has been nicely washed and the scabies is only in the parts of the body above the level of the navel, you need apply the ointment only to the palms of the hands. For the lower half of the body apply it on the soles of the feet. When the scabies is everywhere, anoint both places.

The six ointments used for scabies alone and not for the itch are: Take and mix 1/4 part each of fresh butter and dialthea; two parts of terebinth and 2 dr. of litharge.

Another: 2 parts of white hellebore and 1 part of litharge and mix with rancid (old) lard. Another: Grind the shells of black plums (ie sloe) and mix with rancid lard Another: 1 oz. of soot, 1/2 oz. each of walnut-oil, vinegar and juice of fumitory. Another: juice of lily, white wax, mastic and walnut-oil. Another: juices of scabious and spikenard, white and black hellebore and rancid lard as needed. Shake and stir at length.

Two ointments for itch alone: Finely diced rancid red or yellow bacon fat, soaked in vinegar (ie macerated) for three days, replacing the vinegar twice a day. After decanting the fluid, carefully mash the bacon and add ceruse and litharge, oil of roses and rose-water. Mix gently for a long time. Another: add crushed and macerated bacon fat to a little mortified mercury (see B, below). This ointment is useful for scabies, mal-mort. salt phlegm and all crusted infections.

Medications other than the ointments that are used to treat the itch are fresh water, sea-water and brine (ie man-made) baths, or use a decoction of oxalis. You may add a little vinegar in the bath water. If you use an herbal steam bath, massage the patient while he is in the sweat-box, using a sack containing apium mixed with crushed salt.

III Five Explanations

A. There are three causes for scabies and the itch. 1. The strength or weakness of the repulsive forces of the body. If weak, they cannot force the outward passage of the matter from within the body. The bad matter remains and causes aposthems. and other ailments. A strong repulsive force will rid the skin of all the bad matter and prevent scabies and the itch. That explains the prevalence of those ailments in old persons, and why cures are more difficult. They continue to produce the altered salty humors which they cannot expel through their skin. 2. Spoiled, acidic, salty, etc. foods are causes for the bad humors. 3. A weak digestive force is the third cause.

B. Quick-silver (mercury) which is part of nearly all the compounds used to treat these maladies must not be used before it has been mortified. That process involves vigorously mixing the mercury with saliva and human hair. Or you may use the powdered cuttle bone of squids with saliva. Or you may shake a narrow-necked vial containing the mercury and Roman vitriol and vinegar. Furthermore, you should not mix mercury with medications that have hot actions, because that will create fumes that must be eliminated at once. Use only small amounts of mercury in any of the compounds, especially in medications that are applied near noble structures, on the face, neck, forehead and torso. Use it only on the legs below the knees and on the forearms below the elbows, but not directly on the joints themselves. I have seen many persons who have been badly treated with mercury-containing medicines develop swollen tongues, throats and mouths, and the corruption of dry inflammations within the mouth and gums was followed by the loss of their teeth. Those deprived patients who cannot chew waste away and die not long afterwards. As soon as a surgeon sees the first signs of such complications, he should immediately stop the applications and wash the affected parts with a decoction of anise, camomille and wild mint, and treat the mouth as one usually does for pustules (ie herpetic, aphthous). That may prevent the more serious complications of mercurialism.[141]

C. Scabies is one of the contagious illnesses that induce the recurrences of acute fevers, phthisis, leprosy, delirium, quinsy and anthrax. The term 'gale' has included several kinds of inflammation such as safati, tinea, pox, roseola, purpura, dartre, impetigo, roseola, mal mort, salty phlegm etc.

D. Avicenna described all of these contagious maladies in Bk. I, Fol. 2, Doct. I, Ch. 8. He said that there are certain diseases that pass from one person to another, such as leprosy, scabies, variola, pestilential fevers, cold abscesses and especially those seen in crowded domiciles and where the neighborhood is open to the wind where exposures to such as 'pink-eye', cold teeth [142], phthisis and albarras (ichthyosis) are easy.

Other ailments are inherited via the sperm, such as white albarras, natural baldness, podagra, phthisis and leprosy. The reason that these

---

[141]   The toxic effects of mercury were recognized long before Fracastorius (LDR).

[142]   Cold teeth: Nicaise cites a manuscript which describes the sensitivity to cold in teeth that are carious, and eroded by acidic foods (LDR).

and similar maladies are contagious is that any disease whose matter lies near the skin and that evolves by expulsion, contaminates the air with malignant fumes. Scabies and the Itch and the like are that sort, and therefore they are contagious.

Chapter 15.

On Serpigo (Dartre) and Impetigo[143]

I. Diagnosis. Five Items

A. Descriptions: In respect of these lesions and many others, the authorities contradict each other as to the definitions, the treatments etc. First came the Greeks, including Hippocrates, Galen and Constantine[144], followed by the Arabs, including Avicenna, Rhazes, Serapion et al., and then the Latin writers of Salerno and all the others who until the present have dealt with the questions. I have yet to find two of them in complete agreement about these infections.[145]

---

[143] Nicaise's Notes: *Dartre* is a form of *serpigo*, and often was called the Inconstant Fire, according to Guy de Chauliac (p 416).

Impetigo derives from 'impetus', an eruption, nearly its synonym, but not a complete definition. The Latin authors used the term to define a group of crusted, dry and chronic lesions corresponding to the leicen of the Greeks, sometimes by the *mentagre* of Pliny. The Arabs and the medieval authors agreed with the last (Cahmbard and Guy). (EN).

[144] Constantine The African was the great Salernitan and Cassinian monk who translated from the Greek and Arabic sources during the 11thC (EN).

[145] 'Infections': originally meant to deface, to impregnate, to stain, to color. Using the term Infections of the face, the authors designated certain changes and alterations of the skin and subcutaneous tissues accompanied by changes in color. In some cases, 'infection' may be translated by 'blemish', 'alteration' and 'infection' (inflammation) (EN).

As noted before, the identification of the dermatoses mentioned by Mondeville with lesions found in modern texts presents many problems. Pannus (singular and plural) may be chloasma; serpigo may be ringworm; lentils probably are lentigo, or pigmented nevi, and dartre may be our equally vaguely named 'scurf' (LDR).

It is almost as if they agree to disagree. One cannot take from them a single 'true fact', because what one calls serpigo another calls impetigo and a third says pannus. A fourth names one as a subset of another and insists on the same treatment for both. A fifth says that there are three categories of impetigo, each requiring a different treatment. They deal with morphea and its ilk in the same way, or with even more vague arguments. So it is that we must be sceptical about what any single author or a single practitioner says, something that may contradict or give incomplete information. The subject-matter is arguable and obscure. I hope to go ahead after a broad and easily understandable survey, although I may not be in agreement with authorities and practitioners, or perhaps with the truth.

For the present let us call an *infection* in its everyday sense, a defacing, dark, irregular unnatural defect of the skin and the immediate subcutaneous tissue. It is seen on the surface, anywhere in the body. Some infections are inborn, others are acquired.

Among the inborn are the spots on a fetus that was conceived during a menstrual period.[146] Among the acquired are infections that follow the expulsion of melancholic blood through the skin, such as pannus and lentils that are seen in pregnant women.

There are five categories of infections. First are those in the skin alone and not beneath it, such as dartres, pannus and lentils. Second are those only on the subcutaneous tissues. They include some acquired disorders within the body and can be cured only with evacuations, not with topical applications. The skin itself is not involved, and preserves its clear and translucent attributes. This category includes the residuals of contusions which Nature will cure. The third group involves both the skin and the subcutaneous layer, as do some black and red morpheas and leprosy. The fourth group may involve the bone beneath as well as the skin and subcutaneous tissues, as seen in very slim legs and at the temples and forehead. The fifth kinds of infection lie between the skin and subcutaenous tissues but do not infiltrate (ie attach to) either layer. They are scabies, variola etc. before they become pustules and ulcers. After they ulcerate they involve both layers.

Therefore, serpigo and impetigo are infections in the first category. Serpigo is an irregular spreading snake-like defect; it derives

---

146   Probably congenital hemangiomas (LDR).

from burnt humors. In French, it is called dartre. In impetigo the skin is pale and changed in its consistency and substance.

From these descriptions you can see that they are not identical nor are they completely different. They are identical in what is said to be alike, and different as we define them. Analogies in our definitions mean that the objects are similar.

I will relieve myself for the present of citing what has been written by authorities about each of these lesions. I will summarize all that in the following simply for purposes of comparison. There is no detail about the infections that does not present similarities and differences. Indeed, all infections are similar in that they are the result of weak assimilation and weak expulsive energy, and we define them as similar. Impetigo and dartre are similar in that they never penetrate beneath the skin; they differ in that impetiginous lesions usually are round and are not raised, whereas dartres are elongated and can be felt above the level of the surrounding skin. Furthermore, a dartre sometimes is excoriated and sometimes sheds scurf and flakes, some of which are pale and others are dark, Impetigo has none of those features, Pannus and lentils resemble the foregoing two in that they are only in the skin, but they differ in their colors. The impetigo and dartre do not affect the color whereas pannus is gray and the lentils are reddish. Pannus affects mostly women who are pregnant with girl-children and women who suffer amenorrhea. Lentils and other infections show no preferences. Pannus and lentils appear most often on the hands, the neck and the face, the others are everywhere. Finally, pannus (ie plural) are larger than lentils.

Rosacea and comedo are similar in that they appear in the face.[147] They differ in that rosacea defaces a person with red lesions that are crusted (ie flaky) whereas comedos are nodules which are called verbles (maggots) in France, arising from subcutaneous nodules near the nose. When squeezed they emit material resembling a pâté. If you do not

---

[147] Nicaise commented about the use of 'cossus' here, different from meaning comedo. Here he believed Mondeville used the term analogically, naming a worm that eats leather etc. I think that comedo is the more accurate translation (LDR).

empty them they break down and corrupt the nearby tissues. When they are numerous and hard and at some distance from the nose and they present as inky-black spots, do not remove them or empty them.

Morphea[148] changes the normal color of the skin; we see scattered blemishes, shallow and numerous. The skin is altered as in leprosy. Morphea is akin to algada and albarras, which are the same as al gada and al barras, the al in Arabic is a definite article without a separate meaning. Gada is a form of morphea that does not raise the surface and may even depress it. Sometimes it is accompanied by pustules of one of three kinds of colors: all black, white or purple. The barras is another kind of morphea which affects the skin, the subcutaneous tissues and the underlying bone; it wrinkles the skin along its natural lines. There are two varieties, white and pink, and it is incurable with topicals.

Constantine (*Viaticon*) gave another category of morphea, a white one derived from corrupted phlegm; a black one from melancholy and a livid or purple one from blood and melancholy.

All the gadas are similar in that they affect the skin and a small amount of the subcutaneous tissues. They differ from albarras which can go as deep as the bone. Also, the gadas are produced by a more energetic vital force and have a thinner matter than the albarras, and they can be cured, unlike the incurable albarras which yield only to wide surgical excisions which include the roots.

Although we distinguish other sorts of morphea from albarras, we have not compared them with the others in this chapter. Furthermore, all the other kinds of morphea, no matter which, resemble albarras in that once established they steal their nutrition from their surroundings and transform the region's complexion to resemble their own; just as the opposite, a healthy complexion can improve any poor nourishment that comes its way.

---

[148] Nicaise explained: Morphea derives from the Greek mythological character (morph) who could change his body and face at will. The white morphea were called such because they were white as flour. Bazin believed that was an early phase of leprosy. In our day we call it vitiligo. Many lesions have been called black morphea (EN).

White albarras differs from white gada: the latter can sprout both dark and light hairs whereas the former produces only the light. Albarras differs from all the others in that it is the only one that depresses the level of the skin. Furthermore, it does not bleed when pricked with a needle. On the other hand, white morphea bleeds. Black albarras produces excoriations, but the gadas produce scales, like those of fish or shellfish.

I have taken all this material about the various infections from the works of famous authors and practitioners. Before I proceed further, let me say that I have found little of value for surgeons in that search. That is so because there is so little agreement among them, and there is very little that a surgeon can use in his practice. What good avails a surgeon in the disputes over names when the classes of treatments differ or are the same? Did not Galen (*De Morbo*, Bk. III, Ch.4) say in reference to names in Medicine, that knowledge of things is important, not knowledge of names? How important is it to know that dartre and impetigo are not the same things, especially when we treat both of them with the same ointments? Does it annoy you that certain infections of the skin and the subcutaneous tissues are alike or different, since you usually treat all of them with the same corrosive ointments and a few other things?

The definitions, the effects, the varieties, the causes and the manifestations of impetigo and dartre are sufficiently clear and can be learned from what I have just written.

II. The Treatments: Three Items

A. Evacuations: I am certain that all these ailments and many others that we have mentioned, such as fistulas and cancers, and many, such as aposthems that we will come to later on, all fall within the field of surgery. All of them are caused by internal matter (ie humors) and they cannot be extracted by or consumed by topicals alone. Therefore, if you wish to cure them you must attack their causes and evacuate them. Furthermore, the faulty humors cannot be evacuated unless they are properly digested beforehand. That cannot be accomplished unless you know the causes themselves and their qualities. And that is why we need first to describe the signs by which to recognize all the bad humors and how to tell them apart. When you know the humor at fault, it can be digested and evacuated as supervised by a physician[149], here, as in all things, within the limits of his abilities. In fact, I think it is not beyond the scope of a surgeon who is involved in these cases to know the principles of medicine, and not only in these cases, but whenever a physician is not available.

The type of evacuation depends on the abundance and the congestion of the humors and hinges on two factors, the size of the veins and the strength of the vital force. First, when the veins are full: In such cases usually the humors are plentiful, but not all of them. When the condition is bilious, there may be more bile (ie in the blood) than phlegm, and vice versa. One never encounters pure blood in the veins, so when we say 'bleeding' we mean all the humors. Second: when one humor is more abundant (ie than the blood-carrier) it is harmful by itself, but it does not congest the veins.

The signs that tell us which humor is the offensive one are in the color and the condition of the affected part and of the whole patient. If he is plump, has good color and a pleasant taste in his mouth and he breaks out with pustules, and if he is young and has reddish and cloudy urine, and if he favors good wine and food, and his face and body retain a healthy rosy hue, those are signs that his blood is abundant and healthy.[150] When the patient is languid and depressed and is pallid and if his veins are collapsed and he produces a lot of tasteless and viscid saliva and his urine is watery and thin and his digestion is sluggish and his feces are soft and pale and mucilaginous and greasy, and his appetite is weak and he is intermittently anorectic and he has a habit of eating fish and other moist foods and of drinking

---

[149] We are reminded that disorders of the internal organs were treated by physicians; they ordered the purges, phlebotopmies, clysters etc. The surgeon followed their orders (LDR).

[150] That means that there is no excess of the unhealthy humor in his blood. Blood normally is mostly sanguinous humor mixed with the other three, the product of the hepatic digestion of chyle, and modified by second digestions in other vital organs (LDR).

pure water (ie not mixed with wine or herbs), all those are signs of excessive phlegm. When the patient is emaciated and his veins stand out clearly and his color is a reddish yellow and his taste is bitter and his eyes are yellow and his urine is brown and scanty and he eats spicy vegetables such as onions and garlic and strong spices such as peppers, galanga and other warm and dry substances and he drinks strong wine at home, all those are indicators of a predominance of bile. If the patient is dark in color (brownish or gray) and has a feeble complexion and produces a pale darkish or a scanty brown urine and he eats foods that engender melancholy such as venison, beef, and dry cheeses and cabbage and lentils et al., all these are signs of an increased melancholy.

Then the surgeon, who also should be somewhat a physician, can diagnose the humor or humors that are in excess in the patient. If the veins are full and the humors are abundant, albeit not all of them, he should bleed the patient, that is, if the vital force, the age of the patient, his life-style and other special considerations permit a phlebotomy. Then he will prescribe the proper laxative to purge remaining humors and if, from ignorance or some other reason he had preceded the bleed with a laxative, he follows the second laxation with another small bleed to prevent the activated humors from causing a fever.

If the veins refill with blood in proper balance with the other humors, the initial phlebotomy is all that is needed and need not be followed with laxatives. A bleed evacuates all humors according to their proportions, including the sanguinous. Therefore, after the surgeon carefully considers all the signs and he aims to cleanse the blood, and to improve its complexion and suppress the heat and the sharpness of red bile, and to evacuate some burnt bile and phlegm, he will take about 2 oz. of cassia fruit-shells (not the seeds) and 1 lb. of goat's milk and bring them to a boil. Then he places them in a jar covered by a cloth and sets it in the open over night. At dawn he will filter it and give some of the warmed liquid to drink. Or, he may take three to four drams of powdered epithyme and make a decoction in goat's milk. When you prescribe a laxative tea, it should contain herbs that are suited to the case at hand, such as borage or bugloss, et al. Take 2 1/2 lbs of one and boil it with 1 dr. of spica and mastic and make a large amount of broth. Dose it as we do for evacuating melancholy.

Simple digestive medicines for phlegm are pouliot, calamint, basil, marjoram, wild thyme, mint, savory, abrotonum, tansy, mustard garlic, carrot root, pepper, ginger et al. as well as simple warm and dry spices mixed with equal amounts of honey. A decoction of the last is said to be useful for digesting bile.

The digestive compounds for phlegm contain oxymel or oxymel with squills, or rosat. Dose those and the like with warm water, or use any fresh warm and dry.spices such as ginger-sugar, diatereon pepper et al.

Compounded purges for phlegm are 3 dr. of diaturbith or about a chestnut shell-full of hierapicra, or arthitic pills, golden pills or cochia pills. Another reliable laxative: Take 5 dr. of fresh agaric, 8 dr. of mastic when in season in warm countries—use 6 dr, in cold seasons—2 dr. spikenard, 1 dr. fennel, 1 lb. filtered hive-honey. Mix. Dose two spoonfulls at dawn. It will provoke four stools. It is a very reliable laxative; I often have seen physicians take it without any preliminary medications. When the phlegm is abundant in the head, a common ailment, prescribe this: 1 part saffron, 3 parts of choice myrrh, aloes in an amount equal to all the others. Use enough rosat to form pills. I have seen physicians take three or four pills in the evening, sometimes just after dinner.

The digestives for bile are cool and moist foods and rest. The Simples include violets, roses, acacia, lettuce, endives, purslane, joubarbe et al. The cool seeds of sandalwood, spodium and water— camphor. Also: vinegar and sour verjus. Or: you may make a syrup of the decoctions or clarify them with egg-white. Dose four to six spoonfulls in the morning and evening or take an equal amount of oxymel or a sweet vinegar and warm water.

An evacuant for bile is an electuary of 4 dr. of rose-juice or about 1 dr. of diaprunum.

Digestives for melancholy: The Simples are borage, bugloss, thyme, hepatica, scolopendrum, capers, tamarisk, a good wine, and all warm moist medicines with which you make decoctions and add sugar, as we did for bile. The compounds are oxymel with squill, acetic syrup and rosat.

The evacuant compounds for melancholy are doses of 1/2 to 1 dr. of the Imperial Cathartic; or take 1/2 dr. of senna (diasenna); or 1/2 to 1 dr. of hierarufin; or 3 dr. of hieralogadon; or 4 dr. of theodoricon[151]. Take 6 dr. of epithyme, 1 dr. of mastic or spikenard, grind them gently and boil them in 1 lb. of goat's milk whey. Let it cool overnight in the open. Rewarm it in the morning, filter it and dose it.

2. The Regimen for those who suffer from ailments caused by burnt humors, both the internal and external diseases: After a careful examination and evaluation of everything that I could obtain from the authors of the medical handbooks, and having considered their ideas, and having proposed a regimen that includes five non-natural elements (air, exercise and rest, repletion and evacuation, the digestive and laxative medicines,[152] as they are involved in the complications of the spirit, happiness, sadness, anger, impatience, etc. and in sleep and aging), to consider, when they are lacking, what regimen a good physician should prescribe. There remains for me only to order a regimen containing a proper diet, which is the remaining, the sixth non-natural element. It is the most necessary of all, and I give four reasons: 1. Of all the non-natural elements, the diet is what the patients are most likely to resist. 2. It alone varies from day to day and is necessary every day. We have less need to change any of the five other things—if the air is good, we need no change—etc. 3. Every rational medical or surgical treatment includes a proper diet.

When the burnt humors are too hot and too dry, you must temper them with contraries, that is, with cool and moist foods which will moderate the excessive heat of the faulty humor by overwhelming it with cool foods. And the same holds for moisture.

Every regimen for such cases avoids the harmful and seeks what is useful. Inasmuch as harmful things are outnumbered by the useful,

---

[151] These laxative compounds are Medical rather than Surgical. All are cited in the Antidotary (LDR).

[152] A Reminder: 'Natural' designates things that are inherent within the body. 'Non-natural' things are externals (LDR).

we will set the latter aside for now and we will list the harmful items, taking a shorter route. We will assume that what are not harmful may be considered useful.

The harmful items in the diet: all vegetables (fresh and stored), green and dry legumes, such as fava-beans, peas etc. All purees of vegetables excepting those made from chick-peas and green peas. All tangy vegetables such as onions and garlic, alone or when cooked with other foods. All sour, strong and spicy foods, such as peppers, curcuma and galanga. All potent electuaries such as diatereon pipereon. All salty foods especially when one has not eaten some for a long time, such as lard et al. All fried and broiled foods, especially the charred crusts. Cheeses and foods made with them. Crusty bread and all unleavened bread and whatever contains unleavened dough, such as crisps and pastilles. Beef, fatty game, venison, wild boar, all aquatic birds excepting the small birds that live on beaches and river-banks, that have claws instead of web-feet, and have long narrow beaks. Birds that eat snakes and other venomous animals are harmful, as are storks and peacocks. All salty condiments, juices, broths, and gelatins of fish and meats that are cooked with onions and spices. All ordinary fruits except an occasional sour fruit after a meal, such as quince, medlar and pear, and some that neutralize the bitterness of bile, such as pomegranate, melons oranges and lemons, sour raisins, sour unripe apples, cherries, plums et al. Let us continue with the harmful foods: All beverages that are common in France, except clear wine (white or red) of medium strength, diluted three parts to one with fresh water.

There are some who believe that certain of the infections, as impetigo, some white pannus, albarras, white gadas and all that tend to have a sickly white color, are formed materially from corrupt phlegm. We know that phlegm is cool and moist and as such it is contrary to the warm and dry burnt humors. Inasmuch as the diet prescribed for all patients should be contrary to the causes, the diets for those cases should be hot and dry, just the opposite from what we prescribe in our list. On the other hand, it is not possible to show what is good or harmful for a bilious illness without touching on what is good or bad for a phlegmatic disorder. We turn to Aristotle who said that a discipline and a doctrine of contraries should be adhered to, different for each case. From that viewpoint it is evident that what excites bile suppresses phlegm and vice versa. And what we allow as useful for those who suffer from bile and burnt humors will be contrary and harmful for those who suffer from phlegm. There are some exceptions in the things that are harmful to the sufferers from burnt humors that were not harmful in their primary state (ie before combustion), in their qualities of heat and coolness, dryness and moisture. But because those substances are fatty and gassy, as are the vegetables, or are viscous like unleavened bread, or are hard to digest as are the unhealthy meats of sea-birds, beef et al., we keep all of them away from those who suffer from phlegmatic illnesses. Furthermore, those last-named patients who feel hunger and especially thirst must avoid moist and watery foods, moist fruits and herbs, dairy products, bitter things, some vegetables, fish and everything that is soft and fatty. They may drink hydromel and only small amounts of good wine. Any well-informed person can take from what has been said here and fill in the spaces for his own regimen. What one finds useful here he should know it to be harmful elsewhere, and vice versa.

3. The Topicals: Seven Precautions: a. Do not apply them on chronic infections or on extensive lesions or when it seems that the applications will be difficult, until you have prescribed a suitable general purge in addition to a purge suited to the affected region, whenever that is feasible and when you can observe all our established principles. b. The goal in all these maladies is to rid the patient of all the faulty matter. If simple measures fail, use leeches, cups etc. c. For chronic and widespread disorders and other difficulties, use evacuations and vigorous massage before applying potent topicals. d. All topicals should be applied after the patient has been in a steam-bath or tub bath. e. After the baths, dry him with a brisk toweling. Then apply the topicals. f. Every infection that has been caused by Natural Forces[153] is easy to cure when the signs are favorable, unless it is a complication of another disease. The shedding (ie peeling) caused by a Natural Force that has been activated by sun-burn or by another

---

[153]   An 'infection' itself is the surfacing, that is, the final extrusion of the faulty matter. The 'Natural' expulsive force completes the cure (LDR).

acute disease, and if it is accompanied by bad symptoms, is difficult or impossible to arrest. g. The blood coming from needle-pricks in the skin has a healthier color in patients who have curable infections, if the other signs are favorable. Contrary signs indicate incurability.

The Topicals Themselves

There are many degrees of them to match the many varieties of the infections. If we set aside the lesions that affect the flesh and the bone late in their course and thereby declare themselves to be incurable, excepting the few small isolated lesions that can be extirpated by a surgical means that does not sacrifice a lot of normal tissue, and if we neglect for the moment the lesions which affect only the underlying tissues, such as the black and blue of contused eyelids, and lesions that appear between the skin and the subcutaneous tissues without invading either, such as scabies and early variola whose treatments we will deal with in a later chapter, there remains for us here to evaluate topicals for infections that involve only the skin and the flesh immediately beneath it.

Some of the topicals are weak, others are stronger and some are very potent. When we choose a topical to treat a mild, superficial and unresisting infection which is small and of recent origin, and few in number and involves the skin alone, we should start with weak topicals. If we fail, try another or a third. Only after that series of failures do we go to a somewhat stronger medicine. For a wrinkled, thick, chronic, large and numerous variety we use the stronger topicals from the outset, and apply many of them in succession, as may be necessary, until we reach the most potent of them. For acquired infections that are more than skin-deep, whatever state they are in, use the potent topicals and increase their potency as you need them. Finally, you may use the ultra potent caustics.

The weak topicals: Crush the roots of parietory with a lot of salt and apply. Or: scrape the surface of the affected region and immediately apply the warm, white ashes scraped from the surface of a branch of wood that you have exposed in the flames. Or take 1/4 part each of oil of sweet almonds and fresh butter and 1 1/2 oz. of terebinth and anoint the site. Or take equal parts of dialthea, fresh butter and oil of violets. Melt and stir until you achieve a white ointment. Then rub in a small amount once a day. Or lay some oakum wet with wine, as hot as tolerated on the dartre and replace it every four hours, several times at each session. Or heat some twigs of the broom plant and collect the drippings from the tips and put them on the lesions. Or bind in place a burnt almond or a dry walnut. The lesion will be gone after two days. Or moisten a lesion frequently with the saliva of a man whom has been celibate since childhood. Or cover it with the sap of a cherry or plum tree or the seeds of mustard in vinegar or with the acidified juices of plantain or memitte. Or make a thin paste of fine milled wheaten flour and water, applied repeatedly. Or apply the juice of the wild green plum. Or use the juice of a melon or the ashes of its rind. Or make an inunction of the fat of a bull that was slaughtered when it was in rut.

The strong topicals are: Blood of a hare drawn from its ear or its foot. Apply it when the patient is exposed to sunlight, and leave it until it peels off on its own. Or foment the site with a puree of fava-beans. Or rub in a powder of live sulfur in a cloth sack that is wet with vinegar. Or apply crushed inions or garlic. Or fortify the milder ointment of beef fat described above by adding live sulfur. All of the mild topicals can be fortified by adding small amounts of energetic abstersives such as sulfur, verdigris, sal ammoniac, both types of hellebore, soot and spider web.

The extremely potent topicals are all of the above to which are added large amounts of the substances listed in the previous sections, as well as the other simples and compounds. Strong as they are, they do not erode intact skin. Our list here is taken from various authors' handbooks and from practitioners, and the medications are specially designed to treat these maladies. They include first: Equal amounts of frankincense, hellebore, roasted snails and their shells. Make a powder and knot it in a cloth and rub it on the infection when the patient is in the bath. Or coat the surface with a broth of lime water or the ashes of garlic mixed with honey. Or use sarcocolla mixed with cow's bile. Or mix radish seeds with honey. Or apply either hellebore in vinegar or fumitory cooked in the abdomen of an eviscerated snake.

The most famous and best tested of the topicals are these: oil of wheat thickened with white spider-web that has been cleared of all its contaminants. Place them in a lead jar and heat it in an oven. Another: oil of eggs thickened with some verdigris or sal ammoniac to intensify its actions. Another: an ointment made from 4 oz. of crushed juniper-seeds boiled in water. Filter it and add to the hot filtrate 6 oz.of melted and filtered fresh pork lard and 1 oz. each of wax and terebinth. Keep it over the flame until everything has been melted and dissolved. Cool it with sprinkle water while stirring it vigorously in a mortar, and add 2 oz. of finely ground live sulfur. A famous physician used this ointment and cured a nun, while I looked on. She had endured five years of dartre which had covered her face; it was deforming and full of scales. Many Parisian physicians had given up after failing to cure her.

Another successful ointment: 1/2 oz. each of hellebore (white or black), live sulfur, ink, orpiment, quick-lime, vitriol, alum, oak galls, and ashes of small bones[154], 1 dr. each of mortified mercury and verdegris. make a powder and set aside while you take 3 oz. each of the juices of borage, scabious, fumitory and parietory and bring them to a boil over a low flame while adding 3 oz. each of the lees of old oil and vinegar and 1/2 oz. of liquid resin, and as much wax as needed. When all those have been liquefied, add the powder. The last to be added is the mortified mercury. This ointment has been proved successful in all curable infections, even in those that seemed suited for radical extirpation with caustics.

Another good topical is a rub-in with the juice of spatula (ie stinking iris) and by soaking some oakum pads in the same and bandaging them in place. The ointment of Master Robert Fabré, the king's premier physician is similar; he cured the king of severe impetigo in four days. It seems to work well in all mild infections: Take 1/4 part each of the juices of parietory and scabious, 1/2 part of a very strong vinegar and boil them with 8 oz. of walnuts and 4 dr. each of tartar and soot; add 2 oz. of cinnamon macerated in strong vinegar

---

[154] Nicaise's term is 'clavelée', which is unclear in this context. Variously defined as arch-stones, sheep pox (ie ovinia) and as small bones. The last is Guy se Chauliac,'s usage. Op cit. p. 609 (LDR).

and 2 dr. each of litharge and choice aloes. Another: Marcassite crushed fine with vinegar also works well.

The most powerful local treatments for infections are puncture, massage (ie with corrosives), leeches, incisions, caustics and cauteries. A puncture uses an awl or a scarifying knife. Rough rubbings are used with the topicals noted above, after which you apply an eruptor containing 1 part of cantharides, 8parts of yeast and a small amount of vinegar. Apply it until a blister forms. Leeching also is useful, and so is scarifying the infected skin until blood runs. Another eruptor: 8 parts of quick-lime, 1 part of soot mixed with French soap to make a paste. Another: quick-lime made into a paste with the bark of the green fruits of walnuts, or with water alone. You can cauterize a small bit of infection at a time until the entire discolored infection is gone. Afterwards, treat the region as a wound.

III. Four Explanations

A. How to avoid errors: Predict nothing, nor appear to invite error with false claims that eruptors or similar means of treatment will not cure the infections that require extirpation. I saw some excellent surgeons assisted by physicians who had despaired of a cure of a leg afflicted with dartre. It was completely excoriated after their traditional treatments. Inasmuch as I had not seen the patient at the outset, I came to believe that it was more like morphea than dartre. I had treated an infection on the temple in the same way with an eruptor when the cranium had been involved. The lesion disappeared after a while and the region returned to its former state.

B. When Nature has not been stimulated by some obviously faulty humor to provoke an infection, such as occurs after a sunburn, the physician should not prescribe evacuations and intervene against Nature. She, Herself, not needing medications, can expel the bad matter from the body as a whole as well as from the affected region.

C. I describe four degrees of virtue for the topicals because the poorly trained practitioners may not appreciate how much virtue each has nor be able to determine the resistance to treatment. And, equally important, I describe the order in which the topicals should be applied in weak infections, different from that in the stronger infections, and I insisted that there should be no confusion in that doctrine. It is important to begin treatments of some cases with light remedies, and in others to use more energetic ones at the outset. Constantine in his *Diseases of the Eyes* said we should begin with the most gentle medicines, whereas Avicenna (*Canon*, Bk. IV, Fol. 4, Tr. 3) seemed to contradict him. I explained that contradiction in Notable 7, of the Explanations in Chapter 1 of Doctrine 2, on the treatment of ulcers.

D. I list a large number of topicals for every degree of potency, for three reasons. 1. All medicines are not available everywhere. 2. Even when medicines are available, the most potent and the most expensive may not be used for an impoverished patient. 3. Even if all medicines are available to everyone, no one of them will serve equally well for all patients with the same ailment. Furthermore, a once-effective medicine may lose its effectiveness after a while in the same patient. Therefore, we should have available several medicines for each patient; if one loses its action we can replace it with another. Indeed, we see the medicines that work well on Peter fail on Paul; perhaps it will serve Peter well on one occasion and fail him in another.

There are three causes for those differences. 1. No two individuals have identical complexions, although they may be alike in general. 2. No two medicines have identical complexions. 3. The cause may be celestial, where changes are continual, producing different effects in people. Therefore, the surgeon should abandon vain promises; he should never so praise a medicine for its potency and say that he cannot find a better one if the case demands it.

CHAPTER 16.

MORPHEA AND ALBARRAS

I. Description of Morphea

Enough was written in the previous chapter to know what morphea is, and that it is in the same class as gadas and albarras. and we described what those are. There are three kinds of gadas curable with medications, and there are two kinds of albarras, most cases of which are incurable, excepting those which can be extirpated by surgical means. We described how they differ from all the other infections, and their similarities.

II. Treatments

As noted before, there are three elements: 1. Evacuants that suit the various maladies as well as all others that derive from hot burnt humors, and from cool and phlegmatic humors. 2. A dietary regimen. 3. The topicals that we use for the curable infections, and the instructions for applying them. Here, I will add a few topicals that are suitable for morpheas as reported by physicians. Before listing them, here are a few useful General Principles:

1. Do not bleed the patients you treat for morphea. It is not a contrary action.
2. The chronic white morphea (ie vitiligo) is more difficult to treat than the black.
3. Consider it a good sign when rubbing causes albarras to redden. Failing that bodes ill.
4. In gadas the surface of the skin is level and smooth.
5. Albarras that does not redden when rubbed exhibits a rough and irregular surface.
6. A famous physician recommends a midday drink of the juice of white bryony and a small meal of cooked snake-meat in the treatment of patients with any of the varieties of infections.

The Topicals

Scarify or puncture gadas and cover them with menstrual blood and dust on a powder of cashew nuts. Or rub the lesions with green onions or warmed squill-vinegar.

Or rub the patient in the bath with fig-leaves. Or use a decoction of the roots of water lilies. It is good for morphea and black gadas. Or grind equal amounts of thapsus, seeds of radish, black hellebore, mustard and garance. Mix with vinegar and rub it on while the patient is in a tub bath or sweat-box. This is better for the black gada than for the white. Or: rub on some minced squill. Or use the blood of a black snake as an inunction, as acclaimed by some.

III Explanations

I need not repeat the explanations offered in the preceding chapter. In regard to the flesh of snakes, prepare them for use as follows: Obtain a white snake in the forest, taken from a dry habitat, not aquatic or marshy. Put it in a cool flat container. Swat it with a tree-branch with many stiff twigs, similar to a broom, wet with water. Keep at it for as long a time as it takes to recite four misrere prayers. By then, the snake will be so tired that it cannot move and it will lie in a coil. The servant, taking care to avoid being bitten will grasp the snake behind the head with the thumb and index finger of his left hand and will chop off the head and the tail on a wood-block. After skinning it and eviscerating it, boil it with fumitory and similar plants.

Chapter 17.

Lepers and Leprosy

We shall consider three General Sections

I. Descriptions: Three Parts

A., Definition: Leprosy is a demeaning (shameful) disease that derives from melancholic humors or from corrupted derivatives of them, which are corrupted beyond repair. It affects the entire body in the way a cancer affects a single part. But, as it is for a cancer that involves an entire limb, it can be cured only by destroying or amputating it. And since leprosy invades the entire body, that treatment is impossible.

B. Causes: It can appear before birth and at any time of life. The congenital leper is conceived by a leper, or when a leper has coitus with a pregnant woman, or when she conceives during a menstrual period. Since Jews forbid coitus during menses, leprosy is rare among them. The cases of acquired leprosy after infancy get it from pestilential and infected air or by prolonged use of melancholic foods, or by habitually drinking milk and eating fish at the same meal, or by drinking milk and wine, or by a long and frequent association with lepers, or by coitus with a leprous woman or with a woman just after she had coitus with a leper and his semen remains in her vagina. One should avoid such women. When the urge is strong wait as long as possible (ie after her previous engagement) "Happy are they who distance themselves from danger".

The search for other causes in the medical texts is difficult; leprosy really more a medical than a surgical disease, and since my Book is Surgical, I shall touch on the subject only briefly. However, at times I shall mention the signs and the diagnosis. Furthermore, the surgeon may be able to palliate the cases that lend themselves to such surgical measures as cautery, phlebotomy, cupping, leeching, caustics, sweats, baths and ointments.

C. The Signs of Leprosy: The thickened eyebrows sag; the orbits are rounded (ie wide-eyed); the nostrils flare in front and are pinched behind.[155] The speech is 'nasal'; the color of the skin is livid with streaks of brown and gray; the aspect is fearful with its staring expression. The earlobes are shriveled and are crusted and pustular. There are lumps and vitiligous spots (ie morphea). The worst sign is the reappearance of lesions that were thought to have been cured. The muscles of the web of the thumb are shrunken. The skin of the forehead is taut and glistens. The surfaces of the legs and of the toes show leprous signs. When you try to wash leprous blood-stains from clothing, dark spots persist—a very bad omen. You may find many more signs in books by other authors.

The Treatments

It is feasible to treat leprosy at a very early stage if you are prudent and are an expert physician or surgeon. But once the diagnosis is obvious, a cure is out of the question. Palliation may be possible for long periods of time if you are diligent and wise. What could have

---

[155] A description of a leper's leonine face (LDR).

been a curative measure early in the disease, now serves to palliate it. There are four elements.

A. The Regimen: The diet prescribed for all malnourished persons is good for lepers, and consists of cool moist foods. It is the same diet that we use for victims of internal or external burnt humors, as we described it in the chapter on Dartre and Impetigo. The regimen for the malnourished is purely medical and we leave it to the physicians; you can find it in their handbooks. But all lepers do not have the same complexion, and what is special about one of them is not so for another. All lepers are not of one sort; one may be sanguine, another be more bilious, etc. As Galen noted, and others, too, the doses of the topicals to be applied cannot be fixed for one as for all. Galen, our Prince of Physicians, (*Megatechni*, Ch. 3) wrote that the great artificers and operators have the right and the duty to increase and to decrease and to modify the medications. He added that even when we cannot decide the precise treatments we should indicate that we are about to make a change. He stated further that those who do not know the principles will never be able to change from one medication to another. Damascenus said the same (*Aphorisms*, Part I, in the commentary on the second Aphorism). He cited Aristotle, "Every artisan in his work adds, subtracts, changes, rearranges and does what seems advantageous".

B. The Evacuations: There are many medicinal evacuations which must be preceded by the digestion of the proper foods that we have listed: Laxatives for one, Vomits for another, and Masticatives, Caputpurges[156], sternutatories, sweats, baths, et al. Any of these may be weak or strong to befit the stage of leprosy—which is to say, before the disease is fully confirmed, and the parts are obviously corrupt. They can be given as preventives to persons disposed to the disease, and before they reach the stage we described. After that, the topical medicines may have some palliative value. The physician and the

---

[156]   Caputpurges: Head-purges were General Evacuants, containing the juices or decoctions of marjoram, chelidonium, nasturtium, staphisagre, pyrethrum, nutmeg, long peppers, and small amounts of euphorbia, scammony and rose-water (LDR).

surgeon should not involve themselves in the same case (ie at the same time) unless they cannot resist being drawn to it by the entreaties of concerned persons and by promises of large fees. Even then, they should accede only after they both have declared their diagnoses, which agree that they face a vile and contagious diseaase. Finally, the leper himself may wish to be involved with his physician and to keep him nearby. The physicians who agree, if one really wants to know, are greedy, and despicable, and are to be considered corrupt and revolting.[157]

The weak evacuants are: a morning dose of a filtrate of epithyme and goat's milk whey, taken until a good passage at stool has been obtained. Or use ramich, as praised by some. The stronger evacuants are hierarufin, hieralogadon, the theodoricon[158], as well as colocynth and others made with hellebore. All of them are potentially dangerous (ie if over-dosed). Much safer is the following: an oxymel, a good digestive as well as evacuant, made so: deeply uproot radishes, fennel and parsley and puncture them in many places with an awl and add some slivers of hellebore. Cover them with earth and wait forty days. Dig them up, clean them and dice them and make the oxymel to be dosed according to a physician's orders.

Among other compound laxatives the one that follows has many lay advocates for its ability as a general purge to evacuate and consume the poisonous matter. It is called Serpent's Flesh. When the patient takes it as he should, at the correct time and in the correct amount, he will be completely and perfectly cured! According to its promoters, there is no other cure like it. Here it is: after finding and selecting and dealing with the snake as above, wash its flesh with warm brine and then with wine. Then add some seeds of fennel and dill and some toasted bread and some salt. Boil the snake meat in the soup

---

[157] The'delicate' points at issue are bluntly stated. 1. The physician comes first with his diet and laxatives, etc. Then he steps aside while the surgeon works with his topicals. The physician who hangs on after the Surgeon takes over is a vile and despicable and venal thing! (LDR).

[158] These are Medical favorites taken from Avicenna. Mondeville gives some recipes in the Antidotary (LDR).

until the flesh comes away from the bones. The patient drinks the broth and eats the meat. Another: the flesh and the broth of a capon with added sugar and ginger, or ginger alone, cooked until all the meat is white. Another: an electuary of ginger, nutmeg and sugar. Another: a live snake, the seeds of fennel and dill, some epithyme, polypodium and anise all cooked in a crock of warm wine or wine-must. After it boils and is clarified and decanted, the patient drinks some of it diluted with water and takes no other medicines until he sees how well it works. Two doses a day are enough. Altogether it is a beverage. a medicine and a source of nourishment. If the patient begins to desquammate and have weak spells or faints, those are considered to be favorable signs. Another: fatten hens with flour boiled with snake-meat until their feathers come loose. Eating them is said to produce marvelous results. Another: in an alembic distil the flesh of snakes and some fumitory and save the liquid to be drunk during the seasons when neither ingredient nor anything else are available.

Emetics: Macerate 1 lb. of black hellebore in 4 oz. of an oxymel for two or three days. Remove the hellebore and put in a radish to macerate for the same length of time. The patient eats two or three slices of the radish (ie and vomits). Or crush radish and dose it and an equal amount of the oxymel. Or make a caput-purge with the juice of marjoram and the fluid from crushed live crayfish (riverine or aquatic). Induce emesis with two drops placed in each nostril. Or put in the nostrils a distillate (from an alembic) of the juice squeezed from the herb mercuriale or from marjoram or the inner bark of an elder tree.

A masticatory:[159] pyrethrum and staphisagre crushed and tied in a cloth sack. Sweats and baths: use cool herbs contrary to warm diseased matter, and warm herbs against the cool.

A Gargle: macerate radish-roots in vinegar for three days. Remove the roots and press them and mix their juice with oxymel, used as the gargle.

A Sternutatory: Make it with peppers and white hellebore. Insufflate it into the nostrils to induce a sneeze.

If the physician has no success with all these evacuants he can

---

[159] Masticatory: More like a 'chew' of tobacco, than a stick of chewing gum (LDR)

refer to many other excellent and tested recipes and instructions in the handbooks that describe emetics, head-purges etc.

C. Surgical Evacuations are by phlebotomy, leeches, cupping and scarifications, cauteries, caustics et al. You should bleed only those persons who are disposed to leprosy or those early in the course before they show any facial deformities. Use a vein on the forehead and an hepatic vein in the right arm. Later on, bleeding is to no avail unless the patient has an excess of blood, as with retained menstruation or suffers from hemorrhoids or is short of breath.[160]

The reason for not using phlebotomy or potent laxatives when the leprosy is fully established is well known. Those maneuvers have cooling and drying effects and they weaken the vital force by activating the faulty humors rather than evacuating them.[161] The humors already are at the surface. If for any reason you must bleed, use small veins or apply cups with scarifications between the shoulders or use superficial scratches instead of the incisions with the scaraxation-knife[162] at the shins. Sometimes, when only the face is disfigured you may bleed from the bridge of the nose. Cups are used instead of phlebotomy for reasons already stated, when we want to divert the corrupt humors from the face. Sometimes we place the cups under the chin to attract the humors from the face, or we may use leeches at the same sites. The scaraxations on the shins are used fairly often to evacuate and divert the faulty humors and the infectious and toxic fumes that are emitted by the lesions.

---

[160] Perhaps a description of congestive heart failure (LDR)

[161] The rapid pulse caused by the reduced circulating volume after a large bleed or a potent laxative (both were General Evacuants) was interpreted as due to a stirring (activation) of the humors, which favored their flow away from the diseased region that attracted them, The laxatives attracted the humors to their site of action, that is, to the interior of the body, away from the surface where they could be eliminated. The small bleeds from nearby veins were locally attractive only and did not have the stirring effect (LDR).

[162] 'Scaraxation': Nicaise cites Joubert who claimed that the barbers corrupted the original 'scarification' to indicate a superfical rather than a deeper cut (LDR).

Cauteries are used after preliminary caput-purges and other maneuvers to purge the head and may be indicated at any time. They are especially useful for leprosy that comes from putrefied melancholy and phlegm. They are more successful if used early in the disease.[163]

You should apply the cauteries in five places: 1. At the center of the coronal commissure, as we showed in the chapter on cauteries. 2. At the temple just below the hair-line of the forehead. 3. and 4. Between the sagittal suture and the ears, closer to the midline, just above the parietal bosses (ie horns). 5. At the occiput, just over the cervical fontenelle. Other than the sites at the head, apply three seton cauteries over the spleen, on a forearm and over the liver. Finally, cauterize four spots between the muscles of the arms above the elbows and four spots at the hollows just below the two knees. Use a seton cautery below the occiput at the hair-line. If the patient is dissatisfied, the surgeon should consult the *Albucasis*, in which Rhazes recommends sixty-six places, and says that the more cauteries a leper can tolerate the better off he will be. At any rate, the leper suffers less from the cautery than do healthy persons because their bodies are somewhat numb.

The caustic eruptors are used to divert the leprous matter from the face and are usually placed at the occiput and under the chin. They are much more effective when they are placed alongside cups used without scarification.

D. Topicals: Many and Varied as to types: sweats, baths, massages, ointments, plasters, epithems, lotions, etc. All have been mentioned in previous sections. Others cited here have little effect except to relieve some of the superficial complications and ease some of the serious symptoms of leprosy. I list eleven: facial deformities, pustules, scales, fissures, nasal obstruction, subsided eyebrows, nodules, pale and shiny skin after bathing or sweats, ulcerated gums, morphea, dartre, impetigo.

For facial deformities, after you determine the fault, use frequent applications of this lotion chosen for its lack of conspicuousness after application: Take 1 lb. of rose-water, 1/2 lb of leaves of honeysuckle, 1

---

[163] These are 'point' cauterizations, rather like acupuncture points (LDR).

oz. of live sulfur, 1/2 oz. of squid's bone, 2 dr. of camphor, and pulverize and mix them in a glass jug. Set it in sunlight during several days of hot summer, shaking it well every day. Apply it to the face twice daily. I have been told that it consumes the nodules and lessens the redness. Another: bathe the face with wine or water that contains a decoction of snake-meat. Another: after a sweat or bath, coat the face and body with the warm blood of a hare, or use the oils of eggs and almonds in equal parts. Another: anoint the fat drippings from a roasting hen that had been fed flour cooked with snakes. When the patient eats the chicken and applies the grease, the old skin will be shed and replaced with better. Another: burn the heads and tails of snakes and mix the ashes with washed lime and butter, and anoint the face. Indeed, the only truly effective material for the general health of lepers is snake-meat.

Treat pustules with continuous fumigations for three days, using a decoction of mauves, violets, guimauve et al. Then apply an ointment of rendered lard. Then use a lancet to prick each pustule in three places. Collect the bloody discharge and mix it with a powder of litharge and burnt lead and some mastic. Make a plaster to lay over the pustule. Or dust on a powder of rock salt and red slugs cut open down the middle (? charred); mix the fluid that drips from that with some green mud of a sort that is greasy and bubbly, found in shady damp places. Anoint the pustules to clean and dry them. Or do the same with cimolean mud, cooked, powdered and mixed with vinegar. Or make an inunction with the fatty drippings from damp manure that is fresh and greasy. Or cover the pustules with quick-lime and vinegar.

The scabs are treated with ointments of dialthea alone or mixed with fresh butter and violets, all in equal amounts. Failing that, use the topicals we prescribe for salt-phlegm.

Treat fissures with Rhazes' white ointment that contains thick litharge. Or use an ointment of mastic and flour mixed with chicken-fat, as we use for rhagades.

Treat the sunken nostrils of leprosy by inserting a stent cut to size from the roots of gentian or of the Paulino[164] covered with the yellow

---

[164] 'Paulino': A double error! The correct spelling is 'papilina' a kind of bean of the genista family. It is not a gentian. See Nicaise p.849 (LDR).

(citrin) ointment of litharge or burnt lead. Or use a cloth tent coated with the yellow ointment and sprinkled with powdered litharge, aristolochium and tartar, all in equal amounts. Or insert tents covered with Alexandrian Gold (ie unidentified) or oxycroceum. Every day, wash and moisten the nostrils with warm wine. When that is painful, use rose-water with mauves and bran.

Treat loss of eyebrow-hairs as follows: scarify the area, rub it and apply a leech. Remove it when it is full and burn it at once and save the char. Then anoint the place with this oil: Boil equal amounts each of venus-hair ferns and labdanum in laurel oil. Then dust in this powder: Equal amounts of flayed and eviscerated mole, hazel-nut shells, chestnuts, bees, house-flies, wasps and mouse bile. Char them all. Then add the charred leech and pulverize the mix. Dust it on the brow. Or make an ointment of the powder with oil and wax. Or rub in some juice of abrotonum and onion and some pumice, as used in making parchment. Or burn an eviscerated mole with an equal amount of bees and add an equal amount of flour of nigella and mix with the oil of eggs before anointing. Albertus Magnus (*de Animalibus*, Bk. XXIV) said that the charred skin of a hericium, barbs and all, will cause hair to grow on scars.

The nodules and tuberosities on the faces of lepers are treated so: Make a superficial cruciate incision in the skin over the mass and apply an ointment of cantharides, realgar and dialthea. A practitioner told me that it will erode and rot the node but will not damage the normal tissues. Or burn the skin away with a caustic if the node is caused by a cold humor and if the patient refuses the knife, and consume the interior with a powder of litharge, tartar and other corrosives. Then cicatrize with the yellow (citrin) ointment fortified with litharge. Or fumigate with a decoction of emollioent plants and then apply lard. Or: grasp the node with a hook and cut it out. Collect and thicken with litharge the blood that oozes and reapply it. Repeat the applications until the third day. Then apply cicatrizers—bran-water containing some of the medications mentioned above. Or rub the site vigorously while it bleeds after making a cruciate incision, and dust it with a powder containing equal amounts of litharge, ceruse, orpiment and alum, and then use the cicatrizers. Or thrust two needles (square edged) through-and-through beneath the very roots of the node and pass a tough cord beneath the protruding ends. Tighten the knot daily until the node falls off. The resulting scar will be small.

Treat thin facial skin which is the result of sweats and baths with French soap or a variety of epithyme and galls and vinegar, or with some thick juice of absinthe alone. You may use a combination of all three.

For erosions and ulcers of the gums: rinse and clean the mouth with a decoction of wild mint, dill seeds and flowers of camomille, or with juices or decoctions or waters of such cool herbs as plantains, endive, lettuce, nenuphar, pig-snout, morel, joubarb. Or use a decoction of dove's-foot, plantain and myrtle. Or apply the juice of herb bennett or sanamunda. The latter is popular.

Morphea, impetigo and dartre are treated as in an earlier chapter.

In addition to the eleven complications, and the severe symptoms which are beyond cure by surgical methods, another serious symptom is difficult breathing, which the physician treats as follows: He urges the leper to exercise by walking, running, jumping, shouting and loud singing, all early in the morning. Then the patient drinks as much ewe's milk as he thinks he can digest, and he takes no other food or drink until that is completely digested.

In addition to the problems mentioned early in this chapter we occasionally encounter unanticipated cases. For example, what should we do when a healthy man has had sexual relations with a woman who had just had coitus with a leper and his semen is still in her vagina? In those cases we consider two things: 1. the signs that such relations had occurred with a confirmed leper. 2. The treatment.

A. Apropos the signs: Two items in respect of the disease: 1.The complexion of the infected person, either warm or cool: the different complexions indicate different humoral causes; their signs will influence the treatment. Inasmuch as the treatment should accord with the cause of a disease, so too it should vary with the complexions of the newly infected persons. So, if he is of a warm complexion he will soon feel ill. The sooner it appears in him, the sooner he can seek a remedy. But the disease manifests itself more rapidly because the woman is of a warm complexion, has large organs, a weak (a female) body, thin vital spirits etc. The opposite is true when the complexion of the newly infected one is cool and those of the infector (the woman) and the disease are warm. The effects of the disease will appear even more rapidly. But if both are cool, the disease appears more slowly. Finally, if the complexions are opposite, the middle course is observed. What is warm hastens the progress and what is cool slows it.

When you can determine the signs in the infector, the treatment is more certain and is easier. If you cannot obtain that information you must deduce it from the infected one. So it is: When the infector was of a warm complexion, the victim will notice soon after the coitus an unusual pallid color in his lower parts, and that will spread to the exposed areas. Then he will feel prickly and burning discomfort and perhaps suffer chills, but not always. His face may change from pale to red or vice versa. He will imagine that poisons are spreading under his skin, and it will seem as if ants were crawling on his face. He will suffer from insomnia, and often will have hot flushes.

The signs which indicate that the infector was cool, that is, of a melancholic or phlegmatic complexion, are the changed appearance of the victim on the day of after the coitus: his face will be livid or ashen gray (leaden). The face will become puffy, his limbs will feel heavy to the degree that he finds it difficult to move about as much as may wish to. He will feel cold under his skin and numbness, first of the face and then of the body. The signs of fully expressed leprosy are described in the chapter on scurf (dartre): both in the conduct of the patient and his appearance.

B. The Treatment. 1. A simple treatment advocated by a famous practitioner, for any of the above cases: Immediately wash the penis with vinegar and he will avoid the infection. When his penis burns (ie after coitus with an infected woman) and he has coitus with an uninfected woman before or just after he urinates, she, too, will be infected. 2. A more commonly used treatment, when the above is not suitable, and when the victim and the infector both are of a warm complexion, that is, if both are sanguine or bilious, or one or the other, the regimen, the medications and all the rest tend to be very cold. Bleed the patient from several veins in different parts of the body, taking small amounts at each site, all to be no more than a single large bleed. Take one small bleed every two days for three months. After the initial bleed, dose a digestive of 2 parts of oxysaccharum (oxymel), 1/3 part of hierarufin.[165] On day three, use the sweat box followed by a bath with cool herbs, as we use to treat alopecia. When the patient leaves the bath, he takes a compound of two parts each of garance and saffron and 1 part of theriac with some syrup of fumitory. Follow that routine every three days, with the bleeding or cupping or leeches, scaraxations, caustics, head-purges, masticatories and all the other measures we have mentioned, when any of them seems useful. Twice a week temper some yellow ointment by mixing it in the palm of your hand with some rose-water and rub it on the face, and then some pulp of leaves of plantain. The next morning wash it away with bran-water or similar.

If the infection has a cool complexion, the regimen, the medications and all the rest tend to be tempered. Indeed, although an infection produced by a cool infector in a cool infectee will show the signs of the disease, that occurs much later than when the complexion is warm. Furthermore, the victim does not seek help until much later, and the treatment is delayed. The vapors that impregnated him are thicker and heavier and her passages[166] are narrower. In those cases digest the matter with doses of 2 parts of oxymel and 1 part of syrup of fumitory if it is melancholy or 2 parts of hierapicra and 1 part of hierarufin if it is phlegmatic. On the third day after a sweat and bath that contains such warm herbs as elder, dwarf elder, scabious, fumitory, parietary, inula et al., he takes theriac with the warm juice of fumitory. On subsequent days he is bled every five hours, withdrawing two ounces at each session. He repeats the sweats and tub-baths and the sedatives (ie opiates). Then apply caustics, cups and other evacuants on the diseased spots. Then wash the face with warm wine in which you dissolve one to two dr. of rhubarb or apply yellow ointment fortified with a

---

[165] Hierarufin: a complex laxative, containing colocynth, germander, asafoetida, wild parsley, aristolochia, peppers, cinnamon, saffron, polium, myrrh, honey, etc. See cochia in the Antidotary (LDR).

[166] 'Passages': the privates, the vagina, urethra, etc, which admit air (see Tr.I, Ch.9) and emit vapors (LDR).

powder of Bresillet.[167] Leave it on overnight and wash it away in the morning with bran-water.

Four Explanations

A. The immediate cause for all types of leprosy is a poisonous, horrible and infectious melancholic humor that Nature cannot dispose of (ie by digesting, consuming, combusting, dispelling etc.) in healthy flesh, and that humor provokes the disease. Other humors never are the immediate cause; first they must be transformed into a corrupted melancholic humor. In fact none of them is contrary to Nature or to the basic Life Principles which are warm and moist. The leprous melancholy is corrupted not by the kind of putrefaction that causes suppuration in an aposthem or the putrefaction of liquid humors that cause fevers, but by a dry incinerating putrefaction.

B. All the authorities agree that leprosy must be treated by wetting agents in any acceptable way. They agree also in the use of cauteries as often as the patient can bear it.[168] It may seem that the drying effect of the cautery contradicts our premise. Now, the cauterizations evacuate and consume those poisonous and evil humors. Therefore, they enhance the humectants. And I must add that the humectants recommended by the authorities are sweet and mild, and they repress the bad qualities of salty, nitrous and toxic humors.

C. The fresh bath water itself has no direct action against leprosy, because it rinses and deterges only the skin, whereas the infection is beneath the skin. But if the preceding sweat is made with the correct herbs and you add them to the bath-water as well, they are useful because their detergent effects go deeper than water alone.

---

[167] 'Bresillet': Brazil-wood is an anachronism , see Antidotary (LDR).

[168] This is cauterization by touching designated spots, not to ablate tissue. It is akin to acupuncture (LDR).

D. We use ceruse in many of the ointments to improve the appearance of the face of lepers as well as for other patients, and you should know that it can harm the teeth and wrinkle the facial skin. Also, older women who use it in their face-creams make their wrinkles worse.

CHAPTER 18.

HOW TO INCREASE OR DECREASE THE SIZE OF A LIMB

I. Description

The *Definition* is simple, as one can see and feel the diagnoses.

The Causes:: A.Why is a limb too thin? 1. A constriction will block the entry of its nutrition. 2. A chronically painful joint will cause an arm to lose bulk by disuse. 3. Chronic hypertrophy of the opposite side takes away nourishment from the smaller limb. 4. A wound that enters a joint and has not been properly treated, such as a wounded shoulder or elbow, will cause the shrinkage (ie atrophy) of the entire arm.

B. What makes some limbs too large?[169] 1. The basic and major cause is the dilation of feeble and lax veins which accept all the inflowing humors. Whatever more arrives is received and further weakens the veins. The inflow over-taxes the limb's ability to assimilate it and causes the enlargement. 1.The most proximate (immediate) causes are the humors in the condition of the basic cause; they are the substance of the enlargement. Other causes are remote and relate to one or the other of the preceding two.

II. The Treatments

A. Eight ways to increase the size of a small limb: 1. A fortifying diet similar for treating emaciation and leprosy that is due to burnt humors. 2. Elimination of the cause. Galen said that no ailment

[169] The enlargement is lymphedema or venous congestion (LDR).

can be cured as long as its cause persists. Thus, when a limb is constricted you must remove or loosen the band and relieve whatever pain is there. Apply a mild constriction to the opposite limb to prevent it from accepting nourishment meant for the ailing one. You cannot remedy the late effects of a badly treated wound that has penetrated a joint. 3. Foment the limb until the part shows redness and begins to enlarge. Use decoctions of mauve, guimauve, violets etc. in water. 4. Brief but brisk massages with rough towels to dry the limb after the fomentations. 5. Slapping with green thin willow switches while reciting a Miserere prayer. 6. Apply an ointment made from 8 parts of black tar (pitch) and 1 part of oil, every morning. Cover it with a bandage, and remove it in the evening. Some report enlargement following a treatment described in the *Totum Continens:* a vigorous toweling after a good sleep, in the morning or during the day or night. 7. A plaster of equal parts of naval tar, greek tar and white resin, all melted and applied to the entire limb, or by saturating a large cloth and laying it over the entire limb and bandaging it in place, every morning. Tear it off brusquely in the evening. Then use the fomentations etc. 8.Vigorous exercise will attract nourishment from other parts of the body. To assure proper exercise of the atrophied hand and arm see to it that the patient uses only it in performing the normal daily functions. Also, have him lift weights and pull the bell ropes to ring church bells and carry packages[170]. Bind the good arm to his chest so it cannot assist with the work.

If you use the eight measures in that order for as long as is necessary, every shrunken limb will recover, all or in part, excepting those due to wounded joints.

B. Two ways to reduce enlarged limbs: 1. A diet of contraries against the causes. If the cause is a phlegmatic humor, prescribe the contrary regimen described in the chapter on flaking skin (or dandruff). If the humor is a thick melancholic humor, use the regimen that opposes burnt humors, described in the same chapter. 2. Eliminate the causes, as Galen taught.

---

[170] A physiotherapetic regimen as in a modern 'workout salon' (LDR).

There are two proximate causes: material and dispositive[171] a.
The material causes are: 1. Antecedent causes lead to the malady
and are eliminated by purging with laxatives suited to the cause.
Thus, evacuate phlegm with a a phlegmagogue and evacuate
melancholy with a melancholagogue, as decribed in our chapter on
Impetigo. The concomitant material cause is related to the
antecedent, as well as to the dispositional cause, which is intrinsic in
the limb, are corrected and suppressed in the same way—that is with
the same topicals and the same techniques—excepting that the
topicals for the concomitant material causes differ in the hard and
the soft substances, although the techniques are the same. The
concomitant-material-soft cause of hypertrophy of a limb is a liquid
phlegm (ie edema); the soft limb betrays the cause in a person of
normal color etc (ie not inflamed), as well as other indicators of a
superabundance or phlegm, signs that we already know. Treat this
cause as we described for phlegmatic abscesses, in that chapter. In
brief: saturate a double cloth or some oakum or a sponge in some lye
made from the ashes of oak or grape-vine wood, or use water
containing borax. Wrap it on with a bandage. Moisten it and rewrap
it twice a day. Or you may use a good surgical bandage to hold crushed
iris leaves in place.

The concomitant-material-hard cause is a thick melancholic
blood which you can recognize by the hardness of the enlarged
tissues and by other well-known symptoms of superabundant humors.
Treat with the softeners and emollients that are described in the
Antidotary, and follow with resolutives. b. After all the above, treat
the dispositional causes with constrictive astringent topicals,
bandaged on the limb. After the evacuations and consumptions of
the concomitant matter—as given above—reduce the size of the
relaxed channels and veins and restore the normal virtue of the
affected limb to prevent it from receiving the harmful humors which
have been directed to it.

---

[171] Dispositive: The patient's tendency to be affected; his disposition (LDR).

III. Five Explanations

A. Whatever the dispositive causes, be they lax veins or porous tissues, and whatever the material causes, be they proximate or concomitasnt causes of the hypertrophy, they differ in the way they work. The dispositive cause is inherent in the limb from the very beginning, and is not tied to anything else and is not preceded by any material cause. On the other hand, the concomitant-material causes in the affected limb ones perform side-by-side with the inherent factors.

B. Avicenna (*Canon*, Bk. II, Ch. On Pitch) said that pitch attracts nourishment to limbs and fattens them, especially when it is reapplied frequently after the cloth is forcefully torn away at each change. Rhazes in his *Totum Continens* recommended an ointment of oil to be applied when the patient awakens from sleep, or an ointment of oil and pitch, claiming its excellence.

C. I believe that applying a saturated cloth is better than a direct application of pitch for four reasons: 1. You can warm the pitch only once; that is not the case for the reapplied cloth. 2. It is not easy to remove the pitch from the skin. 3. Absorption from the skin is too rapid. 4. You can use both sides of the cloth when reapplying it.

D. The atrophy of a limb that follows a badly treated wound that had entered a joint means that the wound had suppurated, and the atrophy now is incurable, as I see it. Wounds involving joints are more difficult to treat and take longer to heal, consequently the ends of nerves and veins and other structures are destroyed. The wounds do not heal by primary intention when some tissue has been lost; it is replaced by dense fibrous tissues that cannot be penetrated by nerves and veins. Neither nourishment nor motor function can be freely transmitted. The limb is malnourished and immobile.

E. All the elements of the treatments I have described need not be repeated at every daily or twice-daily dressing. You may eliminate one or another item at any single treatment, such as fomentations one day and flagellation on another, etc. You do not have to be too strict with the routines if the limb is feeble or painful; there is no need to afflict the already afflicted one. I disagree with the 'rule' that says the opposite, that what is bothersome in an element should be relieved by its contrary (ie not by eliminating it).

CHAPTER 19.

SALT-PHLEGM, RHAGADES, NIGHT-SWEATS AND THE NIGHT COMPLAINT[172]

I. Salt Phlegm

A. Definition: This is a vague name for a kind of humor which is discussed in the chapter on The Generation Of Humors. It is applied to a pruritic crusting eruption caused by it.

That eruption is seen most often on the legs and thighs, but appears in other regions. Sometimes it is everywhere on the body, and causes intolerable itching, and sometimes it provokes open sores (ie caused by finger-nails). The scabs may cover an entire limb or parts of it. When the crusts are shed, excoriations and shallow ulcers are exposed.

B. Treatment: We prescribe the diet and the purges used for ailments caused by burnt humors. I often have used the following simple medicines: 1/2 lb. each of green wax, pitch and resin; 1 lb. of oil of hemp-seeds. Mix all and add 3 dr. of mortified mercury.[173] Apply it once daily until cured. If it is successful, it will attract the bad phlegm during the initial three days of treatment to form pustules, which will

---

[172] Here is another cluster of dermatoses: The *salt-phlegm rash* resembles what we called 'winter itch', blamed on dry skin and old age. *Rhagades* is the cracked skin of chapped hands as well as the fissures at the corners of the mouth or herpes simplex, most of them are cold-weather lesions. *Night-sweats and nocturnal complaint* resemble miliaria and *sweat-blisters* are troublesome for mothers of all epochs (LDR).

[173] The mortification of mercury is described in chapter 14, Explanation B (LDR).

dry and heal during the next three days.[174] Another: an ointment of althea alone has been successful in my hands. Another: 1lb. of fresh butter, 1 oz. ground sulfur. When the scabs have fallen away or are about to, they may leave some large and unsightly marks on the skin like mal-mort. For those, apply this well-tested ointment suggested to me by a practitionetr: 1 oz. each of laurel leaves, verdigris, ink, live sulfur and ground couperose; 1/2 lb. of lard, 2 dr. of (mortified) mercury.

II. Fissures (Rhagades, Crevices)

A. Definition: Usually we use the term'rhagades' for fissures around the anus and vulva, and 'crevices' for fissures in other parts, as in the feet and hands (ie chapped).

B. Treatment: The full treatment is used for chronic and multiple lesions, including diets, purges and topicals. However, if only one or a few lesions are to be treated, the topicals alone may suffice. First bathe with the decoctions of softening herbs (as in the preceding chapter). Then put oil of violets or butter in the fissures. Then use a drying ointment: equal parts of oil of roses, duck fat, moist lamb's fat (ie hysop), a mucilage of quince, wheaten flour and tragacanth. If more is needed, use the more potent remedies for fissured lips, as described.

III. Night-Sweats and Night Complaint

The two are almost the same. They come from a vaporous sweat which is the product of digestion of bilious or sanguinous humors. A cold night may constrict the pores of the skin and block the emission of the sweat-humors, which may form tiny blisters. Sometimes the skin becomes rough without blisters. The tiny blisters and rough skin are called the night-complaint because they occur only at night, when cold constricts the pores, never by day when the air is warm.[175] The ailment

---

[174] In other words, the bad humor is delivered to the surface and then eliminated from the body, the goal of treatment (LDR).

[175] Again, the failure to expel through the intact skin is the cause (LDR).

provokes itching, which may be mildly pleasant at first, but soon becomes distressing. The sweat and the complaint are the same disorder.

B. Treatment: The entire course may include phlebotomy, purgation, and the diet prescribed for maladies caused by burnt humors. The patient should not live where he is exposed to cold (ie at night). The topicals are ointments based on oil of roses, such as Galen's ointment. Another excellent one: oil of myrtle, fresh butter, tragacanth and gum arabic. Another: a massage with juices of oxalis or morel. Another: 1 part of hulled barley, 10 parts of vinegar. Cook until the barley dissolves and the consistency of a porridge is reached. That is a good ointment for all conditions caused by hot or burnt humors.

IV Three Explanatory Comments

A. Be aware that mercury itself or in any compounded topical must not be used on or near a principal or noble organ. The risks and the benefits of the substance are described in the chapter on Scabies and the Itch.

B. As to the dark blemishes that persist in the skin after the lesions salt-phlegm, mal-mort etc. have healed, be aware that you never can make them fade, because they are from a pure melancholy plus the dark residues of the other humors. When that humor infiltrates a part of theskin it is impossible to separate either its substance or its dark color of the melancholy.

C. Are night-sweats and night-complaint really the same ailment? It seems to me that sweat is the cause that activates the sick humor to the surface and produces the night-complaint. Any other hypothesis does not ring true. Furthermore, suppressing and consuming the sweat is the cure for the night-complaint. All the authorities agree to that.

CHAPTER 20.

LICE, MITES, TICKS, FLEAS AND STINK-BUGS

I. The Maladies

A.   Lice are treated for four reasons. 1. They cause discomfort. 2.

When there are many of them and the victim cannot restrain himself from scratching at them, often indecently under his clothing, he cannot perform his ordinary activities. 3. They bite and suck so strongly that they make the blood run. 4. Worse yet, they depress the appetite, spoil one's healthy color, deplete the entire body and weaken the vital forces.

The Treatment has three elements. 1. The Dietary Regimen lists things that are suitable and some that are harmful. The suitable items are digestible and unspoiled foods and beverages which prevent decay and tend to have drying effects. The harmful ones are moist and can decay, such as fruits, dairy products and especially dry figs and raisins, chestnuts, chard and spinach. Also harmful are old unwashed clothing. 2. Evacuation of putrid phlegm as we have described. If a physician recommends it you may perform a phlebotomy. 3. The Topicals: First bathe the patient in salt water and follow with fresh water, with several rinses. Wash the hair. Change the clothing. Dust with hot wheaten bread-crumbs on his back and chest to kill the lice. Some recommend spreading terebinth on a cloth and sprinkling powdered staphisagre on the chest and back to kill the lice. For my part, I use the following and have never failed: one walnut shell-full of mercury shaken with the white of an egg until you cannot recognize the mercury. Then soak a piece of linen cloth with the mixture and set it to dry. Then put it directly on the raw surface. It kills the lice in one day and turns them black as if they had been fried. At least, they do not survive. Those that can escape gather in the tail of a cowl or in shoes in such numbers that the common folk believe that they have been put under a spell.

B. Nits: Treat them with a shampoo of strong vinegar and rub the ends of the hairs with broom plants, salvia and mint boiled in vinegar or with flour of lupin boiled in vinegar. C. Mites (sarcoptes, scarabs): These are tiny animals that nest in cavities or erosions between the hair and the skin, especially on the hands of idle persons. Treat them by washing with the juice of ground-ivy or macerated aloes in vinegar. Heat the liquids and use them as a fumigant as well as a fomentation.[176]

C. Are night-sweats and night-complaint really the same ailment? It seems to me that sweat is the cause that activates the sick humor to the surface and produces the night-complaint. Any other hypothesis does not ring true. Futhermore, suppressing and consuming the sweat is the cure for the night-complaint. All the authorities agree to that.

D. Fleas [177]: These collect around a bit of fig-tree wood covered with beef fat. Another: place a bit of shiny armor in a corner of the bed; the fleas will gather on it. In the morning discard the piece. Anoint a flea-infested dog with oil. The fleas will swell up and burst. Other measures are described in Ch. 2, Doctrine II, Treatise II, and in the explanation for Notable VI.

E. Ticks: These are small, flat, round animals with many legs. They attach themselves and do not move about and are difficult to pull away when they are alive. They commonly inhabit the chest and the axillae. Treat them by applying an oil and ashes. I believe that any remedy that kills lice will also kill the ticks.

F. Cossi (blackheads, comedos) which in common French are called verbles.[178] These are firm nodules that are fixed in the skin and subcutaneous tissue near the nose and press against it. They empty a white matter resembling bits of a patê. If you do not empty them they may become inflamed. Treat them by washing with a gruel of bran. Then sprinkle with this powder: equal parts of olibanum, galbanum, serpentaria and squid-bone (ie cuttle). Then wash the place with egg-white and wheaten flour mixed with water. Another: mix 2 parts of French soap and 1 part of black pepper and apply it. Another: moisten

---

[176] 'Sarcoptes':" Nicaise notes the failure to blame the mites for scabies, here or in Ch. 14 of Doct I. (LDR).

[177] Here the fleas are nocturnal pests (LDR).

[178] These are discussed here with the insects because the squeezed comedo exuded a pasty form that resembled a maggot. Hence 'verbles'. See Ch. 15, 'rosacea' (LDR).

the places and rub it with the following: Filter finely powdered live sulfur through a linen cloth. Then take 1 oz. mixed with a good rose-water and pour it into a glass jar and set it in the sunlight during the dog-days. Shake it well every day.

G. Punaise (stink-bugs)[179] I described them and their riddance in Chapter 2 of Doctrine II of Treatise II, Another way to kill them or get rid of them is to smoke them out by burning the leaves or nuts of cypress trees or use the smoke of crushed rue. You may wash them away with a decoction of absinthe or rue.

II Two explanatory Points

A. Lice and other imperfect animals of that sort arise from putrefaction in this way: When a body contains a bad kind of humnor which Nature cannot control or reject promptly, it repels it from noble organs as far as it can. When the humor is thin and lacks resistance and the vital heat is potent, the ailment is entirely resolved, or it is transformed into sweat. Or the opposite may happen. The humidity is not compliant and the vital heat is impotent andyhe humor either putrifies and causes scaling lesions, or it may capture some vital spirit (ie heat). Then the heat may overcome the humidity and engender animals like lice etc. They are lively and they move about. If no heat is captured the humor forms scurf (furfur) or lentils or pannus (ie not alive).[180]

B. Skin has three laminae or thin layers (ie pellicles). The lice and other similar animals are engendered between the two outer laminae in such a way that the vital heat acts on the substance which forms their carapaces and hardens it. It becomes a shell in which the animal collects humid vapors of its own and retains its own vital spirit.[181]

---

[179] Murgantia histrionica (LDR).

[180] The varieties of scales included pannus, lentils, dandruff, dartre, seborrheic keratoses, flakes, etc. (LDR).

[181] This explanation of spontaneous generation is very different from the Egyptian concepts about dung-beetles. Note that Mondeville makes no mention of it when he includes scarabs with 'similar animals' in item C in Part I (LDR).

CHAPTER 21.

BURNS AND BLISTERS CAUSED BY FIRE AND HOT OIL OR WATER

I The Lesions

A. Burns: The term defines the lesion. The Treatment has two aims: 1. Prevent blistering with simple and compounded topicals. The Simples are vinegar, all kinds of clay such as bol armenie, gypsum, terra sigillata, argilla, poitiers-clay, and others that are found along river-banks as dry mud that is greasy and viscous and resemblesa thin poitiers-clay; morel, joubarbe, purslane, umbilicus-venus, plantain, lettuce, shepherd's crook, oxalis, knotweed, and many other cool herbs. We use their pulp and their juices or the distillates of the juices.

The Compounds are ointments, plasters, and cool lotions made from the Simples and others. This is a commonly used ointment: Oil of roses with egg-whites or yolks or whole eggs. Thicken it with something like barley-meal or ceruse. All medications are applied cool and serve as coolants, or if warm to begin with are cooled with snow before being applied, and must be renewed frequently to overcome the local heat at the burn. 2. To treat blisters after they appear, use simples and compounds. The Simples are oils of roses, nenuphar, myrtle and other cool and oily and earthen (ie clay) substances as listed above. The Compounds are Rhazes' white ointment, an ointment of ceruse, an oil of egg-yolks[182] and roses, the juice of endives and similar plants, all with added barley-meal and raw egg-whites. Or use an ointment of oil of roses, wax and lime which has been washed or soaked in water for a day. This last ointment and the other nonspecific medications act easily and rapidly and they are entirely adequate as the sole topicals.

---

[182] See Antidotary for the recipe (LDR).

Blisters can be a primary lesion as well as a complication of a burn. The treatment (ie of both kinds) comprises four elements. 1. The Dietary Regimen and 2. Evacuations are those that we prescribe for ailments due to burnt humors, an overabundance of bile or a plethora of phlegm. 3. The Topicals for both the primary and the secondary blisters, whether they are full of liquid, have intact skin or are ruptured: An ointment of oil of roses, ceruse, litharge. Another: 1 part of white wax, 4 parts of oil of roses, and enough dissolved ceruse to thicken it. Then add egg-white and stir. 4. The surgical procedure consists of lifting off the bleb without injury to the underlying intact skin. If the blister is large, simply perforate it in two or more places and use the ointments described in the Antidotary.

I consider it foolish and unjustifiable to repeat over and over the recipes for ointments which we use for various purposes. Furthermore, all the medicines used for treating burn-blisters are there to be used for all kinds of blisters, and vice versa.

II Three Explanations

A. Although I have made statements that contradict other authors and their handbooks, experience shows us that when you burn your finger and you immerse it immediately in cold water, the pain increases. Perhaps the cold enters and keeps the heat inside, In contrast, if you keep the burned finger warm near a fire and allow it to cool gradually, it can exhale its heat and the pain subsides. That is why your cooling medicines are to be used after they are slightly tempered.

B. Medicines that prevent blistering should be very cool and wet. But after the blisters have formed or when the burn is large, the medicines can be less cold. Besides, they should be abstersive and mildly astringent, but not biting.

C. Lime is useful in treating burns, although it may be warm. Therefore, first thoroughly wash it in a lot of fresh water or macerate it for a day. Then mix the purified stuff with enough oil of roses to eliminate all of its corrosive quality.

CHAPTER 22.

ROSEOLA, VARIOLA AND BLATTES[183]
(ALL MEAN THE SAME)

I. The Maladies and Their Treatrment

A. Descriptions: Some claim that roseola and variola are different in that roseola derives from corrupt melancholy and variola from other corrupt humors. Other authors say that they are the same, except (ie for names) at the onset when they are red (ie call them roseola) and later on when they are variegate (ie call them variola). It may well be that the are both, the same and different, and most writers put them in the same chapters of their handbooks and treat both as infections with tiny pustules. They form just beneath the epidermis, from corrupt humors, the roseola from melancholy, the variola from the others.

There are five General Causes: 1. Menstrual blood that has been trapped since birth in parts of the fetus.[184] 2. Conception of the fetus during the woman's menstrual period. 3. The inclusion in her diet of fish, dairy foods, green and moist foods and other soft and easily decayed foods at a time when she does not menstruate (ie pregnancy). 4. Corrupt and pestilential air. 5. An incomplete crisis during a sanguinous fever, and one of the four corrupt humors predominates and causes the disease.

---

[183] Mondeville does not commit himself with differential diagnoses because the surgeon dealt only with the esposed lesions of small-pox. The disease, well-known even in ancient times, was named variola by a Bishop Marius in France during the 6th C. Rhazes was the first to clearly distinguish it from the other pustular exanthems. Mondeville describes milder cases of small pox, and perhaps confuses them with varicella. He gave little heed to devastating lethal epidemics which did not occur in southern Europe in his era. Perhaps the endemic prevalence conferred some protective immunity in those populations, even against the introduction of foreign strains by the returning crusaders. (LDR).

[184] Hence it was a childhood exanthem, perhaps varicellla. (LDR).

The lesions form as follows: Before they appear the patient feels an irritation in his nostrils and sees lights and has nightmares; his face reddens and he feels prickles everywhere and he has backaches. All those symptoms are a prodrome lasting three or four days. Then tiny spots appear, at first the size of needle points then of grains of millet or heads of ants as they enlarge and form pustules which dry and flake off.

The treatment includes four items:

1. Five general rules: a. Do not provide cold foods or beverages which make the faulty humors curdle. b. Do not apply cold or anything cooling on the affected skin. c. Avoid repercussives, because they will direct the toxic matter from the surface inward towards the noble organs. d. Use no cool topicals as oils, or medicines that block the pores, for obvious reasons. e. Avoid cold air.

2. The evacuations will be phlebotomies and the cleansing of the blood as described in the chapter on bleedings and on impetigos etc. Follow those rules to the letter. We comment here that you should never bleed after the eruptions appear, do it only before that phase if you can recognize the prodromal signs as described above. Also, do not prescribe laxatives that are too potent, because these illness themselves loosen the bowels, and you need have no concern for constipation.

3. The diet tends to cool and dry foods. Offer barley-broth and gruel, dry raisins and figs, herbal juices and acidic fruits such as oxalis, green grapes, pomegranates, oranges et al. For other suggestions consult the medical handbooks. Really, this is not a surgical problem beyond the preparation and the application of the topicals and the fact that the common folk often consult us when the eruption appears.

4. The manual procedures never open the pustules until they are completely 'ripe'. Then use a needle or gold lancet. Sometimes you may use a chisel and trim away some slough or skin at the margins to prevent them from closing over as a new pustule.

5. The Topicals are used to treat the lesions and to lessen the serious complications. They are extractives and desiccatives

used in an effort to avoid unsightly scars, and to get rid of those that remain.

The *Extractives* which treat the lesions themselves.are as follows: Wrap the patient in a red cloth. Another: boil several handfuls of fennel and wild celery and a lot of lentils in water. After it is cool, dip cloths in it and squeeze them damp and wrap the patient in them.

The *Desiccatives* : Open the ripe pustules and dust the body with this powder: Equal parts of flours of fava-beans, lentils, chick-peas, lupins, vetch-seeds; litharge, ceruse and aloes. Also dust it on small pads which will comfort the patient (ie self-application). To Prevent Ugly Scars: Prevent the patient from scratching and keep him away from pork products while he is sick. If he disobeys and eats them in secret, the scars will be worse. To treat scars after the fact, use the lard of a donkey cooked with oil of roses. Another: equal amounts of litharge, cachimia and ceruse. Wash them all and take the pulverized ashes of reeds and mussel-shells and add enough oil of lilacs and wax to make an ointment.

The topicals that prevent or ameliorate serious complications are many and of many sorts. Some protect the eyes, eyelids and face. Apply a collyrium of saffron and rose-water, frequently. Another: rose-water in which sumac has macerated for a long time. Another: collyria of coriander, sumac, rose-water and egg-white. The topical to treat the nose is made of verjus (ie sour grapes), rose-water and vinegar. For the throat use a gargle with the warm juice or the wine of mulberries, blackberries and barley-water. You help the lungs with electuaries of diapapaver, diatragacanth et al. Protect the stomach with troches of sumac, coriander and spodium, mixed with the juice of oxalis et al. Or use a decoction of saffron and lentils.

II. Two Explanations

A. The common folk have so much faith in the surgical potions, that patients often request that they be prescribed. Here is one that is potent and is successful if it is properly composed and used by an educated surgeon. Take equal amounts of dry figs, macerated tragacanth, the milky juice of fennel seeds and lentils and a bit of saffron. Crush them and boil at length. Dose it as a decoction and it will bolster the vital forces, comfort the spirit, sweeten and dilate the passages of the chest, open and clean and improve the work of the liver and support the heart. Well-meaning uneducated women since ancient times have dosed their children when they are smitten with these maladies with a decoction of saffron and lentils. Some have used the water of a decoction of figs with a little saffron. Others use hot water and honey and yet others make a tea of the juice of scabieuse.

B. Do not err in feeding the patients. The foods should be fatty during the first week in order to thicken the blood and to prevent an additional combustion of the sanguinous humor and cause a relapse. That diet includes jujubes, verjus, sumac, berberis and the juice of astringent fruits such as crab-apples, quince, medlars, et al.

CHAPTER 23.

WARTS AND PORREAUX[185]

I The Lesions

A. Description: Warts and Porreaux are firm fleshy excrescences on the skin anywhere in the body, but more often on the hands and feet than elsewhere. They differ in that porreaux derive from melancholic blood and are firmer and have fissured surfaces, somewhat granular as a result of the gritty nature of the melancholy. Warts are formed from thick phlegmatic blood and are softer than the others.

B. Treatment: The Dietary Regimen for Warts: Abstain from foods and beverages that engender thick phlegm. For Porreaux: Avoid foods and beverages that engender melancholy. The other items of the regimen are the same as those which we prescribed for dartre.

The Purges are the same for both hunors.

The Topicals: Some are light and gentle; others are energetic and some are very potent. The gentle topicals recommended in the handbooks are oils of myrtle and roses; acacia, bol d-armenie, ashes of grape-vine wood or willow tree bark, one or the other is mixed with vinegar. Others: juice of green walnut fruit-husks with powdered dry

---

[185] Porreux may be seborrheic keratoses (LDR).

goat-feces. Or use meadow marigolds that grow among the grape vines at Paris, crushed with salt. Or crush agrimony and salt. Or crush bryony with rancid lard. O a massage with purslane; or cover the warts with tree-moss and let the leaves rot. The warts will decay at the same rate. Or touch the warts with chick-peas, then toss them behind the patient's back. Or scrape out the insides of the roots of a plant that is called the rays of Larchamp in France (ie a strong radish, perhaps horse-radish) and apply the scrapings to the wart and cover it all with a thin round slice of the same root. That treatment is called the corn of the toes. Or grind equal amounts of the leaves of rue, yarrow and Robert's herb. Rub it on the wart and put some more of it on the wart and cover it with a bandage. After three or four days the wart will be consumed. Or use this simple and well-tested remedy which is quite reliable: Put red slugs and some salt in a lead pot and let them sit for four or five days. Then you will find some liquid with the slugs in it. Touch the warts or porreaux with the liquid, three or four times a day for six days; at each episode touch the lesion three or four times. The excrescences will come away with their roots, without discomfort to the patient. Those are all the gentle remedies we will need.

Stronger local treatments advocated in the handbooks: Peel off the wart or open a blister with a finger nail or knife and introduce the juice of a fig or of euphorbia, some cataputia, powdered asphodel, couperose, verdigris or other mild corrosive.

The very potent topicals are eruptors as decribed in the chapter on morphea and other strong corrosives such as quick-lime, arsenic (sublimated or not), realgar, honey of anacardus et al.

The most radical measures are cutting away the wart with its roots, or tying a strong thread at the base, or burning it off with a hot cautery or with hot melted live sulfur.

We need not describe the techniques for using the topicals, they are simple enough. When you apply strong caustics do this: Apply the caustic for two hours, then remove it and lift off the crust. Then introduce sublimated arsenic et al. into the defect, just enough to be useful. Let it remain for as long as deemed necessary.

The technique for radical excision consists of cutting it away and applying caustic on the defect. Do the same to the defect left behind after ligating the base of a wart. When you use a hot nut cautery, place it in a perforation of a plate that exposes only the wart. If you use the burning sulfur, place some powdered live sulfur in the perforation of the plate and ignite it. Keep it burning until the lesion is destroyed.

II Two Explanations

A. A large wart may breed many small ones around itself. But it contains nothing in itself that can produce other warts. When a large wart is formed it attracts a large amount of melancholic blood, more than enough for its own development. The excess of humors is expelled by the wart's own Natural force, into the neighboring soft, thin and porous skin. It is not reasonable to claim that the tiny warts are formed by the same blood that engendered the large one; we have seen the little ones where there was no sign of blood. Therefore, I think my explanation is correct.

B. I argue in opposition to the authors of the handbooks who advocate the use of drying topicals. The substance of the warts and porreaux already is dry and the treatment should use contraries. Their dry topicals are not contrary to dry warts. Indeed, dry topicals are either constrictive astringents, which are not the intention of the handbooks, or dry resolutives, consumptives and occasional carburants. Certainly the last ones are the only ones that cure, and that is our common goal.

CHAPTER 24.

SIMPLE TUMORS (SWELLINGS)

I. Descriptions

At times a tumor is a simple enlargement of a tissue, soft to the touch, caused by an inflation or a vapor mixed with a viscous humor which cannot be properly consumed because the local natural heat is inadequate for the task. Also, the external pores may be blocked.

This definition is clear enough. Inflation is the category; and simple and soft are the differentials. We identify the material and the efiicient causes and the predispositions. This definition befits a tumor and excludes all the other swellings that may or may not go on to suppurate: warts, nodes, joint-swellings, phlegmatic and melancholic aposthems, etc.

A careful examination comes first, to determine whether or not it is a medical or a surgical tumor. The common folk tend to believe that all swellings are surgical, and they go only to the surgeons because the lesion can be seen and felt.[186] If we mention some other swellings in this chapter it will be a matter of convenience not due to ignorance. We aim to serve the interests of the surgeons who seeks information about treating tumors, and we will discuss all kinds of tumors in that setting.

II. Treatments: Two Kinds

A. Evacuations, three varieties: 1. By laxation. 2. By vomits. 3. By others, including massages, baths, abstinence, exercise and rest; not much to choose from for the surgeon.

B. Topicals: Those which we use for all kinds of tumors, in all parts of the body except the liver, the stomach and the spleen. Boil some laurel berries and powdered cumin until the consistency is right. Apply it after a general evacuation. Another: crushed absinthe and mauve boiled in water. Another: oakum pads soaked with the warmed juice of mullein. Another: a lumpy porridge of barley-meal and the juice of berula. Another: a fomentation with a decoction of sage. Another: leaves of crushed joubarbe fried in oil and butter. There are many others of that ilk.

Special Topicals, according to the various sites: For a tumor of the ear, both insert warm oil or vinegar in the ear and cover it. Or boil fava-beans in vinegar and direct the vapor into the ear. Then apply an ointment of camomille around the ear.

---

[186] Recall that the formal arrangement during th 13thC gave surgeons responsibility for external ailments only. Mondeville excludes internal neoplasms and limits his 'tumors' to edema, hematoma and inflamation. He includes tympanties, priapism and ascites (LDR).

For a tumor of the jaws, use white marrubium crushed with salt and applied warm.

For tumors of the tongue and palate: first bleed from a saphenous vein if the general conditions permit it. If you cannot bleed or if it is not successful, prescribe cochia pills and set a cloth wet with the juice of lettuce or a similar herb around the tongue. Another: use a gargle. Another: put bits of head-purges in the nostrils to deflect the matter from a tumor on the same side.

For a tumor of the uvula, first bleed from a vein in the hand between the index finger and thumb and from a vein under the tongue and apply a cup at the occiput. Some practitioners put their feet against the shoulders of the supine patient and pull his occipital hair.

For a tumor of the neck, bleed from the same vein on the hand as well as from one on the forehead (ie temple) and apply a warm oil of camomille on the tumor, or use raw hemp fibers or oakum soaked in a mild lye or water. Or cover the tumor with ashes as hot as the patient will tolerate.

For a tumor of the arm, after a bleed, apply wheaten bread crumbs boiled in wine. Or: apply 1 part of oil of roses with 1/2 part of dialthea.

For tumors of the hands and feet: begin by warming them and apply a sponge soaked in cold water and vinegar or in Pontine wine. If the tumor is cool, use a sponge wet with lye.

For tumors or enlargements of the breasts, use prolonged compresses of hemlock, cumin and frankincense, all crushed and macerated in vinegar. Or: use bread crumbs or fava bean flour with warm juice of apium.

For tumors of the stomach, liver and the spleen, use oils of mint, absinthe and mastic applied warm over the affected organ. For a stomach tumor apply a sponge wet with warm vinegar and water or some oil of laurel, spica et.al. Here, I limit our treatments to tumors in the abdominal wall in front of any of the three organs mentioned and not to tumors of the organs themselves. Those tumors are the responsibility of the Physicians. They are treated first and principally with oral medicines rather than with external applications of ointments, plasters etc. Finally, cauteries are useful when all else fails.[187] Inasmuch as those organs have feeble complexions and perform vital functions for the entire body, you must be careful not to apply topicals on the abdominal wall over them which harm them and impair their functions. For a tumor over the liver apply a plaster of absinthe and mauves fried in oil of mastic or similar. For a tumor of the spleen, use a plaster of mauves and absinthe cooked in a good wine with added tamarisk tree-bark.

For a tumor of the abdominal wall, first use a clyster to get rid of the gas and then lay a large cup, without scarification, centered over the umbilicus. I saw a four-year-old child whose belly was so swollen to the extreme that it appeared it was ascites. He recovered after five days of treatment solely with an inunction of mastic.

For a tumor of the penis (priapism) and elsewhere in the region evacuate by vomiting, then use massages, exercise, baths and a plaster of pigeon droppings, barley-meal and vinegar. If heat predominates, use Galen's ointment which is made with melted white wax and oil of roses. Cool it before applying by immersing the jar in cold water or snow. If cold dominates, apply solvents (ie dissolutives) and thinners (ie attenuatives), consumptives of gas, and such desiccatives as rue, cumin, ache, dill, anise, parsley, laurel berries, et al.

For tumors of the testicles in old men use clays with vinegar.

For tumors in the hips, thighs, legs and feet of phthisical and doddery old men who are edematous, dyspeptic, liverish and who are bothered by the swellings,[188] use fomentations of lye made from grape-vine-wood ashes, and apply cloths and bandages wet with the same lye. Or boil some leaves of elder, box-elder and laurel trees with pigeons' and goats' droppings. Foment first and then apply the wet cloths and bandages. Or you may use the roots of radishes with sheep-feces and salt. Or you may rub lightly but at length with the oil of roses and vinegar to which you have added a small amount of salt and powdered nitre. Or you may use pigeon droppings mixed with a strong vinegar applied on a cloth. For a tumor limited to the lower

---

[187] When the surgeon may be called upon by a reluctant Physician (LDR).

[188] This doleful recitation is Mondeville's tongue-in-cheek description of himself (LDR).

leg, use this especially famous application, a cloth wet with the liquid distilled from the juice of morel. For a simple swelling of a foot, use the leaves of elder trees or of parietory, crushed and fried in oil or butter. For a swelling that remains after a podagra has subsided, anoint it with rue cooked with wine and rancid oil. A swollen foot caused by marching or hard labor (ie digging) is relieved with plantains ground with vinegar. Or use a fomentation of water and warm ashes which has been put on a cloth and wrapped with a bandage.

III. Five Explanations

A.   Every humor can be associated with a specific aposthem. Indeed, the very term aposthem, according to all the authorities, describes a non-natural tumor in a part of a body. But a tumor that does not have the potential to suppurate is not an aposthem. A tumor is a kind of deforming malformation, and we all know it as such. Therefore, this chapter is not placed with that on aposthems but near the chapters on skin-blemishes.[189]

B.   Galen (*Megatechni*, Bk. XV, Ch. 3) established the difference between a phlegmatic aposthem and a simple tumor, and said that the difference is seen in their treatments. Phlegmatic aposthems are soft and can be indented by a finger, and the hollow soon refills, and the contours are restored. A simple tumor, to the contrary, is caused by an inflation that does not yield to the finger and show a dent.

C.   Galen (ibid,), in the case of a penile tumor, said that a vomit works better than a laxative, whereas, for a tumor of the tongue, the reverse is true. From these comments, we can infer the valuable Notable that a diversive evacuation from a distance is better than one from close by. I discussed that precept in Notable II of the Introduction to

---

[189]   Nicaise's Note: Mondeville called this the Doctrine on Embellishments and wrote it without a clearly defined order. He appended it to chapters on cutaneous ailments and to other chapters that could better have been placed elsewhere. Unlike Mondeville, Guy de Chauliac placed his chapters on cauteries bleeding, cupping and leeches at the beginning of his antidotary, and placed his chapter on simple tumors in his Doctrine on aposthems (EN).

Part 7 of Chapter 1 of Doctrine I of Treatise II, where we considered the issues of prevention and treatments of warm aposthems and dyscrasias.

D. In this chapter, where we summarize the treatments, I have been brief, because there will be seven more chapters completely devoted to special tumors, and to the treatments of certain tumors by Physicians or Surgeons or by both.

E. We often need new sponges when we resolve tumors, and we cannot always obtain them at the time of need, and we must reuse old ones. In such a case, be diligent in washing them many times with egg-yolks and French soap.

Here Ends Doctrine I of Treatise III

VOLUME II

PART II

TREATISE III

DOCTRINE II

CHAPTER 1.

THE GENERATION OF HUMORS

I The Process of Generation

Food is chewed in the mouth until it is ready for digestion in the stomach. Then it is attracted into the upper esophagus and expelled below into the stomach. There it is mixed with the beverages into a mass which is nicely and entirely enclosed by the gastric wall which warms it and digests it on contact—its normal action. After that has been accomplished, the stomach exerts a second function and it separates the bulk of the impure matter from the lesser quantity of useful pure matter. The stomach discharges the impure feces and it is attracted into the intestines where it descends. That matter is not entirely fecal, because some chyle escapes with it. The fecal mass traverses the four parts of the intestine, the duodenum, the ileum and the jejunum and the cecum.

The chyle is attracted from within the intestine into the mesenteric veins and carried by them to the portal vein and thence into the liver. To sum up: the first digestion begins in the mouth and is mostly completed in the stomach. However, the small amount of chyle that was admixed with the feces is not completely digested and absorbed until it reaches the cecum.

The chyle, the product of the first digestion, now separate from the feces, has been brought to the interior of the liver by various expulsive and attractive forces. There it is boiled and digested again, just as the must of crushed grapes is acted on in a vat or a jug.

The cooked chyle then becomes the four humors.[190] The first is a white froth which rises to the top. According to Avicenna (Bk. I, Fol. 1, Doct. 4, Ch. 1) and Haly (*Techni, Treatise on Causes,* Ch. 38) it is the phlegm. However, Lanfranchi claimed that the top layer is bile. The second liquid substance formed by the heat of the second digestion is red bile. The third humor, formed by the heat in the core of the chylous mass is pureblood. The fourth is the residue of the rest, deposited beneath them; it is the melancholy. These four humors differ in terms of their humidity and other qualities. Phlegm is cool, moist and white and it corresponds to water. Bile is warm, dry and brown. It corresponds to fire. Blood is warm and moist and corresponds with air. Melancholy is cool, dry and black and corresponds with earth.[191]

II The Separation of Humors

After a second digestion of the humors, God and Nature, who do nothing in vain as explained in our Anatomy, separate them and each, excepting phlegm, has its own reservoir where it can remain uncontaminated and unmixed. Thus, pure melancholy, which is the lees of the humors, is attracted by a divine agency from the liver to the spleen. Pure bile is similarly attracted into the gall bladder. Pure

---

[190] See Tr. II, Doct. I, Ch. 3 on Phlebotomy, The Examination of the Clot (LDR).

[191] The magical and mystical properties of the number four were accepted without question in the medieval epoch as they had been in every era since Pythagoras. The correspondence of humors to earth, air, fire and water was unavoidable. See VF Hopper, Medieval Number Symbolism , Columbia Univ. Press, 1938. p.144 and ff.(LDR).

blood is taken into the veins. Phlegm and watery urine are separated from blood as it leaves the liver, but not completely, and some phlegm is carried with blood in the veins, as are portions of the other humors. The water to form urine makes the blood more fluid so it can penetrate with ease, and all together the three humors accompany blood and participate in the nutrition of the body and all of its parts. Therefore, a phlegmatic member is nourished with phlegm, and the same for all the others. Lacking that, the non-sanguinous structures would not be nourished: remember, nutrition is made by similars. So it is that the remaining urinary water in the blood nourishes the kidneys when it reaches them.

Phlegm has no special reservoir. Most of it is found in the joints, the brain and the lungs, to lubricate the organs that are continuously in action. Some of it remains in the veins to be available when needed, and it nourishes as a companion of the blood.

III The Special Character of Each Humor

Each of the four humors can be either Natural or Non-Natural. It is Natural if it conserves its original qualities, has not set itself apart and has not been altered. Those limits are somewhat loosely drawn, leaving room for variances; for example, a humor can be more or less warm or dry and yet remain Natural. I cite Avicenna for his statement (*Song of Songs*, where he spoke of blood which is heated in the heart). He wrote, "Blood in the heart is warmer and drier than in the liver."

Now we must describe the differences between the humors. All of them, together or apart, contribute to the matter of the aposthems in the body. (see Chart of Humors)

Melancholy is of two kinds: 1. The Natural can vary, be more or less cool or dry within the limits of its Natural complexion. But if it exceeds those limits or if it is admixed with Natural Melancholy is produced in five ways, yielding five varieties. a. From burnt blood. b. From burnt bile. c. From burnt phlegm. d. From burnt natural melancholy which had become

A TABLE OF HUMORS DERIVED FROM CHYLE

(a copy from Nicaise's General Introduction in Volume I)

PHLEGM

NATURAL PHLEGM   NON-NATURAL PHLEGM
                                  Aequous
                                  Mucilaginoue
                                  Vitreous
                                  Gypseous
                                  Salty
                                  Sweet
                                  Pontine (two types)
                                  Acidic (two types)

BILE

NATURAL BILE   NON-NATURAL BILE
                                  Citrine
                                  Vitelline
                                  Prasine
                                  Aeruginous
                                  Burnt (three types)

BLOOD

NATURAL BLOOD   NON-NATURAL BLOOD
                                  Five Types

MELANCHOLY

NATURAL MELANCHOLY   NON-NATURAL MELANCHOLY
                                  Five Types

more liquid and bitter. e. From burnt natural melancholy which had remained thick and less bitter. As stated above, Natural Melancholy is the lees, the residue of the mass of humors produced in the liver, just as tartar is the lees of wine. Non-natural Melancholy, to the contrary, is the result of a combustion the ash of the humoral mass produced by excessive digestive heat in the liver.

Blood also has two varieties. Natural Blood is a red humor—clear, sweet, free from bad odors or flavors. Its complexion is not strictly limited, but it cannot exceed the limits and remain natural. Non-Natural Blood has exceeded the limits but it continues to have the complexion of blood. It has any of five forms. a. The chyle is gently heated beyond that needed by the liver to make Natural Blood; it is less red and its complexion is less warm than that of Natural Blood. b. It definitely has been over-cooked and it acquires qualities that are opposite to the Natural. c. It is thicker and heavier, or, d. It is too watery. e. Natural Blood that has been mixed with some other humors which change it and its complexion.

Bile also has two types. Natural Bile is clear and transparent; it is light in weight; it is yellow or nearly so. It forms the membrane on the surface of a blood-clot. Its limits are not as strictly defined as they are for the others. When the bounds are overstepped for any reason, it becomes Non-Natural Bile. Again, there are five varieties: a. The Citrine is weak, and derives from a small amount of natural bile. b. The Vitelline has the color of egg-yolk and derives from an admixture with a small amount of phlegm. c. The Prasine gets its name from its resemblance to the juice of green onions or of emeralds. It is very strong, bitter and acidic and very tart. One often vomits it at the onset of a tertiary fever. d. Aeruginous Bile is reddish, the color of rust. e. Burnt Bile includes three types: 1. Combusted Natural Bile by itself and nothing more, and, not enough to cause a separation of its thinner and thicker contents or to incinerate itself. When a separation occurs, it becomes burnt melancholy. 2. A small amount of burnt bile is burnt with a large amount of natural bile. My term 'small amount' distinguishes it from a larger amount of burnt bile, which is combusted to form melancholy. 3. Natural blood that has just entered combustion forms burnt bile. Take note that completely combusted blood becomes non-natural melancholy and cannot be considered a burnt bile.

Phlegm has two varieties. 1. Natural Phlegm is a cool, moist, white humor with a flavor somewhere between sweet and flat. It floats with the urinary water atop the other humors in a dish containing clotted blood after a phlebotomy. Here, too, there are latitude and limits.

2. Non-Natural Phlegm has eight types, four showing different matter and four having different flavors. The four with different consistencies are: a. Aqueous Phlegm is obviously liquid, barely cooked, translucent, sticky and viscous. b. Mucilaginous Phlegm is not homogenous, but is lumpy like clotted milk, both liquid and semi-solid. c. Vitreous Phlegm derives from the aqueous, which changes over time when exposed to cold. d. Gypsum-like Phlegm is thick and as firm as plaster without any residual heat. In many people it appears on the face as bits of salt between the skin and the underlying flesh.

The four Non-Natural Phlegms that differ in their flavors are: a. Salt-Phlegm which tends to be more dry and warm than the others. It consists mostly of tasteless phlegm and a small amount of burnt bile, thus it's flavor is burnt and salty. b. Sweet Phlegm is almost the same as Natural Phlegm; there are two types. In one the process of combustion is shorter than that for blood. The other has a little blood as a sweetener and a larger amount of bile. c. Pontine Phlegm derives from a liquid phlegm which has acquired the (pontine) flavor of any unripe fruit. There are two kinds: in one cold has congealed the matter which is the source of the flavor. In the other a small amount of pontine-flavored melancholy has been added. d. Acidic Phlegm began as a liquid and became acidic as it thickened. One of the two kinds was boiled (ie bubbled) and putrefied and then cooled to make it acidic. The other type is a mixture of a large amount of phlegm with a small amount of bitter melancholy, just as milk is soured in two ways. It will sour in time simply if left to itself, or when you add some coagulant (ie rennet) that will make it ferment, and then you allow it to cool.

IV Eight Explanations

A. Avicenna, Haly and Averroes (*Commentary on Avicenna's Song of Songs*), all deal at length with this question, as do many illustrious physicians, to explain the Natural things. We can review the matter briefly and not repeat what they wrote except to deal explicitly with the different surgical implications.

B. A knowledge of how things form, etc. is necessary for an operating surgeon, because all aposthems and great numbers of other maladies derive from the four humors, and from water and wind, from each alone or in combinations, natural and non-natural,. A surgeon cannot cure a malady unless he knows its cause, as we can infer from Avicenna (Bk. III, Fol. I, Ch. On The General Regimen For Fevers) who said, "Know this: you cannot cure a fever if you do not know what came before it, and you cannot know that was the cause unless you know about the cause." That is clear, and it is all that needs to be known. Everything in existence has a cause and given the cause the effect is known, You can suppress the cause and thereby suppress the effect. Therefore, the malady should be known by its cause, and when you wish to cure a malady you must seek the cause, and suppress it. Galen said (*de Ingenio*, Bk. VIII, Ch. 9), "There is no doubt that we cannot cure an illness so long as the causes persists."

C. Some writers disagree with Galen's and Avicenna's limits of four humors, and they claim that there is only one, which is blood, and that the others are merely superfluidities separated from blood during digestion, just as foam and tartar are separated from wine, and of themselves are not wine. They explain that blood is the sole nutriment of the entire body and as such it is the sole humor, stripped of any others. If that is the case, a melancholic structure would receive no nutrition, because all parts of the body are nourished only by their similars. While it is true that the blood nourishes the body, the blood in the veins consists of the sanguinous humor along with some of all the other humors. Indeed, we must say that blood nourishes everything with its sweet major component, and also with the non-sweet other substances, which are the humors.

D. I disagree with Avicenna and Haly as to the phlegmatic nature of the foamy scum for this rteason: When you remove blood by venesection, it coagulates and you can define only four parts. The deposit at the bottom is melancholy, the firm red film over the coagulum is bilious and what is between is sanguinous. They say that amounts to only three humors, and there must be four, and the foamy scum floating on top must be phlegmatic. On the other hand, in my argument with Avicenna and Haly, I claim that bile is lighter than phlegm, and we should expect that it would occupy the uppermost tier.

E. One may ask how chyle, a uniform substance, can be digested by the uniform heat (ie vitality) of the liver into four very different substances. I reply that chyle itself consists of various things, coming from the various parts of the stomach into various parts of the liver, and that explains the different sources for the four humors.

F. Take note of what Avicenna wrote (Bk. I, Fol. I, Treat. 5, Ch. 1), "A humor is a fluid part of the body, formed from the first mixture of foods, and foods are formed from the admixture of elements, and the parts of the body are formed from the first mixture of the humors."

G. Phlegm alone can provide nourishment whenever that is necessary. Bile and melancholy cannot provide it, because the those humors are more combusted than blood, and they cannot be reduced to blood. In fact, whatever has been more-cooked cannot become something that is less-cooked. What has been less-combusted, such as phlegm, can be subjected to additional combustion to any degree.

H. Here I add Explanations about the separation of the humors. If all the humors arrived en masse and remained undivided at the sites they must nourish, they would be rejected, especially by melancholic, bilious and phlegmatic structures. The mass of humors would be expelled from the blood into the subcutaneous tissues, and we all would become phlegmatic, yellow-pink in color or dark-skinned or even worse.

CHAPTER 2.

THE GENERAL TREATMENT OF APOSTHEMS

(Excluding Particular Lesions And Their Causes)

A. Definitions: An aposthem is a swelling or an enlargement beyond the normal dimensions of a member. All the rural surgeons and all the illiterates in the world believe that there can be no aposthem without pus and only if the swelling is large. That is not what Avicenna said (Bk. I, Fol. 2, Ch. On Compound Maladies), "small swellings are small aposthems, as the larger ones are large aposthems."

B. Varieties: Some derive from a single humor, some from water, some from wind. The single humoral cause may be blood, bile, phlegm or melancholy. The humoral cause may simply be an overabundance, or the bad qualities of a natural or a non-natural humor. A single pure humor, blood for example, may be a cause or there may be a mixture of humors, such as blood mixed with bile and phlegm. In addition to the internal (ie humoral) causes, there may be such external sources for an aposthem as a blow or a fall. Some are caused by congestion, others by diversion, and some from both mechanisms. Some are common, others are seldom seen. Some are painful, even in extreme degrees, others not. Some involve principal organs, others may be near them, and others arise at a distance. Some appear in fleshy tissues only. Some arise in persons with healthy complexions, others not so, occurring in feeble as well as robust bodies. See chart on p. 802.

The Aposthems According To Mondeville

(copy of p. 93 in Nicaise's Introduction in Vol. I.)

Formed From Humors:
    From Natural Humors (4)   Bloody   Phlegmon
                            Bilious   Erysipelas
                            Phlegmatic   Edema
                            Melancholic   Scirrhous

From Several Natural
    Humors (9)   Blood and Bile
                                  Phlegmonous-erysipeloid
                                  Erysipelous-phlegmatic
                    Blood and Phlegm (thin)   Intermediate abscess
                    Blood and phlegm (thick)   Glands and soft masses
                    Blood,phlegm, melancholy   Scrofulas
                    Bile and Phlegm   Nonsuppurating abscess
                    Phlegm and melancholy (thick)   Glands and hard nodes
                                            Hard scrofulas
                    All four humors   Carbuncles, Anthrax

From one unnatural
    Humor (8)   Non-malignant
                            Blood   Thin Blood and heat
                                        Thick blood
                                                  Bastard Erysipelas
                                                  Boils
                            Phlegm   Thin and soft
                                                  Goiter and Testudo
                                    Gypsum-like
                                                  Hard nodes

                    Malignant
                            Burnt Bile   Thin and liquid
                                        Thick Bile
                                                  Persian fever
                                                  Formiculation

                          Burnt Melan-
                              cholic bile.   Hematoma (prune)

From Several Unnatural
    Humors (4)   Nonmalignant   Glands and Hard Scrofulas

                        Malignant   Miliaria and Herpes

From Various Humors   Cancerous abscesses and ulcer
                                              Watery and gaseous abscesses
                                              Nevi and Warts
                                              Simple tumors

C. Causes: Before we go on to new material, let us remember that the aposthems, may come from congestion or diversion or from both. Those that are congested are due to a The superfluidity of nutrition, which, for some reason, the member cannot convert into its own substance. Then we say it is due to congestion, an accumulation. The defect is the excessive inflow or the lack of ability of the part to consume or expel the accumulated humors.

Another type of aposthem is caused by humors diverted from another part of the body, which the member cannot resist for any cause. That is called an aposthem of diversion or delegation.

Other causes: The recipient structure may be weak and it may lack expulsive force. Its anatomical site may lie caudad to the source of the humor, or the source may be cephalad. The cause may be the quantity or the severity of the bad humors, or the size of the veins that carry it into the member, or the smallness of the veins that can carry it off, or the distance between the source and the recipient, or the sponginess of the sick region as compared to normal glands.

We would serve no purpose to name all of the secondary factors here. You will find them in Treatise II, Doctrine II, Ch.1 on Ulcers. In fact, you will find it very useful to know all the causes for aposthems and how they are generated. You will find more on those subjects in the Explanatory Comments (ie at the end of this chapter).

D. Symptoms (signs): Avicenna (Bk. I, Fol. 2 Doct. 2, Ch. 1) wrote in regard of the symptoms of aposthems in general, that the senses of sight and touch give the observer the knowledge and reveal the obvious about an aposthem, and later he added very little about the general subject of aposthems. A recital of his special comments will fill a long and tedious essay. It is better for us here just to mention that he offers those items in his Bk. IV, Fols 3,4,5.

My reason is clear: an aposthem is not an abstract concept, but is an observable natural phenomenon, presenting in such forms as phlegmon, erysipelas, etc. and the signs are those of each in particular. From what we have said about the descriptions (ie definitions) the varieties and the causes we can infer some of the signs by which we can diagnose the true facts by feeling and looking.

A surgeon who treats aposthems should know the definitions and the parts, the varieties and the special conditions, the causes and an infinite number of other things, including the symptoms. In fact, among all those things to know, he must recognize how each of them pertains to his case at hand, because for each there is a different operation. That is our single generalization that is applicable to the entire doctrine on aposthems and similar maladies.

II Treatments In General

A. Preventive Measures: We have dealt with this question in general and have defined its content in the Introduction to Doctrine II of Treatise II. There are three sorts of preventive treatments that correspond to the three stages of an aposthem. First are what we do at the earliest hint that an aposthem is on its way, before we see a full-blown sign. We can arrest it at once. The second prevents the further development of the lesion now in progress. The third stops the evolution of the aposthem before it suppurates.

The first has four elements: 1. A diet that contains some of all six non-natural things. 2. Abstinence. The regimen should be known by every properly accredited physician [192] who will obey the principle of contraries. 3. Proper medical and surgical evacuations, properly timed and in the correct succession, as we explained them in Part 5 of Ch.1 of Doct. I of Treatise II, and in the Explanations that followed. 4. Preventive treatments after contusions and other external accidents. Those do not require general diversive evacuations. Indeed, we should know to divert only from the affected site. If you control the body as we say here, no aposthem should evolve.

The second level of preventions has three elements: 1. A correct dietary regimen. 2. Evacuations as described in Doct. I. If that source is inadequate, you may supplement it with the medical texts. 3.

---

[192] Nicaise explains Mondeville's frequent use of 'medecin ou chirurgien suffisant" as a reference to the certification of competence of the practitioner by the local lieensing juries (LDR).

Furthermore, you may learn from your own experiences with three types of topicals: defensives applied around the diseased site; repercussives applied atop the aposthem wherever that is feasible, as described in the same chapter. Then apply resolutives over the aposthem. In some cases you use them one at a time. At other times you mix resolutives with the repercussives.

In the third stage of prevention you try to avoid suppuration in an established aposthem: diet and evacuations as described and topicals, including resolutives as we will prescribe later on for each type of lesion. All of them are described in the Antidotary.

The principles of preventive treatments were set down by Galen (*Megatechni*, Bk.IX, Ch. 4) and by Haly (*Techni, Treatment of Causes*, Chs. 34 and 38). In the latter, read 'preservative treatment' where he wrote 'preservative cause'. There a surgeon may read everything he may need.

B. Curative Treatments: Eighteen General Principles.

1. The internal cause for an aposthem either is a superabundance or a corruption of the humor or a water or a wind, each alone or together with the others.

2. Every aposthem either is repercussed (ie aborted) or becomes a solid mass or suppurates and becomes an abscess. Avicenna declared that only when an aposthem has suppurated does it merit the name 'abscess'(Bk.IV, Fol.3, Ch.3).

3. If you are assiduous, you can prevent suppuration in an aposthem. When suppuration occurs a fever will accompany it and you must try to avoid it as a dangerous complication

4. The surgeon should make a careful examination of the aposthem to determine if the cause is internal or external. If an internal cause is over-abundance, he should begin his treatments with an evacuation. If the internal cause is only a mild corruption of humors, a simple single medication will be enough. That is why Avicenna (Bk. I, Fol.4 Ch. 1) said, "All alterations of complexions need not be treated by contraries, that is, by an evacuation. Often, a good diet will suffice to correct a repletion or a deterioration of the complexion. Or you may

not need topicals after a successful evacuation. Indeed, you need not repercuss a single matter that is already on the move, or even more than one." Galen said (*Techni, Treatise on Causes*, Ch. 34), "If you want to repel matter that is on the move, an already full body may not accept it. Rather, you may observe the opposite happen after the application of a repercussive, even though you may have obtained a brief relief of local discomfort. The local tissues may harden and cause more suffering than before. Also, you may do harm with all resolutives, excepting camomille; they may attract more than they resolve. So said Galen. Elsewhere (*Megatechni*, Bk. II, Ch.3), he said, "Every dissolutive is warm, and all warm topicals attract unless they are tempered. You err by applying a maturative which, by its heat, often attracts the liquid matter which was ready to leave the mass. That is when it will increase the pain and exacerbate the aposthem."

The subject of purgations is discussed in Tr. II, Doct. I, Ch. 1, and more will be found in Tr.III, Doct. III.

5. A large aposthem in a replete body is treated in four ways: a. A general purge. b. A diverting purge. c. A repercussive applied over the lesion and a defensive around it. d. A resolutive applied when the repercussive fails. These are Galen's measures, as cited above, where he wrote both as a physician and as a surgeon. He added, "when the preceding measures are inadequate, us maturatives." But that is contrary to the primary goal (to avoid suppuration) as stated in Rule #2. Nevertheless, when the aposthem is ripe, open it and empty it.

6. Do not prescribe harsh local evacuations for plethoric persons, such as scarification of the affected part, except to drain pus. The purpose is to avoid the attraction of more humors to the site of the aposthem, which the painful procedure will induce. Galen said so (ibid.) Also, when an aposthem at the anus appears in a plethoric patient, do not use laxatives (Galen *De Ingenio*, Bk. III, Ch. 4) but use emetics. When the aposthem is above the umbilicus, use laxatives and never use topicals that

are too hot lest they, too, attract from the rest of the body to the site of the lesion. However, Avicenna (*Colliget,* Bk. VII) amended that rule by saying, "When the phase of increase of the aposthem has peaked, you may perform evacuative procedures near it."

7. The surgeon should try to initiate his treatment of an aposthem when it begins in a plethoric subject with general and diversive purges, according to our general protocol. In a non-plethoric patient, do not purge. Follow with simple repercussives except in the following nineteen cases: a. When the faulty matter is heavy and compact and will not submit to the usual repercussives, they may do more harm than good. b. When the matter is cold and resistant. c. When congestion is the cause. d. When the matter is too abundant to be affected by the repercussives. e. When the patient is a pregnant woman, the repelled matter may harm the fetus. f. When the aposthem involves a noble structure. g. When the aposthem is deep, as in the buttock (hip). h. When the aposthem has made the patient critically ill, repercussives may be lethal. i. When the patient is recovering, the residue of matter may be dealt with by Nature alone. j. All the authorities agree that you should not use repercussives when the aposthem involves an emunctory. Rather, use cups or other attractive measures (see below). k. When the patient is an infant. l. When the patient is a feeble oldster. m, The same holds if the aposthem is near a vital organ. n. When the matter is hard, it cannot be dispelled and repercussives make it harder. o. When the matter is intensely inflamed, impelling it toward vital organs is perilous. p. The same is true if the matter is venomous. q. When the aposthem is over a joint, you may push the matter into the joint itself and corrupt the nerves and the ligaments of the region. r. When it is near the anus, an attempted repercussion will lead to a fistula. s. When the aposthem has an external cause, drainage of the 'ripe' corrupt matter through the overlying skin, is better than an attempt to dispel it inward.

The Ninteen Cases are framed in this mnemonic verse:

> Size, cold, congestion, abundance and pregnancy,
> Noble-site, critical illness and convalescence,
> Tenderness, infancy, old-age, proximity to vital organs,
> Hardness, violence, virulent matter, proximity to joints and
>   anus,
> External causes.
> In all these, do not repercuss
> Nor purge before or after, but
> In all other cases, after purging a plethoric body
> And not purging a depleted one,
> You should repercuss.[193]

8. No defensives are needed to treat an aposthem caused by an external agent on a healthy body. Topical repercussives and anodynes (comforters) will suffice. However, Avicenna (*Medications for Aposthems*) said we should apply resolutives and emollients at the onset. When the patient is plethoric, first prescribe a purge and the other measures we described in the chapter on Contusions in Treatise II, Doctrine I.

9. Apply cool topicals on warm aposthems after the purges, thus using contraries. Galen (*Aphorisms,* Part 2) said that the repercussives in those cases should be very astringent, such as cimolea et al. For a cool aposthem, if you diagnose it, the repercussives should be the same but diluted with warm resolutives, or with simple astringents which also have some resolutive qualities, such as absinthe and squinanthus.

10. Do not allow vigorous exercize of a limb that contains an aposthem, nor let it rest in a dependent position. Therefore,

---

[193] The original: "Grossities, frigus, congestio, grossesse: Nobilitas membri, crisis, post hancque, resurgens; Et sentina puer. senex et prope nobile dura; Et furiosa virus in juncturis et in ano; Causa forensis. In his casibus nunquam retropellas . Nec antequam tu purges nec postea sed tu Casibus in reliquis purgato corpore pleno. Aut non purgato non pleno pellere debes.(LDR).

the patient with the affected leg should not walk nor sit with his leg down. Someone with an affected arm should wear a sling (Galen, *De Ingenio,* Bk. V. Ch. 4)

11. When the aposthem is deep in the tissues or presents an indurated mass or exhibits a thick (edematous) overlying skin you will need potent medications with proper contrary qualities. An example is an aposthem in the foot or in the breast.

12. Never perform an evacuation caudad to the level of the aposthem when the torso is above the site of evacuation and of the aposthem, such as bleeding from a foot when the aposthem is in the thigh. No matter the success in attracting the matter from the abscess, you will pull more matter from the body toward the aposthem. The humors that are drawn past the mass to the site of the evacuation may not be eliminated and will return to the aposthem and enlarge it. Eventually that accumulation will suppurate.

13. For an aposthem at a distance from the torso, such as in the hand or foot, when it is in the stage of onset or increase, urge the patient to exercise the opposite limb vigorously while it wears a snug band and grasps a weight (ie a stone). (Avicenna, Bk. III, Fol.4, Ch. 25). In that way one may displace or eliminate the aposthem or at least reduce it by withdrawing the faulty humors which are flowing to the aposthem when it is just beginning to form.

14. When the aposthem begins at an emunctory, strongly attract the matter that is flowing to it, using cups when other measures are unsuccessful.

15. When the aposthem is fully mature (suppurated) the pain will decrease as will the redness and the insomnia, after you incise it, after if it erupts on its own. Follow the procedures we have taught and observe all the precautions that weset forth. Estimate the volume if the contents and their nature and the site, and assess the patient's general condition. Is he a delicate city boy or a feeble oldster or child? Is he debilitated? What are the astrological signs? Is the moon exposed or covered? Is it

on fire? Are you at the end of the field of Libra or at the beginning of Scorpio, by 12 deg.? Are you in the sign for the affected part of the body. An astrologer can answer those questions if you cannot find them in *The Planetary Almanac* or in the small treatise called *Circa Instans*. [194]

The drainage is better accomplished by incision for two reasons: 1. Any undrained pus will continue to corrode (Galen, *Megatechni*, Bk.III, Ch. 3). 2. If you wait too long, the abscess will rupture on its own and form a round ulcer that is more difficult to cure than a nice oval wound created by a longitudinal incision.

16. Do not use moist emollients after the incision. Instead use desiccatives, as you do when treating ulcers. Beware more of an ulcer than an aposthem. That complication is worse than the original lesion (Avicenna, ibid.)

17. Do not drain an aposthem or abscess before it is fully mature, except in six cases, because the signs of maturity will vary according to the type of pus, as you will observe later. The reasons for this dictate are: a. the part of the body that you drain may not be able to complete the maturation. b. a premature drainage (ie before the pus is completely liquid) is more painful, and will deplete the vital forces and prolong the duration of drainage.

The six cases in whom you may drain an unripe abscess are: a. When the pus seems to corrupt and to inflame the part, it is better to open the aposthem for incomplete drainage. b.When the aposthem is near a noble organ, you drain to prevent the spread. c. For the same reason, when the aposthem is near an emunctory. d.When the aposthem involves a joint and the pus can spread to involve the ligaments of the joint. e. When the lesion is at the anus, where it can form a fistula into an immodest structure. f. When the pus is adherent (viscous) and thick. The abscess may persist for a year before completely suppurating.

---

[194]   Nicaise cited Pagel who claimed this to be work of Matthew Platearius. Salerno, 12thC (LDR).

18. Open the aposthem as soon as it is ripe and drain the pus, following our standard procedures, except in three cases, as Galen explained (*Techni, Treatment of Causes,* Ch. 33), "You should remove whatever is un-natural." I discussed that at length in Tr. II, Doct. I, in the chapter on the General Treatment of Wounds. These are the exceptions: a. When the aposthem is filled with burnt humors, as in anthrax when the pus is thick and viscous and as tenacious as collagen from tendons or leather, and it will not drain through your small incision. b. When the patient's vitality is completely depleted, any incision can be lethal, and that will be blamed on the surgeon. c. When the surgeon should receive his fee before he sets to work. The adage is 'be paid while the patient suffers, because after the abscess is opened and emptied, and the pain and the fever are gone, and the rest of the malady has subsided, the patient will have no further respect or money for the naïve surgeon.

To supplement these General Principles, the surgeon should refer to certain rules in Treatise II, Doctrine II on the treatment of ulcers, from which he will learn what he needs for his practice.

From the foregoing sections on Definitions and Explanations, Varieties, Causes, Symptoms and General Prnciples and other relevant items the well-taught surgeon can infer nearly all the basic routines, which consist of four parts: 1. general evacuations (purges), and particular diversive evacuations. 2. a suitable dietary regimen. 3. local surgical topicals. 4. manual operations.

In respect of the foregoing surgical measures there are five Notables:

A. Every curable aposthem goes through four phases: onset, expansion, stable-state and decline (termination). In Stage I the accumulating matter begins to enlarge the site and some impairment of local functions may be noticed. Nature has not yet begun to react. The Stage of Augmentation follows and the advance may be slow. The enlargement is observed and lasts until the increase peaks, during which period the natural force weakens. The Stable Phase is when the enlargement has halted, there being neither more nor less of the matter, and Nature acts on it in some way. The Final Phase extends from the moment the aposthem begins to shrink and its destructive properties decrease until it is completely cured or it is transformed into another lesion, such as an ulcer or a fistula.

B. The duration of each of the four phases can vary. The phase of onset has a beginning, a middle and an end. Sometimes an aposthem may remain dormant, a barely noticeable mass, for a long time. The same is true for the second phase. At times such phlegmatic aposthems as goiters, loups[195], nodes, etc. enlarge continuously during many years, even a lifetime. So, too, the phase of augmentation may have several stages, and also for the stable state and that of decline.

If at the onset of a warm aposthem, as most authorities agree, you apply cool, dry and astringent topicals and they are not completely effective, and the aposthem enlarges, you should add mostly repercussives and some small amount of resolutives. Therefore, the longer the period of onset before the second phase, the less you will need cool, dry and astringent topicals; use less of the resolutives and more of the repercussives. But if the second phase comes soon after the onset, mix less resolutives with the repercussives. And the longer the second phase lasts the add more of the resolutives to make them more potent (ie energetic), until they predominate. The same principle holds for the duration of the stable state, and also for stage four. You mix your topicals according to the need, adding some and reducing the others.

C. The stages and the variations of each must be considered by the surgeon who wants to follow our protocol to the letter. However, some aposthems get well in spite of the surgeon's disregard of the stages. Success in those cases is a matter of luck rather than of skill, not because his treatments were of the best kind.

D. These Notables apply only to aposthems that do not suppurate as desired, and to those that do suppurate but do not go through the four stages that we discussed earlier in this chapter.

---

[195] Loups: edematous folds of skin and panniculus, resembling a wolf's neck (LDR).

When by chance or after an improper treatment has failed, then you must promote suppuration and the maturation of the aposthem, as we will describe.

After the aposthem is ripe, open it, observing all the rules and taking all the precautions discussed in the chapter on the treatment of ulcers. After the drainage use your mondificatives in the abscess, then the desiccatives and incarnatives. After new tissue (ie granulations) appears use your cicatrizers as prescribed in Tr. II, Doct. I, Ch.1, Part 7, on the treatment of wounds. Follow the instructions for each type of aposthem.

The topical medicaments suitable for cases in the present chapter and in other chapters that deal with the particular types of aposthems will be described in the Antidotary. Refer to them there and keep in mind that this is only a summary.

E. In addition to the manual operations noted above, there is another based both on good theory and a long experience, that is a wonderful cure for all aposthems which you expect will suppurate. This is something the old-time surgeons may not have known, or they kept it secret because, being so simple, it earned little in their practices. When we think an aposthem is about to suppurate, lay on a plaster of the leaves of mauve not stripped of their veins but with the stems removed. Cook them on embers (ie charcoal) and gently mash them before wrapping them with wet(water) oakum. Apply the plaster continuously for a full day and night. Then apply three to five leeches, to cover an area about the same size as the aposthem, not to exceed the vitality of a patient of any age. Repeat the measures at the spot where Nature seems to point the pus. Then foment the place with warm water. Dry it gently and apply some similarly crushed and cooked (in olive oil) pear leaves. Atop them lay pads of dry hemp to retain the heat. Continue the cycle of treatments which last three days. On the fourth day reapply the plaster of mauve, then the leeches and pear-leaves. Repeat all of it until the tenth day, at which time the mass will have dissolved or it will have matured. The abscess will be smaller than the original aposthem, and may have erupted on its own, without the need for a corrosive or a knife. If an incision is needed, let it be done as we have instructed.

After the incision, insert a drain covered with lard to keep open the wound and prevent the edges from adhering and sealing. The slippery drain facilitates drainage. Then lay on a compress of wheaten flour, water and oil, all cooked to the consistency of a bread-dough. Change the compress daily but do not replace the drain (ie at the first change). Continue until the abscess is empty.

Palliative Treatments

This will be a General Summary of my own experiences with many cases, and as garnered from the texts of many authors. As I wrote in the Introduction to Tr. II, Doct. II, I state again that palliative measures for aposthems are suitable for three types of lesions:

1. When it is obvious that the aposthem is incurable, such as a previously undetected cancer, or a completely dry aposthem that fills a breast or an eye[196]. 2. When the patient is incapable of withstanding any kind of curative treatment, even when the aposthem itself may be curable. 3. When the treatment of an aposthem, chronic hemorrhoids as an example, may provoke one or more worse maladies such as leprosy, dropsy or madness.

Other items of palliation will be discussed later as [art of the treatments of particular aposthems in chapters devoted to them.

III. Eleven Explanations of Preceding Topics

1. Take note of Avicenna's statement (Bk. IV, Fol. 3, Doct.1, Ch.1), that the words aposthem, 'dubelech', tumor, lump (ie eminence), elevation, enlargement and abnormal (unnatural) swelling all are terms that describe the same feature, which belongs to every aposthem. Also, it is observed in every abscess, pustule, pimple, etc.

2. An abscess differs from other aposthems and other lesions that share the property of unnatural enlargement; it contains pus, whether or not it drains. An abscess defines a warm

---

[196]   Perhaps a necrotic ulcer (LDR).

apposthem or one that was complicated by inflammation, only after it has suppurated and not before. Avicenna, in the same chapter cited above, added,"when a strong throbbing pain and a persistent heat are felt, you can be sure that an abscess is forming in an apposthem. Until those signs appear, the mass remains as an apposthem."

3. According to Avicenna (ibid.) pustules and pimples are different: A pimple is a small mass in which the matter is immediately under the skin and not in the deeper tissues, and it is not toxic. In France we call it a small bubo (ie bubette). However. a pustule may contain venomous pus that can eat away the surrounding tissues.

4. Avicenna (Bk. IV, Fol. 2, Doct.1, Ch. 5) indicated that a warm apposthem signifies not only a sanguinous or bilious source, but any in which the original matter is warm or that became warm during suppuration. Furthermore, the signs of an apposthem vary according to the various matters that produce it, be it phlegmatic, melancholic et al.

5. Take note of the terms primitive, extrinsic, external and antecedent in one group, and of another group of terms including intrinsic, interior, congestive and congregative. The terms within each group are synonymous. So, too, are the terms derivation and delegation.

6. The four items which follow qualify the values and the meanings of the terms in the preceding paragraph. Galen (*On Simple Medicaments,* Vol. I, Doct. 2, Ch.1) wrote, "An error in the terminology that defines causes can lead to great harm." In Doct. IV, Ch. 2 he added, "Physicians confound and corrupt names, and not only names but the science of names and things."

7. Avicenna (ibid.) wrote that an apposthem is a faulty compound, which is to say that there are many kinds of bad things about it. Its matters are bad and its complexion is bad, because the fact of an apposthem presumes a faulty complexion. Its altered contour is bad, no apposthem is without some change in shape and appearance. And it is a separation (ie dissolution) of

continuity of normal tissues, because the infiltrating aposthem pushes them apart.[197]

8. An aposthem is a faulty assemblage, common and official; common because it arises from similar tissues, be they bones, nerves or flesh; official because it develops on a previously normal assemblage of tissues, such as a hand or foot. It is common, because it can involve any structure except the heart, where unique noble qualities forbid an aposthem.

9. According to Avicenna (ibid.), one can see something resembling an aposthem in a bone. Indeed, whatever receives nourishment can be over-nourished, even such as bone, and the very substance of bones is such an overabundance (ie callus).

10. Galen (*Techni, Treatment of Causes,* Ch. 34.) explained the pathogenesis of aposthems as follows: Humors flow to a part of the body where they are the proper nutriments for the part. It is distended, and its small veins now are full of matter and become visible, such as one sees in the conjunctiva of a swollen eye. Some of the matter escapes from the veins and infiltrates the spaces in the tissues, where it cannot be controlled by Nature.

   Galen went on to explain suppuration (*De Interioribus,* Ch.IV, and *Commentaries on the Aphorisms,* Part 2.) And I have explained the nature of pus, virus, putrefaction, sordidness, flakes, crusts, the underlying conditions, the material causes and the effective causes etc. in the Explanations and the Notable IV in the chapter on ulcers.

11. There are three meanings for the term General Malady: a. When we say that a malady involves everything at the same time, such as a fever. b. When we may mean that the faulty matter is the malady itself, as in the terms consemblable, official and common maladies, such as in aposthems, scabies, itch, dartre (scurf) etc. c. When we say a malady is general because it can occur in any part of the body, such as a wound or an ulcer.

---

[197]   Hence it is like a wound (LDR).

CHAPTER 3.

THE TREATMENT OF VARIOUS APOSTHEMS
CAUSED BY A SINGLE, SIMPLE, NATURAL HUMOR:
ONE OF FOUR KINDS[198]

I. The Sanguinous

A. Diagnosis, Three Items: 1. *Pathogenesis:* In the chapter on the generation of the humors we learned that each of the humors is Natural or Non-Natural. We also saw that each has loosely defined limits. For example, the natural humor may be warmer or colder in degrees; so, too, for the non-natural one, etc. Then, in the General Chapter On Aposthems we learned what an aposthem is, its varieties, its causes, its symptoms and the basis-in-common for all of them. From that information we can infer the mode of formation of particular kinds of aposthems, which now is our topic. A surgeon must know by heart everything in the preceding two chapters; then he can take from the information there nearly all he needs to know in his treatment of particular aposthems. Therefore, we can rather quickly pass on to the special treatments. 2. *The Causes* of sanguinous aposthems are internal or external, and you can supplement what is given here by referring to the chapters on ulcers and on aposthems in general. 3. *The Symptoms (signs)* of the sanguinous aposthems: They are light red in color, similar to that of normal blood. Pain is caused by the infiltration of the humor which separates the normal tissues. The accumulation (ie the flooding) in the depths causes throbbing pain worsened by the conflict between the natural heat and the non-natural. One can feel its heat at the surface. At times an ephemeral fever causes the humor to boil, especially when it approaches the stage of suppuration. The region expands and hardens and it is tender. 4. *The Treatment* for this and for nearly all similar aposthems from the time one first is aware of them, all other things being equal, has three elements, each

---

[198] The four: From blood comes phlegmon; from bile comes erysipelas; from phlegm comes edema; from melancholy come the scirrhous (hard) (EN).

proper for the stage of development. You treat a replete patient who can tolerate evacuations differently than you treat a replete patient who will not accept them or a depleted patient. The replete patient should be purged, followed by repercussive topicals. The replete patient who is intolerant—this excludes children, oldsters and debilitated persons—should not have an evacuation nor the application of repercussives, because they will not repercuss. Recall what Galen said (*Techni, Treatise on Causes*), that a repleted body cannot accept what is sent to it (ie repercussed) from the aposthem. A patient who is not already replete, on the other hand, may not tolerate a purge but will tolerate repercussives.

B. The Treatments include three elements.

1. A good dietary regimen is what we prescribe for all ailments caused by warm humors, as we described it in the chapter on scurf (dartre).

2. The evacuations are those we use for aposthems and similar conditions according to the status of the affected patient, selecting the one that suits his complexion. If he is partly replete and has a small aposthem a small bleed will be sufficient and a blood-cleaning may not be needed. First use a diverting general phlebotomy if the patient's general condition permits it. If the aposthem persists unchanged, bleed from a nearby vein. If more is needed apply a cup over a scarification close by. The following evacuations are suitable for a fully replete patient with a large aposthem. When the preceding measures are ineffectual, go to a potent detergent of blood (ie of sanguine humor) as described in the chapter on scurf, and, if it seems proper, apply a cup over a scarification on the aposthem itself. I described the method in Part V of the Ch. 1 of Doct. II, Tr. II, on the use of potions and evacuations in wounded patients, and in Tr. II, Doct. I, Ch. 3 on phlebotomy.

3. The topicals follow the evacuations in patients who tolerated them. But the topicals come first when the evacuations are eschewed. In those cases use repercussives and resolutives; they are good in all kinds of these cases. When one type is not

enough, go to the next types: maturatives, aperitives (ie pointers), mondificatives, regeneratives and cicatrixatives, one after the other, when the preceding one no longer is effective. The repercussives that follow evacu:ations and the blood-cleaners are used only when the case requires them, and they must be applied at the onset of warm aposthems, as Galen authorized (*Megatechni,* Bk. XII, Ch. 1). "Our goal in such warm aposthems is to deflect the matter that is flowing to the lesion and to dissolve what has coagulated, acting differently, case by case."

For bilious aposthems, begin early with potent cool repercussives, at least when the aposthem is not in a nerve-bearing region. During the phase of enlargement use even more potent repercussives, exceeding the amounts of added resolutives. In the phase of declineuse pure resolutives when the aposthem begins to shrink. If instead the lesion continues to enlarge and shows no signs of erupting, go to the maturatives and aperitives. After it ripens and drains, use the mondificatives followed by incarnatives and cicatrixatives.

For the sanguinous aposthems use mild repercussives at the onset, neither alone (ie 'pure") nor cool in action. During the phase of enlargement mix in resolutives which are more energetic than the repercussives. During the stable stage use equal amounts of emollients and resolutives. When the stage of decline is about to begin, use pure resolutives. To reduce the induration, foment with warm water and apply unmixed emollients. If the treatment fails during the second phase and painful enlargement continues, use your maturatives as in the cases of bilious aposthems.

For the last-named cases and others like them, open them in any way the surgeon deems best, using aperitives or incisions as we taught in Rule 16 in the General Chapter and in the chapter on incisions, abiding by the sixteen precautions. Open the abscess at the most dependent site where the skin is thinnest. The incision should be about as long as the width of an olive or myrtle leaf. Make it longer for the larger aposthems and consider the tolerance of the feeble patient as well as the amount of pus to be drained, and do not empty it all at once. After the drainage treat the abscess as you do a virulent ulcer; fill it with oakum for three days or a cloth on which you have spread a paste of raw egg-yolks and wheaten flour. Then apply strong detergents or mild corrosives. Then use incarnatives and follow them with cicatrizers; keeping in your mind's eye the seventh principle in the General Chapter on aposthems, which deals with the adjustments of the repercussive topicals, and where we listed the nineteen cases in which we must not use them. Indeed, no chapter dealing with any particular aposthem can be complete of itself. You must always rely on the lessons learned by experience based on the general principles. Also, refer to our Antidotary where with Divine assistance[199] I dealt at length in special chapters devoted to the simple and compound repercussives for all sorts of surgical disorders. We have examined their qualities, degrees and effects, their modes of action, case by case, the timing etc. And we did the same with all the categories of surgical topicals.

II. Bilious Aposthems

A bilious aposthem is called erysipelas because it adheres to hairy skin. The name comes from *heiresis,* and *pelas,* which means skin.[200] Because the bad matter is always just under the skin it isolates the hairs and destroys their roots. This discussion will have two parts.

1. A brief *Description*: This aposthem derives from superfluous natural bile, as we described in the general chapter.
2. The *Symptoms*: Its distinguishing characteristics are its moderate hardness, between that of a firm sanguinous and a soft phlegmonous mass. It has the shape of a pointed pine-cone as it rises from the surrounding skin. Because bile is a

---

[199] The Lord kept him alive, to finish the Antidotary and Doct. II of Tr. III (LDR).

[200] Nicaise commented on Mondeville's faulty etymology. Erysipelas cames from 'ery', meaning red, and 'pelas' meaning skin (EN).

thin humor it is easily displaced toward the surface beneath the skin. Its color is a reddish orange (ie saffron), resembling bile itself. The dryness and warmth of bile in a large aposthem causes much inflammation in the region as well as fever. The pain rarely throbs because the lesion is at the surface and does not penetrate. Because the matter is so irritating it offends the hand that touches it.

3.   The *Treatment*: The diet and the evacuations are what we use for scurf. At the onset we use repercussives cooler than those for phlegmon. The aposthem is slow to suppurate on its own, but it responds readily to maturative topicals and it resolves easily. Once it is ripe, it discharges itself without external intervention. After it drains, treat it as we have instructed in the preceding chapters, except that you should use cooler topicals than those for sanguinous aposthems. More information on the subject is available in the General Chapter and in the chapter on sanguinous humors.

III. Phlegmonous Aposthems, Called Edema

1.   *Description*: This derives from a superabundance or superfluidity of natural phlegm. It forms in the same way as the bilious aposthem.

2.   *Signs*: It is recognizable for its pale color (ie whitish), that of phlegm itself. It is soft because the phlegm is soft and liquid. Pressure indents it, and the contour is restored when you remove the finger. It is cool to the touch as is its matter. It is barely painful if at all, because there is no conflict of heats except when it begins to suppurate. That is when the pain begins, and it persists until the pus drains. In fact, all cool and indolent aposthems become warm and tender when the begin to mature. Only then are they to be called abscesses, as purulent aposthems.

3.   *Treatments*: After proper diets and evacuations and a policy of abstinence, as we use for scurf, we apply the repercussive topicals used for warm aposthems. However, here we always

mix them with equal or nearly equal amounts of resolutives. Then we follow with maturatives etc., as we taught for the pure cool repercussives at the beginning of the section on warm aposthems. Therefore, we use repercussives and resolutives together, and we may add medications that have those as secondary properties, such as absinthe, artemesia (avrotonin), schoenanthus, stoechas, both centauries, gentian et al. Use two or more together. Crush them and boil them in the water which you will use for prolonged fomentations. Apply the mashed pulp as a compress, or use a lye made from wood-ashes of grape vines or oak in which you soak a sponge that is large enough to cover the entire lesion. Hold it in place with a good bandage wrapped as tightly as the patient will tolerate. Do not remove it until the lesion is cured, but wet it thoroughly with the lye, through to the skin, every day. Be diligent and perservere until the entire phlegmatic mass disappears. Another: apply a well-tested plaster composed of salt, millet-seeds and lard. However, if a little blood has been mixed with the phlegm, or if the purely phlegmonous aposthem is old, or when you have already used maturatives—in those three cases the phlegmon may have been corrupted and have been transformed into pus, and has begun to be hot and painful—open the lesion with a knife or use aperitives and eruptors or the cautery. Do not wait for a spontaneous eruption. You could await 'pointing' for ten years, during which time it may corrupt the flesh but not the overlying skin.[201] It is more stubborn (ie hard) than the flesh, and the matter in the aposthem is heavy, viscous, cold and it resists your treatment (ie is obtuse). I will provide many medications suitable for these cases in our Antidotary.

IV. Melancholic Aposthems

These derive from Natural Melancholy and are called scirrhous or sclerotic.

---

[201] A 'cold' abscess, eroding the back, bones and muscles, etc. (LDR).

1. *Description*: This lesion is formed from a superabundance of natural melancholy. It forms in the same way as the phlegmonous kind.

2. *The Symptoms*: The mass has a very hard texture due to the dryness and heaviness of the matter. It is not very painful because it is not warm, and the matter is so densely compressed that it is insensitive. Its color tends to be livid. Although there may be a lot of it, the heavy matter keeps below the surface and shows itself less at the level of the skin. It differs from phlegmatic aposthems which are whitish and soft, and often has an external cause. The melancholic aposthem is livid, hard and rarely has an external cause. It differs from a cancerous mass which derives from a non-natural putrefied melancholy that has acquired heat in the way we described in the chapter on treating natural phlegmatic aposthems.

3. *The Treatments:* After a suitable evacuation, dietary regimen and abstinences, all of which were contrary to the cause of the lesion, we use topicals in sequence or in combinations, including emollients, delayants and resolutives, in that order if not admixed. In fact, if the emollients do not come first, the others will extract the thin matter and will thicken what remains and cause the aposthem to become cancerous.[202] Or, if you apply the topicals when they are too hot, they will agitate the matter and attract even more of the inflammable stuff. Similarly, if you use maturatives, you will corrupt the matter and provoke suppuration. Let the surgeon beware that in those three occasions a purely melancholic humor becomes a cancer, and let him reduce the aposthem with such emollients as mauve, violets, mercuriale, brancus ursinus, figs, dry raisins, seeds of fenugreek and flax, wax, common oil, the lees of all emollient oils, all marrows, fresh lard (the less salty the better), some gums, such as ammoniac, bdellium and serapinum et al. The delayants (ie liquifiers) make the matter thin. They include hyssop, thyme, savory, camomille, absinthe, avrotonin, schoenanthum, calaminmt et al. The resolutives include

---

[202] When the emollients will have no useful role (LDR).

camomille, melilot, pellitory, wild mauve, cabbage leaves and seeds, aneth, anise, nettles, bugloss, borax, spikenard, elder, dwarf-elder, bran, all diuretic seeds, moist lanolin, butter, terebinth, many kinds of gums and all herbs that are warm up the 2nd deg. Note that many of the foregoing have two or three virtues (ie properties).

With the simple medicines the surgeon makes his compounds to suit his case. The melancholic aposthem needs no maturatives, mondificatives or others which we have not listed. If in error you unnecessarily have applied a maturative and it has provoked suppuration, you must then recognize it as a cancer and deal with it as such, as we taught in Tr. II, Doct. II, in the chapter on cancerous ulcers. You may look into the Antidotary for other medications suitable for that lesion.

Six Explanatory Comments

A. Some of you may wonder why good healthy blood can form an aposthem. I answer that even healthy blood can be compressed when the blood vessels are congested, and the blood acquires bad properties which corrupt it and convert it to pus. That does not occur in normal situations. Avicenna (Bk. III, Fol. 4, Ch. 1) said, re a sanguinous aposthems, "It derives from laudable blood which turned bad by being too thick or too thin."

B. Natural blood is red, clear (ie bright surface) and free of bad odors and flavors. It is at fault only when it is superabundant and forms phlegm. In that case its quality deteriorates and thins and forms a false (ie bastard) erysipelas. When it deteriorates by thickening, it forms furuncles. When it is too thin and too cool it forms a small cartbuncle which the French surgeons call 'echarbongle'.

C. It is said that all sanguinous aposthems are alike insofar as they are aposthems in all ssizes down to the very smallest, and that they should all be treated alike, with the same medicines and other methods. But when they differ in certain ways, they need different measures. The same is true for patients; their individual qualities differ. The treatment of an aposthem in a replete patient will not be the same for a non-replete patient. So, too, the surgeons differ. One will try to resolve the aposthem, and he may be successful. Another surgeon may try to mature a very similar or identical aposthem, and then he will mondify the suppurated lesion. The differences, small as they may be, one or many, and involving any or all things, will lead to different treatments.

D.   Diverse evacuations from distant regions are more useful early in the course of an aposthem than those performed near the lesion.

E.   All knowledgeable therapists are aware that a sanguinous aposthem in a replete patient calls for a bleed or a blood-cleaner, one or the other, whereas in a non-replete patient neither are called for.

F.   We recommended a larger incision to drain a cool aposthem than needed for a warm one. Yet now I must explain why that is not always the case. 1. For small abscesses where we are not concerned about major evacuations. 2. For all aposthems with undrained corrosive pus which you must not allow to accumulate. 3. You should have no concern for the size of the drainage incision in a robust person who can withstand a wide open drainage. 4.If the patient is debilitated and the pus in a large aposthem is not corrosive, you should make a small incision. A large one will imperil the victim.

CHAPTER 4.

APOSTHEMS DERIVED FROM A MIXTURE OF NATURAL HUMORS

They are 1. Erysipeloid Phlegmon. 2. Phlegmonous Erysipelas. 3. Noncystic Aposthem, called Phlegmonous Edema or Edematous Phlegmon, which is intermediate between phlegmon and edema. 4. Nodes and Soft Glands. 5. Glands, Scaling, Fistulous and Nodose Scrofules. 6. Non-suppurating Aposthems At Joints, derived from bile and phlegm. 7. Nodes and Hard Glands. 8. Very hard Scrofules. 9. Anthrax and Carbuncles

This chapter will deal with the formation, symptoms and treatments of each type, and end with Explanatory comments.

I. Introduction

Everyone knows how seldom we encounter a single humor and accordingly how rare are aposthems with single causes. Why, then, did we devote so much time and space to those few? I answer that it is not for their sakes alone, but because the more common compound aposthems exhibit the conglomeration of signs and symptoms of the simples, which we can recognize and use as guides for treatment.

The various types of compound aposthems are formed in the following ways, from several or from all the natural humors. 1. Blood predominates along with some bile to form the erysipeloid phlegmon. Avicenna (Bk. I, Fol. 2, Doct. 1. Ch.1 on compound maladies) said, "A compound malady takes the name of its predominant humor." Really, what we say for number 1, also holds for number 2 (phlegmonous erysipelas). Their signs are sanguinous, as described before, somewhat attenuated by the bilious. The same holds for the treatment for both. When natural bile predominates over natural blood we call it phlegmonous erysipelas and the bilious signs predominate. 3. When thin natural phlegm is mixed with blood, an intermediate aposthem forms, in character somewhere between phlegmon and soft noncystic edema which matures easily. Its symptoms and the treatments are composite. 4. When a thicker phlegm combines with blood, they form nodes and soft glands. 5. Blood, phlegm and melancholy combine to form scrofules and glands in which the sanguinous and phlegmonous elements can undergo suppuration, but not so the melancholy that results in fistulas and crusted lesions which shed scales. The crusts act as opercula and hide pus and virus beneath them. 6. Bile mixed with phlegm that it has liquefied form a seldom seen aposthem that cannot form pus. They usually involve painful joints. The treatments use remedies for bilious and phlegmonous aposthems. To treat the painful joints, refer to Ch. 43 in Doct. III of this Treatise.[203] 7. Thick phlegm predominates over natural melancholy and they produce soft nodes which are harder than the others (ie in # 6.) 8. When Melancholy predominates in a mixture with natural phlegm, very hard scrofules can form. 9. The four Natural Humors as well as the Non-natural ones are the sources for anthrax.

A. Descriptions

First let us take note that some aposthems are enclosed in sacs; those are cysts that resemble chickens' gizzards. All cystic aposthems contain cool thick matter, as in nodes, turtles, goiters, glands and scrofules, and the matter in some of them may undergo suppuration. And among the latter some mature rapidly, as do turtles and soft nodes. Some are slow to suppurate, as do the hard nodes. Those that do not form pus are formed from gypseous phlegm and melancholy. In general, the cystic aposthems are slower to mature than the others, while those that occur in children, especially in girls, break down faster than in adults.

Nodes and glands are similar because they derive from the same sources. They differ in that nodes form on harder surfaces such as the wrists and the forehead and glands always form in soft regions such as the neck, the armpits and the groins. The node always is single and a gland may or may not be single, whereas scrofules are multple; they simply are clusters of glands. The word scrofule comes from scrofa, a sow. That animal always gives birth to clusters of young, never to only one.[204]

II The Treatment

In general, the treatments for all aposthems are nearly the same, with additional special measures for scrofulas. They include:

---

[203] This is another reference to the Treatise for which Mondeville lived to write only the detailed rubrics (LDR).

[204] Nodes are what we call ganglia; glands are lymphnodes (LDR).

A. Evacuations: use a phlebotomy when the circumstances allow it. The laxatives are what we prescribe for phlegm in the chapter on scurf (dartre). Or we make a powder containing equal amounts of turbith, ginger and sugar. Dose 2 dr. Do not use emetics which will eliminate the laxatives. Scarifications, leeches and the like are avoided because they simply cause pain and do no evacuate the matter enclosed in a cyst.

B. Diet. Favor such easily digested foods as capons, chicks, forest and meadow-birds, partridges, larks, blackbirds, pheasants, hard-boiled eggs, castrated rabbits, lambs, veal, kids. The drier the cooking the better. Put croutons and dry pastes in the soups. Salted and highly spiced foods are good. Sprinkle salt and spices in the soups and over the meats. Drink good, clear and undiluted wine in small amounts, such as claret. Or drink a special hydromel.

Avoid such harmful foods as these, and do not overindulge in food or drink, especially the moist and watery foods, including vegetable juices, purees etc. Avoid difficult-to-digest foods such as beef and game such as venison and wild boar. Aquatic birds are harmful as are such unripe moist fruits as pears and apples and all astringent fruits as medlars, sorbs and coctanes. Avoid cheeses, cabbage, milk, lettuce and all raw vegetables. Avoid such tangy vegetables as onions and garlic and all legumes, including fava beans, peas (except chick-peas), and all such unpleasant foods as lards and cold water. Do not go to bed with a full stomach. Avoid unsalted pork, all fish that have no scales (except river crawfish), and fish with soft and thick flesh. Avoid thick and cloudy fresh red wine.

In sum, the goal of the dietary regimen is to keep the patient a little hungry and thirsty. He should sleep with his head elevated and avoid harm caused by a too-tight cord in the cowl or bonnet (ie a night-cap).[205]

C. The Topicals and Manual Operations: Four Parts

1. The Treatment of Nodes: One kind resembles a soft fig dangling from a soft thin pedicle. Simply ligate the pedicle with a thread or transect the pedicle and apply a caustic to the stump. When the pedicle is broad you must use one of three methods. a. Take a

---

[205] Because the neck was a common site for glands and scrofules (LDR).

lead disc at least as large as as the node, and prepare two oakum pads of the same size. Immerse the pads in salted egg-white thickened with sugar-alum. Put one pad on the node and cover it with the lead plate. Place the second pad atop the plate and wrap the assembly snuggly with a two inch-wide bandage, as tightly as the patient can accept for eight days, to be removed only for urgent reasons. When that is necessary, replace it and keep it until the node is destroyed or the treatment fails. b. For soft and recent nodes: Put the limb on a strong stool with the node exposed on top. Then swat it forcefully with a ball or a dish or a paddle. The node will disappear. I have seen it work without reccurence. c. Here we perform the smash as above and then apply the lead plate and oakum pads. When used with a proper bandage the method never fails.

Those measures do not succeed for glands and Scrofules. They lack pedicles to be divided and they do not develop on hard (ie bony) surfaces that would favor the 'swat' treatment or the tightly bound compresses.

2. The Topicals used for Glands and Scrofules are not suitable for nodes. It consists of applications of the approved surgical resolutives, such as Rhazes' diachylon and Mesuë's ointment, or a paste of dry goat-turds mixed with oxymel. Or you may use them when moist, cooked with honey and vinegar. Or make an ointment of quick-lime and fresh or old hog-lard. Or use cow-dung with vinegar, or the ashes of cabbages with liquid pitch and old lard. The following is a better topical than the others: 1 part of incinerated fig-tree wood and 1/2 part of the roots of guimauve in vinegar. More similar topicals are described in the Antidotary.

3. Topical Treatments for Nodes, Glands and Scrofules.: a. Repercussives are not to be used, because these aposthems derive from dense, cool and compact matter which does not lend itself to repercussion. b. If you wish to resolve the matter with emollients and resolutives, use the emollients first, or with mixtures of the two. c. If any or all the treatments described before are unsuccessful, or if any treatment has not been tried because the patient had no surgeon in attendance, or if he had one who could not or would not be involved in the treatments, the only measure that remains is a total extirpation by

surgical excision. To do it safely and to avoid complications—there is not rescue for a dead patient—first study our instructions for control of hemorrhage which are in Tr. II, Doct. I, Ch. 1, and also go to Tr. III, Doct. I, Ch.1 on the surgical methods for incisions in various conditions. In the central part, we discuss incisions for removing various excrescences, and list seventeen precautions to be observed. Here I will add five more: 1. Do not remove normal glands from the region of the emunctories. 2. You cannot safely make incisions in the neck. 3. If there are many separated scrofules never try to cut them out. 4. The same holds for a single very large scrofule. 5. The same holds for a scrofule that is firmly adherent to a large artery, vein or nerve.

After a properly performed excision suture the wound and dress it with the white royal ointment (ie Rhazes') described in the Antidotary, or with compresses wet with warm wine as we described in Tr. II. Doct. I, Ch.1. When an aposthem or excrescence shows signs of maturation or if the patient wants it to suppurate, you may help that along with the following topical: broil three onions and three eggs. When the eggs are hard, peel the shells (ie now charred) with some adherent egg-white. Mash the rest and add an equal amount of pork lard and apply it. Or you may use one of the maturatives we will describe, taking note that if a scrofule is only beginning to suppurate, delay your incision until the maturation is complete. Then open it and mondify it by introducing a wad of oakum soaked in vinegar, and kept in place for three days. Or use an oil of roses thickened with sugar-alum. Mondify until the cavity is empty. If any any harmful matter remains, use more vigorous detergents such as Apostles' ointment, powder of asphodels or others. And, if some of the cyst remains, introduce butter or oil of roses mixed with melted butter. You will destroy the remnant if you persist patiently for a long time. Then use the regeneratives and consolidatives.

4. Special Treatments for Scrofules: They are also useful for glands. In addition to the common treatments in everyday use we have two kinds of medicines, oral and topical. First let me make it clear that Scrofula, the Royal Disease—so-called because a king can cure it—and Chirade all are synonyms. *Scrofula* comes from the word for sow. The term *chirade* comes from the Greek word *choiros* which means pig.

In Scotland it is called 'worm', because when the patient is covered with honey and set in the sun, worms leave the patient's head (ie the scrofulous neck) attracted by the honey which they eat. They could not otherwise be extracted without breaking open the scrofules, and thereby making them incurable.[206]

The oral remedies suitable for scrofula are potions and powders: a. The Potions: the juice of piloselle or the powder of round aristolochia, one dr. of either, every morning. Another: a compound potion of agrimony, potentilla, leaves of olive trees, filipendula, large garance, tansy and red cabbage. Mix all with wine and honey. Another: a root of scrophularia mashed in wine exposed to the night air. In the morning filter it and dose it. To evacuate scrofulous matter through urine, use the juice of the diuretic herb quadrangula (ie figwort) applied on the scrofule. The scrophularia are still called millemorbia and quadrangula, although they are not identical. Quadrangula has many branches and its leaves are large and have (ie four) white tips, something I have yet to see on millemorbia. Scrophularia has a single stem, about one cubit in length, and its leaves are similar to those of the large nettle and its roots have nodules like a scrofula and that, plus its curative powers against scrofula are why its is called scrophularia.[207] b. The electuary is made from the ashes of incinerated sponges and the bones of squids cooked with honey. The patient takes a dose the size of a walnut in the morning and evening. There are many different electuaries. c. Powders are used to make the electuaries. One of the best which supplements the diet in the foregoing list has been successful in many cases treated by me, and I have not had a failure. Make it with equal amounts of the roots of filipendula, potentilla and agrimony and use it when the scrofules are draining, as ulcers. When they are not open and draining add the roots of the large garance to the powder, about 1/2 as much as each of the other substances. The twice-daily dose is about as much as three fingers will dip up, taken with a small amount of a good light wine.

---

[206] Probably maggots in the draining mass (LDR).

[207] The 'f' in scrofula and the 'ph' in scrophularia are the English usage. Nicaise spelled both words with an 'f' (LDR).

Topical medicines for treating scrofules

Some are applied directly on the surface, while the others are not. The latter should be used after the purges and the oral medicines. These topicals are dripped (three drops) into the ear on the same side as the lesion (ie in the neck): Oil of bitter almonds or warm oil of petroleum (the Holy Oil called benite). Use only one drop in children. In both, use the drops two or three times a week. A better oil: Take 1 measure each of crushed thapsia, radish-root and pouliot. Put them in petroleum oil for three days and then heat them until it is boiled out. Add more oil when heating until the roots are soft. Mash them and save the liquid to drip into the patient's ear, one drop at a time. Continue with the treatment until the scrofules in the neck are cured. If the lesions are in the groins or armpits, use the liquid as an inunction. Water-of-life [208] instilled in the ear also may be used.

The topicals for drect application on scrofules are either artificial or empirical. The artificial ones have been described above. The empiricals include: roots of plantain in a sack suspended from the neck. Another: cook a snake in a stone jug in an oven. Then boil the flesh with vinegar and honey and anoint the scrofula. If that causes too much local inflammation (ie heat), add a large amount of coriander. Another: use this to treat ulcerated scrofules; Take the powder of filipendula etc. with which we made the potion described above. Dust the powder itself on the lesion once or twice a day, along with the potion by mouth. If the patient will not take the potion, the powder alone will not suffice. Then go to the usual detergents as ointments and plasters such as the Apostles' Ointment, or a powder of asphodels, or Mesûe's Egyptian ointment, or others that we describe. Or you may dust the lesion with a powder made from dried human feces, three or four times a day. Or, as a certain person told me, cut the cord that joins the upper lip to the gum underneath it, and that will cure the scrofule.[209]

---

[208] Mondeville (Nicaise) does not identify the term. I assume it refers to Holy Water rather than the modern 'eau de vie' as grape brandy (LDR).

[209] One may guess that such tongue-in-cheek entries in Mondeville's text are clues to his general attitude toward the empirical measures which he attributes only to unidentified 'certain' persons (LDR).

II. Eleven Explanatory Items

A. Glands and similar aposthems do not appear in nerves or firm structures because of their sources. Hard structures do not accept glands. On the other hand, they do appear near the emunctories which are spongy, loose and receptive.

B. All the contents of glands increase, both in old and young people, although they seldom actually begin in the elderly who have tougher tissues. Glands favor children more than the older persons because the young are more gluttonous and plumper.

C. Glands are easier to dissolve during the early (ie two) months after they first appear, if they are treated artfully, but not afterwards. The cyst's capsule does not disappear, and it persists as something tough and dry. The illness will not recur (ie after a successful early treatment) if the patient follows a good diet. When he lapses, the malady will relapse.

D. Glands near emunctories should not be completely extirpated because they are normally there to receive the superfluidities of the principle organs.[210]

E. All our predecessors used cruciate incisions to excise cystic aposthems. The few exceptions among the surgeons were the skillful ones, working only in the head, who made curved-flap

---

[210] Nicaise's Note: The emunctories of the ancients seem to be the lymph nodes. It is strange to encounter an observant surgeon, a fourteenth century precursor, who says that their total ablation is contraindicated because the superfluidities then would not be eliminated but would accumulate. That conception reminds us today of the consequences, recently demonstrated, of total thyroidectomy. While it is true that the latter is not an emunctory, it is a necessary contributor to the formation of normal blood. Nevertheless, the two concepts are related. This is another reason why we should know what our predecessors knew, and how that will influence our present-day investigations. (EN)

But how much more interesting it is to us today, more than a century after Nicaise's misconstruction! We see that Mondeville correctly described the consequence of some radical regional lymphadenectomies, especially at the groin, as burdensome lymphedema in the distal limb. Nicaise's analogy to the effects of thyroidectomy is unacceptable today (LDR).

incisions shaped like shields. Although none of them actually removed skin, the scars always were ugly. We moderns make incisions in the way we taught in our chapter on incisions. The results are scars that are only slightly unsightly.

F. For nodes, the treatment by contusion alone is the better way when it is feasible. After the blow apply a compressive bandage. Those who treat by incison and drainage of the cyst must use maturatives, incisions, detergents etc. That routine is hard on the patient as well as the surgeon, because, once the matter is drained and the remainder is corroded, and any small bit remains, the aposthem persists until you make a larger incision, at least large enough to permit the introduction of corrosives. But the worst of all the treatments is that which destroys the skin and flesh and opens the aposthem and leaves the cyst wall behind.

G. I prefer to abandon a node on the head if it adheres strongly to a cranial suture or to the bones. My reason is self explanatory.

H. Some treat nodes with slim and long pedicles by cutting them off and applying corrosives. Others use the actual cautery. I think it is more reasonable to use the cautery alone than to use the knife, because you can easily cauterize remnants of the root after removing the node. It serves little purpose to do more when less is good enough.

I. When you have to loosen a bandage before the anticipated time, do not remove it completely. Simply retie it as well as you can and secure the plies of a good bandage with a needle and strong thread. Indeed, the rebandaged injury will heal nicely under a good bandage if it is held in place for enough time.

J. If the patient with a scrofule has a low brow and narrow temples and relatively small jaws, you may find it difficult or impossible to apply a compressive bandage to dispel the humors downward from the vertex (the route taken by catarrh).

K. No matter who may have said it first, nearly all the authorities since have agreed that emetics are useful in treating scrofules in the neck. But that treatment seems irrational to me, because, according our general principles, diversive general evacuations (ie from a distant site) are proper for aposthems before they mature. Galen said so (*Techni, Treatment of Causes*, Ch. 34.), and he said it again more succinctly in the *De Ingenio, Bk XIII*, Ch. 4), citing examples, as "when an aposthem is near the anus, never evacuate with laxatives."

CHAPTER 5.

THE TREATMENT OF CARBUNCLES AND ANTHRAX, FORMED FROM ALL THE NATURAL AND NON-NATURAL HUMORS

I. Descriptions: Three Parts

A. *Definitions*: A Carbuncle or Anthrax is an aposthem formed from a mixture of all the natural and non-natural humors, all of which have been transformed into the venomous matter that corrupts the local region.

B. *Pathogenesis:* Sometimes the causative humors retain their natural qualities and do not bring on the worst signs that come from the conflict between the contrary qualities in the humors, and the result is called a carbuncle. In other cases, the same humors are far from their natural states and they corrupt. That happens when corrupt and malignant melancholy and burnt and salt-phlegm produce their most dire offenses. The result is an aposthem called a true anthrax. Although carbuncle and anthrax are similar, they differ in two ways: the condition of their matter and the symptoms that they provoke.

The diseases are more common in some countries, as in Catalonia, Apulia and Asia Minor, where they exist in the pestilential and corrupt air. They often come after bathing, after coitus and after exercise when the stomach is full.

Both conditions begin as a carbuncle and the anthrax may follow. The carbuncle sometimes derives from humors that exhibit a very malignant natural state, and that is the kind that goes on to become anthrax. Frequently carbuncles form on theback, in the breasts and in the neck.

C. *The Symptoms of Carbuncles*: They are red, brown or yellow at the onset. They appear as conical mounds which enlarge rapidly, causing pain and fever. When they mature as carbuncles the pus is white, and it resembles the tissue nearby. As an example, when it is near damaged nerves it is very viscous. The lay folk call it the root of evil. The carbuncle is firm and very warm, and the accompaning fever varies in intensity.

The Symptoms of Anthrax: At the carbuncular stage they are the same in kind but more intense and include the various colors of the rainbow. The apex may show blebs resembling blisters after a burn. You may see a dark ashen-gray pustule under the skin which seems to be firmly fixed at its base. When the anthrax empties, the opening is black. The patient is anxious, restless, nauseated (vomits), has palpitations and he may faint. The lay practitioners call it a deep sleep which forebodes death. Other lethal omens are lesions appearing under the breasts, especially on the left side, or in the axillae or throat, or when it is black, or when a black blister is sunken (ie retracted), or when the patient vomits, suffers palpitations and faints.

The signs that bode well for recovery are: When the symptoms of carbuncle alone do not deteriorate. If you open an oven-dried pig's gall-bladder and place a patch of it atop the pustule and it adheres, that is a good sign. Another: if you apply dissolved galbanum and the pain subsides, the anthrax is curable; but if the pain persists, the patient will die. Another: apply a salted raw egg-yolk over night. If the pustule enlarges before morning, it is curable. Another test: Hippocrates said that when their is little or no local pain, the lesion is incurable (*Aphorisms*, Part 2)./ Galen agreed (*MegatechniI*, Bk. IX, Ch. 3), "When the local pain is very intense, the lesion is easily cured, when it is mild, the cure is slow to come.

II. Treatments

A. *General.* I agree with the authorities: Galen (*Megatechni*, Bk. IX, Ch. 4, and *De Ingenio*, Bk. IV, Ch. 7); Avicenna (Bk. III, Fol. 3, Tr. 2); and such famous practitioners as Theodoric, Lanfranchi, Bruno and many others, that the treatment consists of three categories of things.

Evacuations: First evacuate blood by phlebotomy and cupping.

Then go to laxatives and clysters. The venesection should be on the opposite side if the patient has vomited or fainted or if the apex of the mass is black or livid. Bleed until the patient faints or use cups. If the conditions permit it, on the following day bleed from the same side as the lesion, to avoid attracting the venom into the body.[211]

The gentler clysters are used at the onset of both the carbuncles and the anthrax. The medical laxatives and clysters are used throughout the course of the illnesses but not every day. They include decoctions of such fruits as prunes, of sweet-smelling flowers as violets and borages, of cleansers of blood such as bugloss, of cordials of roses and oxalis. You may sweeten the decoctions with sugar. The clysters use emollients such as mauves.

Oral Medications: include two matters, Antidotes and Diets

A. *Antidotes* against venom, especially the Grand Theriac: Use it with four caveats:

1. *Category One has four items:* a. Test the theriac before dosing it. Give somebody a potent laxative, as hellebore. Then give him some theriac. It is perfect if it blocks the effects of the laxative. You may find many other tests for theriac in the medical practica. b. If the patient is constipated, use a laxative first. c. If the patient does not evacuate, wait at least twelve hours before repeating it. d. If you use a theriac as an antidote against a poison taken by mouth, first prescribe an emetic. But if the poison is on the surface do not use the emetic; it will attract the venom into the body.

2. *The Dosage*: Avicenna in his small book on theriac prescribed a two penny-weight dose with some watered wine. Rabbi Moses Maimonides used between 1/2 and 1 1/2 dr. Bernard of

---

[211] The usual protocol: a general evacuation from the opposite side, followed by the less diversive bleed. Here Mondeville is inconsistent in recommending a bleed before a purge (LDR).

Gordon used between 1/2 and 2 dr. We use smaller doses in warm weather. Thus, the different dosages may be a matter of different geography or of different doctrines. In brief, let your doses be prescribed in consideration of where you live, the weather, the age, the complexion etc. and especially, your assessment of the patient's Vital Force.

3. *The Method* of Use of Theriac. It should always be given with a suitable liquid, such as a small amount of warm wine when the patient has no fever. If he is febrile avoid the wine, with three exceptions to that rule: a. If the fever is pestilential (ie epidemic). b. If the patient has swallowed poison. c. When the patient has incurred a poisonous bite or sting. Unless there is some other reason, use a decoction of tormentilla, scabious or black jacea or oxalis or other substances that you can also use in cases who are not febrile when the patient requests them (ie instead of wine). Theriac is not to be used in cool and soft persons, as we stated in Tr. II, Doct. II, Ch. 2, on treating poisonous bites and stings, and in Notable V of the Explanations.

4. *After the Theriac.* The patient should abstain from all else for twelve hours, to prevent overheating or over cooling.

Give the theriac as soon as you see the anthrax, and repeat the dose once or twice. Some patients dread it and some physicians and surgeons hesitate to order it for them in face of that response. In some northern and western countries theriac is not so well known, but it is viewed with suspicion as an antidote against venomous snakes and other animals, which are rarely encountered there. In such resistant cases we can substitute for the theriac and prescribe a powder made from the roots of tansy, about as much as you can dip up with three fingers, or we can use 1/2 oz. of tormentilla-roots. Both of them have been used as often as theriac.

The victims of anthrax can and should frequently use such sedative (comfortive) electuaries as those which contain diamargaritum, sugar, roses and powdered emeralds.

B. *Diets:* The foods should include porridges of bread crumbs wet with water or ssimple decoctions, barley-gruel, pomegranates, lettuce, purslane, zucchini cooked in vinegar. A debilitated patient should eat the meats of kids, chicken boiled with herbs, verjus. and vinegar. He may drink pomegranate-wine or verjus diluted in half with fresh water or vinegar diluted in fourths.

Topical Medications, Three Types

A. Applications on the Aposthem: Two Categories.

1. The Rationals [212] are of four types: a. To apply at the onset and during the phase of enlargement. They include the cool and dry and mild resolutives, cool without being repercussive You should not mix your resolutives lest you cause the venomous matter to be combusted further. The useful topicals include cooked lentils and arnoglossus (plantains). Here is a doggerel verse: "Lens, arnoglossa, cum gallis, panis porcinus, Exsiccant prunas possuntque resolvere virus."

(Lentils, plantains, galls, and sowbread (ciclamen) and plums can dry and resolve the virus)

A good theriac applied at this time dries the anthrax as itself is dried. b. Rational topicals for the stable phase are maturatives. One recipe takes 2 oz. of dry figs, 1/2 oz. of mustard seeds mixed with rosat and flour of fenugreek and flaxseeds. Another combines the foregoing with equal amounts of yeast. If the suppuration is slow to develop, foment the anthrax with warm brine to dilute the matter. Then scarify and gently apply a cup. Renew it as needed. Or you may drip some melting candle-wax on the summit of the mass or touch it with a cautery. c. Rational topicals for the stage of deliquescence after the anthrax has burst and drained its thin bloody liquid. Apply a surgical porridge made with honey, apium, and a mixture of wheaten flour and powdered roses, all cooked together. d. Useful at any stage of anthrax are slowly simmered acidic (ie unripe) pomegranates in vinegar which you mash and spread on a cloth and apply as a plaster.

---

[212] Rational meant according to the doctrines taught in the schools by the physicians. Empirical meant the methods use by the common folk lay practitioners, based on their own experiences (LDR).

Or you may apply salted raw egg-yolks, or a lily-root boiled in water and mashed.

2. The Empirical Topicals are useful in all stages of anthrax and for all venomous pustules formed from corrupted non-natural humors which have turned black, and when manual interventions no longer are useful. There are five topicals: a. Mashed scabious with or without lard. One practitioner insists that they be mashed only between stones. b. The small consolida which is called the glass-plant in France and vetrimola in Tuscany. Mash it for use. c. Tormentilla. d. Other types of consolida. e. Crushed dog-wood (punaise).

B. Applications around the anthrax. These can be used at all stages. They are cool, astringent and sweetly scented. The best defensives contain bol d'armenie, rosat and vinegar. Many other have been described.

C. Topicals to be applied at a distance from the mass: They are cordials [213], scents and air-sprays to reduce the stench in the patient's quarters. The Cordials are: 2 oz. of red sandalwood, 3 oz. of red roses, 1/2 lb. of barley meal, 2 dr. of camphor. Grind all and add lard to make a paste with added rose-water and a sprinkling of vinegar. Lay it on the chest if the anthrax is on the back, or the reverse. Here and there in the practitioners handbooks are many such plasters. The Scents are all cool, such as roses, violets, willow-flowers, plum blossoms, oak-buds, camphor etc. The Air-Sweeteners include cold water, fresh herbs, leaves of willow trees, plum trees and oak trees, grape leaves, roses, violets and water lilies.

Continue our treatments of the aposthems with a succession of all types of medicaments—regeneratives, cicatrizers—as long as the patient lives. It is the same routine of treatments we use for all venomous pustules that derive from corrupt, venomous non-natural humors. Those diseases include Persian fire, formicula, pruna, miliaria and herpes, which we will describe in the next chapter. In all those cases you should observe the general rules: Treat day and night and renew the topicals at least five times in that period. They are most

---

[213] Cordial: Something to make the environment more pleasant. Not a liqueur (LDR).

effective when they are first applied, when they have absorbed less of the venom and therefore benefit from frequent changes. Another rule: during the course of treatments do not change your medications, cordials, etc, but simply reapply fresh amounts until all the complications have subsided. All the while, frequently offer comfortive electuaries. Prescribe beverages that cleanse the blood once a day, but not every day.

III Eleven Explanatory Comments

A. Although Avicenna and many other authorities have discussed at length and completely the treatment of aposthems other than anthrax, which are formed from corrupt, burnt, venomous non-natural humors, and have devoted long chapters to their causes, symptoms and treatments, it seems to me that all the materials for each type are so similar that it is not necessary to repeat all of it for every variety. I doubt that there is a surgeon anywhere who can give a complete and detailed discussion and tell me or any other intelligent colleague the precise differences between Persian fire, formicula etc. And the same is true for other kinds of aposthems, and the surgeon is not so miserable that he cannot give a fancy name to the aposthem that he is treating. Really, it is important to give a terrible name to an aposthem only to collect a fee from a foreigner. The diversity of names does not entail the same diversity among the aposthems, or of the manual procedures. Every kind of aposthem may be bad, worse or the very worst, and the surgeon must make an assessment with his eyes and his fingers, and from the complications and from the patient's symptoms. Thus informed, he must select the most appropriate measures. Inasmuch as there are great similarities, he can consider them (ie carbuncle and anthrax) as one, as Galen said (ibid.), "Admirable Nature with the help from all the associated things can supply what is lacking." It seems to me that all aposthems of this type, other than carbuncles and anthrax, but alike in their malignant qualities, should be treated as we treat those two.

B. All aposthems derived from burnt humors that are at the stage of suppuration and contain either a watery red pus resembling a liquid that has been used to wash raw flesh—a penetrating virus—or a thick, viscous and tenacious pus that resembles the collagen from skin or nerves, that adheres to the inner surface of an abscess, like rootlets of plants that grasp the soil. When the aposthem is small, as are the formicula, the rootlets are seen only at their undersurfaces. When the aposthem is large, there are several large separated roots. I think that the thin fluid pus is a recent arrival in the aposthem and has not yet been combusted and that it derives from the clear part of the humor. The thicker pus has been there for a longer time and had more opportunity for combustion. The evidence for my hypothesis is this: If you drain off the thick pus in the evening, by morning more will have formed anew; the same amount has formed during eight hours[214]. Therefore, I conclude that you should not drain such by incision unless you have to let blood, too. Indeed, the thin pus can find its own way out, whereas the thick matter needs more time to mature before it drains. At the time of drainage you find the pus adherent to the lining of the abscess, as the rootlets of plants cling to the soil.

C. Anthrax never derives from natural humors, although some authors disagree. Galen (*De Ingenio*, Bk. IX, Ch. 7) was the first to say this, "Anthrax and similar aposthems derive from thin matter which is very warm and very harmful. Certainly no natural humors are like that. Besides, when treating anthrax, everybody uses theriac which is designed as an antivenin. Therefore, the pus of anthrax is venomous, and since natural humors are not venomous, anthrax comes from non-natural humors." Lanfranchi agreed (*Compendium*,) stating, "The

---

[214] Mondeville is not clear here. The two kinds of pus from different sources are distinct and remain so. Further maturation of each type continues independently. The thick pus does not come from the thin.(LDR).

treatment of Anthrax and other destructive venomous matter . . . . etc." Furthermore, Avicenna (Bk. III, Fol. 3, Tr. I) assigned the treatment of anthrax to the category of aposthems corrupted by corrupt humors. It is clear that natural humors are not part of it. Finally, when someone fails to prescribe theriac and denies that anthrax is derived from venomous matter, it still must be clear that natural humors do not ever cause such malignant symptoms. Well! You be the judge.

D. According to Galen (*Megatechni*, Bk. IX, Ch.4) these aposthems sometimes are not surmounted by pustules, even at the onset, and sometimes a single or several pustules appear, usually during the second phase and where the patient has scratched himself to relieve itching. The surgeon should not be misled by the presence or absence of pustules.

E. Avicenna (Fol. 4, Ch. 29, on cauterization) and Rhazes (*Albucasis*, Ch.2) wrote that the body should not be touched by cold iron (ie the knife) but only with a hot iron (ie cautery). Heat improves the complexion of the region, conferring benefits wherever it is used and doing no harm.

F. In some countries the victims of anthrax who are always nodding off to sleep, day and night, are put in rooms across the road from public squares where the loud music of trumpets and drums leads them to dance as if at wedding parties. There one can see them insensible, dancing to death without realizing what was happening.

G. When our recommended herbs and spices are not obtainable or when you see little benefit in them, the best maturatives are made with a mixture of yeast and oil.

H. Slice the roots of iris in rounds and cook them in a pot that holds twice their volume, and add about a lb. of oil and 1/4 lb. of wax to make an ointment which is an excellent attractant of venom from pustules.

I. The (*Totum Continens* of Rhazes states that a carbuncle frequently is accompanied by chills and fever which disturb the patient at night. It is called the Included (ie concluding) Fever. Its matter has eroded into the muscles under the skin and into veins.[215] It attacks the patient when he is goes to bed and encounters cold sheets.

J.   Some say that you should prescribe theriac in these and similar conditions. I disagree, because theriac is warm, and so is anthrax. Therefore, etc . . . Everyone says the opposite; therefore, etc . . . I must say that you should prescribe it when it seems to be the proper agent to restore the human complexion and to prevail against all sorts of venoms, whether they be warm or cool, when the venom is contrary to a person's complexion. With the same properties it will bind up loose bowels or loosen them when constipated. Common sense tells us that when theriac is inappropriate in these cases and in others it will not be because of its heat or coolness. Averroes said the same and Avicenna agreed (*On The Heatr's Strength*). They wrote, "Many physicians are afraid to prescribe 1 dr. of theriac and yet they will give 1/2 oz. of diacumin or other substances, not knowing that . . . etc."

K.   For a complete discussion of the treatment of anthrax and abscesses (pustules) I refer you to Averroes' book on Theriac and to Avicenna (ibid.) and to Tr. II, Doct. II, Chs. 2 and 4 of this *Surgery*, and to other practica where you will find descriptions of unusual cases.

CHAPTER 6.

TREATMENT OF EIGHT KINDS OF SIMPLE APOSTHEMS DERIVED FROM SINGLE SIMPLE NON-NATURAL HUMORS

A surgeon needs to know only the following generalities concerning the aposthems formed from non-natural humors.[216] Some always derive from single humors, whereas others come from mixtures of them. Some others may derive from one or several. In the group

---

[215]   This observation describes what we now call bacteremia, pyemia and sepsis. The medieval surgeon's concept held that the cold bed clothes caused the patient to shiver. (LDR)

[216]   See p. 217 (LDR).

that always come from a single source, the humor may be benign or venomous (ie malignant). Among the non-venomous aposthems some are purely sanguinous and others are phlegmonous. Among the sanguinous are those from the blood that is more liquid and warm than natural, and that kind of aposthem is called the bastard or non-genuine erysipelas. Another sort derives from blood that is thicker than natural; it is called a furuncle.

Some of the aposthems from corrupted non-natural phlegm derive from the more liquid type, and they are soft and are called goiters and turtles. I will discuss goiters in Chapter 19 of Doct. III of Tr. III where we deal with lesions of the head (ie never written). Others form from gypseous phlegm; they are the hard nodes and their treatment has been described in the preceding chapter.

The three aposthems that come from single corrupt, venomous (malignant) and burnt humors derive from bile or melancholy. From thin burnt non-natural bile comes Persian fire. From the thicker burnt bile we get the formicula. Burnt melancholy produces the prune. Thus, we have eight kinds of simple aposthems formed from single non-natural humors, of which five come from benign humors—the bastard erysipleas, furuncles, goiters, turtles and nodes. Three come from venomous humors—Persian fire, formicula and prunes.

The Treatments: Two Parts

A. Two of the Five Non-malignant Aposthems: (ie three are discussed in other chapters).

1. Bastard (false) erysipelas[217], a brief review. The aposthem derives from blood that is warmer and thinner than normal, resembling natural bile in those qualities. Its signs: The contents move under the skin, ready to be expelled. It is fiery red. When we indent it, the finger creates a pale spot where you displace the fluid, and the color returns when you remove the finger. The place is warm and the patient feels the warmth. Treat it as we described for phlegmatic erysipelas.

---

[217] Hemangioma or hematoma (LDR).

2. Furuncle: Some derive from thick blood, and they are the more malignant ones, and are called pig-boils by the common folk, because they resemble the malignant pustules, albeit they are less serious. We treat them as we do phlegmonous anthrax. In France we usually use a phlebotomy before prescribing a regimen of abstinence and the application of diachylon plasters or extractive plasters [218]. Apply them fresh every day.

B. Treatments for the three malignant Aposthems:

1. Persian Fire. Here we see clusters of pustules surrounded by a yellowish-red zone. The pus is watery red, resembling a lotion after it has irrigated raw flesh. The affected skin seems to have been seared by an open flame. Avicenna (Bk. IV, Fol. 3, Tr. 1) said that Persian fire, the aposthems caused by burns [219] and prunes are terms that describe corrupt pustules with blisters. Later in the same chapter he said that you can use any of those terms indiscriminately, because they are so much alike.

2. Formicula. This aposthem is formed from thick corrupt bile and bears a single pustule. It burns and forms a crusts and is hot to the touch. It undermines and is corrosive. Avicenna (ibid.) said, "These are aposthems that appear in the skin and are no larger than ants" Then he added, "If you do not begin your treatment with a phlebotomy, as you should, and you proceed dirfectly to treat with topicals, they will heal locally, only to crop up somewhere else."

3. The Prune is a pernicious aposthem formed from burnt melancholy appearing as a livid or black pustule. With its complications it is especially serious when it is near noble structures.

---

[218] As used for retained arrowheads (LDR).

[219] A blistered mass after a burn, a so-called charbon (charcoal) or braise. Guy de Chauliac (p. 712) identified braise and Persian fire with carbuncle. The 'prune' remains imprecise. See Explanations at the end of this chapter, where Mondeville throws up his hands after a vain effort to resolve the confusion of terms.(LDR).

The treatment of the three is what we use for carbuncles and anthrax, as we described it. After they open and drain we use desiccatives as for ulcers. Avicenna (Tr. II, Doct. 2, Ch. 1, "medicines for ulcers") did the same., We described the technique in Tr. II, Doct. II, Ch. 2. Later in his discussion (ibid) Avicenna cited Galen for his treatment of formicula and prune, recommending the placement of the open end of a hollow reed over lesion and the other end set afire, an 'ambule', which extracts and exhales the venomous matter.

II. Explanations

I have searched all the authoritative texts and the practica that I could find and I have perused them thoroughly, only to discover that no two of them agree in all points as to causes, diagnoses, symptoms and treatments of these apothems and abscesses. What I call formicula another surgeon calls it a prune or a Persian fire or a charbon. Furthermore, no one has offered a treatment that some other author or practitioner has not disagreed with or called disgraceful.

Therefore, since all of these lesions are formed from the same or nearly the same matter as anthrax—corrupt, burnt, malignant and venomous—we hardly do wrong by going to the treatments for carbuncle and anthrax. And everything we have said about treating Persian fire, formicula and prune may be applied to miliaria and herpes, which we put in the same category.

CHAPTER 7.

THE TREATMENT OF SOME APOSTHEMS DERIVED FROM MIXTURES OF NON-NATURAL HUMORS

I. The Aposthems: Two Parts

A. Causes and Pathogenesis. Some come from benign humors and others from malignant, corrupt, burnt, and venomous humors. The first group come from vitreous and hard gypseus phlegm and melancholy. They include glands and very hard scrofules, nodes and scaling aposthems. Their treatments were described in earlier chapters.

The second group also has two types. One comes from a burnt phlegm produced by very mild heat. It is miliaria. The other type derives from heavy bile and liquid burnt melancholy. These are herpes (called 'wolf'), cancer and red erysipelas, which is called Our Lady's Disease in France, St. Anthony's in Italy and Burgundy, St. Lawrence's in Normandy, and many other regions have their own names for it.

B. The Symptoms for each: 1. Miliaria, commonly called Wild-Fire in Normandy, presents as many tiny white pustules, like millet-seeds, which give it is name. It is accompanied by heat of a less degree than formicula and the other aposthems in that group. The pus in the lesions is white and irritating. Treat it as if it is a mild carbuncle: evacuations and diet followed by ointments such as the populeon, or Rhazes' white or others like them.

2. Herpes esthiomene, means that it eats itself. It involves a member that already is gangrenous and it is not apparent before destruction of the limb begins.[220] It manifests itself as an advancing black discoloration and a horrible stench, that of cadavers. The herpes is unrecognizable before the skin is gangrenous. When you press that you feel it as a sack filled with flour and you can touch the underlying bone.

The herpes forms in five ways: a. From thick bile and thin burnt melancholy which are the most corrosive humors, and they destroy and eat away the affected region. b. From the product of faulty treatment with humectant topicals of any of the malignant pustular ailments described above, as formicula et al. Indeed, you should treat all ulcers with desiccants. c. The hands of the extremities (hands and feet) may become gangrenous when they are exposed to extremely moist cold. d. More commonly we see gangrene in fractured extremities that have been bound up too tightly, especially when the bone and the surrounding soft tissues are comminuted and the wound is treated with heavy topicals which smother it and prevent the escape of noxious vapors. e. When one delays the treatment of any venomous lesion.

There are many other causes for herpes (ie gangrene), but we encounter these five most frequently.

---

[220] Nicaise's note : The term Herpes chiefly represented the various types of gangrene (EN).

The treatment for herpes begins with purges as vigorous as the patient's vital forces can withstand, to eliminate the burnt humors, Then apply defensives all around the lesion and use the hot cautery to burn the zone between the gangrene and the healthy tissues. Repeat the process until the gangrene has sloughed away. Some surgeons apply sublimated arsenic on the lesion when it first appears. Theodoric (*Major Surgery*, Part 4, Ch. 9) recommended using the arsenic on the zone of demarcation in plethoric patients, not waiting for purgation. He claimed complete cures, and he added that, although the arsenic causes severe distress whenever it is used, it eases the pain of the lesion itself by destroying it and preventing its advance. On the other hand, Rhazes (*Albucasis*, Ch.1) said that no corrosive medication can be compared with the virtue of the red hot cautery. After the herpes has been destroyed, use your regeneratives, consolidatives et al. One good prescription contains equal amounts of fresh honey and salt. Heat them together on a metal plate until the residue can be powdered. Sprinkle the powder on a pad moistened with vinegar. After the lesion has been washed with vinegar and dried apply the pad.

II. Explanations

You will find the material in Treatise II, Doct. II, Ch. On Cancerous Ulcers. In general, whatever is useful for treating cancers is applicable to herpes, and vice versa.

CHAPTER 8.

THE TREATMENT OF CANCEROUS APOSTHEMES DERIVED FROM ONE OR SEVERAL NON-NATURAL HUMORS

I. Descriptions

A. Definitions: A cancerous aposthem is formed from a corrupt burnt melancholy, which itself is derived from bilious matter or from a humor that has been reduced by combustion to the status of a melancholy. The causative humor has accumulated in some part of the body. The illiterate surgeons call it 'pourficus', which means 'a perfect fig'. They divide the class into two types, the true or intact (not open) called the simple pourficus, and the open type called boiled pourficus.

Even when it is open, a cancerous apposthem is not an ulcerated cancer, because an apposthem is a bulging mass whereas a cancerous ulcer is an excavation.

B. Pathogenesis: It is formed in one of two ways. 1. It begins sui generis and does not develop from a preceding apposthem. Then it is called a simple cancer. Galen (*De Ingenio,* Bk. IX, Ch. 6) said that it is cancer from its birth. It appears at the surface as do new sprouts in a garden, at a stage when even an experienced gardener cannot tell one plant from another. Avicenna said the same (Bk. IV. Fol. 3, Tr. 2), "At first its identity is hidden, etc." So it is; one cannot recognize the mass as a cancer at its onset unless he is an expert, a surgical savant. It grows most often in glandular regions, because the firm melancholic matter can spread more easily in spongy places than in dense ones. 2. The second type begins as a benign apposthem. It becomes cancerous in two ways: a. From a solid apposthem derived from natural melancholy on which someone has applied a caustic (ie burning) material. b. A warm apposthem has attracted additional corrupt matter. This is the type of apposthem that is formed from a corrupt, burnt non-natural melancholy. It evolves in either of two ways: The burnt melancholic source may not yet have putrefied. Then the onset is slow and gradual, and it causes little discomfort unless it ulcerates, by some mischance. There is little pain until the matter in the apposthem putrefies. The second way: The source is already burnt and putrefied, and there is pain from the very onset, even when there is no ulcer. Often an ulcer forms in the usual course of the lesion, sooner or later, varying with the nature of the matter and the part or the region in which the cancer develops.

Other relevant items are discussed in the Explanatory Note VI in Tr. II, Doct. II, Ch. 4, on cancerous ulcers. In Note VIII of that chapter I explained that there are two kinds of non-natural melancholy, the putrefied and the non-putrefied. The former includes three varieties:

1. that which is in large veins near principal organs and provokes a continuous quartan fever. 2. that which is in medium-size and small veins and causes an intermittent or quartan fever, simple or double, according to the amount of matter present in them; 3. that which is in capillary venules and causes a cancerous aposthem. You will find more related data in the Explanations in the same chapter, and you should read it again to learn much that you should know about this subject.

C. The Symptoms of a cancerous aposthem differ from cases of hard aposthems derived from natural melancholy and from other sources. As we have stated, the cancerous aposthem cannot be recognized at its outset. As it takes form from the burnt melancholy it causes throbbing pain that increases with its enlargement. It is surrounded by tender and engorged black veins. On the other hand, natural melancholy is the lees of other natural (unburnt) humors. Its aposthem is not tender or painful; it does not throb. It grows slowly and exhibits no engorged veins.

II. Treatments. Three Categories

A. *Prevention*: You prevent a nonmalignant aposthem from degenerating into a cancer by diligent attention to all the preventive, curative and palliative measures used in treating the non-cancerous aposthem. If you commit no errors, no benign aposthem will be transformed into a cancer. Therefore, the treatments to prevent simple cancerous aposthems are the cool and moist diets used for treating scurf and the other ailments caused by burnt humors.

B. *Curative*: 1. The dietary regimen and 2. The evacuations are those described in the chapters on cancerous ulcers and scurf. 3. The Surgical Maneuvers and the Topicals:

A surgical drainage may be made with by incision (ie knife or cautery) or with corrosives.

Incision has ten elements: 1. A radical total extirpation, as we described it in the chapter on incisions. 2. Bleedings from small incisions at the periphery of the aposthem, to eliminate the dark and infectious blood.—*The following eight items are not relevant when the patient will not accept the knife or the cautery.*—3 (or 1). Use the hot cautery to char the surface after the cold-knife excision or the burn away the roots and the entire cancer, as we described it in the chapter on cauteries, or after you have used an eruptor as we described it in the chapter on scurf, where we discussed the most potent topicals. 4 (or 2). Put defensives around the cancer as we described it in the chapters on ulcers. 5 (or 3). Use cool topicals on the eschar to suppress the inflammation caused by the cautery or the eruptors. They are described in the Antidotary. 6 (or 4). After #5 succeeds, promote the sloughing of the eschar with suppuratives, such as hog lard, other animal fats, oils or butter, alone or mixed. Use emollients, such as mauves, violets, brancus ursinus or powdered seeds of fenugreek and flax. 7 (or 5). Follow with detergenst. 8 (or 6). Then use the desiccatives. 9 (or 7). Follow with regeneratives and 10 (or 8) consolidatives. All of the remedies and the methods of use are described in Tr. II, Doct. II, Ch. 10, and in the Antidotary.

Another scheme of treatment which may be equally effective has three parts.

1. A good diet. 2. Evacuations. 3. Local measures as follows:

a. Use resolutives to destroy the cancer (see the chapter on cancerous ulcers and the Antidotary).

b. If the resolutives are not completely effective, at least try to prevent further enlargement by diet and suitable topicals.

c. If you cannot prevent enlargement, prevent ulceration with those measures.

d. When ulceration occurs, use appropriate palliatives of sufficient potency, as we taught in the chapter on cancerous ulcers, also to be found in the Antidotary.

C. *Palliative Treatments*: are the same as those for cancerous ulcers, as they are completely and adequately described in the designated chapter.

III. Two Explanatory Notes

A. Refer to the chapter on the treatment of cancers, especially to Explanatory Note I where we cite the authorities by chapter and word, because cancerous ulcers, cancerous aposthems clearly are related as to names and treatments.

B. There are two kinds of natural melancholy. One is the true melancholy, the lees of the entire mass of humors. The other is the combusted (ie the ash-residue) of blood, bile or phlegm. Both are called natural melancholy. But, if they are combusted, one of them is twice burned, and is more harmful than that which was burned only once. From the very beginning of combustion both of these melancholies are no longer classed as natural by the physicians who cannot agree about the differences. We more practical surgeons are not involved in such profound investigations. We concern ourselves less with what is in books. We consider that cancer is the worst and, all things being the same, that it is formed from the worst matter, which is the twice-burnt melancholy. That is what we strive to eliminate, more rapidly and more intelligently.

CHAPTER 9.

PURE WATER-APOSTHEMS: OCCASIONALLY
AN APOSTHEM FILLS WITH PURE WATER.[221]

I. Description:

A. How it happens. The water accumulates either because there is too much of it, or because the body's vitality is too weak to eliminate it, or because the normal exit paths are obstructed,[222]

---

[221] Mondeville is vague. Probably he refers to local edema not related to a contusion (LDR).

[222] Normal skin contains pores through which vapors from within the body can be attracted and expelled. That concept was the basis for and explains the use of the medicaments called attractants (LDR).

or because the patient cannot pass enough urine. The stagnant water all over the body drifts and collects in the feeble region and forms a mass.

B.   The Symptoms: When a small collection is just under it, the skin is taut and shiny. When the aposthem is large, bimanual percussion can detect an impulse transmitted from one hand to the other.

C.   Treatments: The most suitable dressing encloses a firm ball of oakum as the center of compression.

D.   The topicals should be applied over the entire mass as well as over a rim of surrounding normal tissue.

So it is that this aposthem is treated as we do the phlegmatic ones. Purges are not necessary, and, as we explained in the chapter on impetigo, the topicals should be very dry, contrary to the moist contents of the mass. The dressing is similar to what we use to reduce abnormal hypertrophy of a limb.

II. Explanations

Consider Avicenna's action (Bk. IV, Fol. 3, Tr. 2) when he placed the instructions for this aposthem with those for the phlegmatic ones, and nowhere else. And he did not give it a special name.

CHAPTER 10.

WIND-APOSTHEM

This mass is produced by pure wind and nothing else.

I. Descriptions

The wind that forms aposthems anywhere in the body comes like the smoke rising from a heap of burning logs, the result of intense heat. Indeed, a smoldering ember gives off no smoke; the heat must be strong enough to consume the wood and produce the fumes. However, the fumes in the body have been generated by a feeble heat, and they collect in some region and produce a gas-filled mass. The commonest sites are at the openings of the abdomen, into the groins[223] and in other membranous regions.

B. The Symptoms: There is no discoloration. The mass resists indentation by one's finger, and on percussion it resounds. The only discomfort is from stretched tissues, not true pain. It resembles the color of phlegmatic, aqueous and melancholic humors whenever they do not discolor the skin. It is unlike the phlegmatic aposthem in its firmness, and it is unlike the melancholic, which also resists digital pressure, in not being tender. They differ in 'feel' as a stone differs from an inflated leather sack.

II. Treatments

A. A good dietary regimen is designed to consume gas. The alimentary elements include garlic, cumin, anise, dill, pomegranates, fennel, et al. Also taken by mouth are electuaries of berries of laurel, of diacumin, diahysop, diacalamint, anise-candy and ginger-candy. It avoids gassy vegetables and other foods. It avoids cold foods and beverages as well as exposure of the body to cold.

The externals are topicals that must be able to penetrate and work deep to the skin in order to resolve the gas, as are plasters of cabbage, yeast and wine. Sometimes we use cups without scarification. The topicals must be completely cooled before they are applied. Another topical: a plaster of wax and oil of myrtle dusted with a powder of hyssop. Apply them when their intrinsic warmth can act and for as long as they retain it.[224] Other topicals may be found in the chapter on hydrops and in the Antidotary.

---

[223] A so-called wind-hernia in the groin (LDR).

[224] The intrinsic warmth is part of the complexion of the medicine, and does not mean that it has been heated (LDR).

III. Explanations

A. If at all possible, lessen your own difficulties and ease the patient's suffering. Some recommend that you disperse the wind to a less hartmful region when the apposthem first appears. That may be possible in some cases. 1. When the apposthem is in a vital structure, such as the penis. 2. When the apposthem is near an important organ, as the ear. 3. When it is on a noble (ie delicate) place such as the face. 4. When the first premonitory sign appears in an organ as delicate as the eye. 5. When it appears in a debilitated patient or when he has had the apposthem for a long time. 6. When it appears in and threatens a dangerous place, such as a joint.

B. Six factors favor the displacement of wind. 1. When the more favorable region is just beneath the wind-apposthem, as the leg is beneath the knee. 2. When the region is adjacent to the apposthem, as the pubic region is next to the groins. 3. When the regions are functionally related, as are the breasts and the penis. 4. When the 'new' region is feebler and cannot resist what comes from the stronger apposthem. 5. The openings into the new region are larger. 6. Some Topicals that act to displace gas are: 3 parts of peeled garlic and 1 part each of lily-roots and sage-leaves, mashed and placed where you want it. If the patient will drink the juices recovered from the crushed plants, the results will be better. Another: Crush tansy and apply it, and wet the route between the apposthem and the new region with the juice. Another: Crush equal amounts of ground-ivy (ie clematis) and hyoscyamus and apply a walnut-size amount. Renew it three or four times daily. All of these measures seem to be reasonable and safe, and they are not costly, and therefore, they may be beneficial for the patient. An honest surgeon will try them, but he should not continue them simply to increase his fees. You can imagine some chap telling his well-to-do patient, one who always complains of the slightest discomfort, "Sir, there is a good reason for your discomfort; you have an apposthem which already is spreading within your body. However, its location threatens your life and I advise you to let me try to disperse it, etc. Then he provokes an apposthem in a place that was entirely normal. Then he collects the money and the glory, and he is exalted by the philosophers and the prophets.

CHAPTER 11.

APOSTHEMS ON THE SURFACE OF THE CRANIUM EXCLUDING TURTLES AND NODES

In brief, simply treat these lumps as we treat all other aposthems, with four additions.

A.   Do not use repercussives.

B.   Do not wait for a complete suppuration. The reason for A. and B. is the proximity of the aposthem to a noble organ (ie the brain).

C.   Open it for drainage with a triangular flap, with a wide base and two cuts coming to a point, shaped lke a shield, or an inverted arabic number 7 (see the algorism) or an inverted letter V.[225] That is the only kind of incision to use on the scalp; both cuts are straight and follow the hair lines. Furthermore, the scalp is thick and tough. A simple linear incision will not favor an easy evacuation of the matter.

D.   After the incision, use detergents introduced on pads of charpie wet with oil of roses and sugar-alum. Persist until the liquid part of the pus is gone. Then clean out the more solid and viscous and adherent residue with the apostles' ointment or the like, as we used them for chronic ulcers.

I have been brief here, eschewing the tripartite arrangement of the other chapters.[226]

---

[225]   Another indication that Mondeville directed his remarks to semiliterate students or those who had no knowledge of Arabic numerals (LDR).

[226]   Again the reader will notice a sense of urgency as Mondeville approached the end of this Doctrine II, sensing his own decline and mortality, yet eager to complete his master work. He takes short cuts and he frequently refers his students to earlier chapters, whereas in the earlier Doctrines he did not hesitate to repeat himself, often verbatim (LDR).

CHAPTER 12.

APOSTHEMS AT THE BASE OF THE EAR IN AND AROUND THE EXTERNAL MEATUS

I. Descriptions

These aposthems usually come from warm matter, only occasionally from cool, and, as for all aposthems, the differences noted in our General Chapter are recognized. The exceptions here arise from the continuous enlargement of the warm type, causing pain and fever, even delirium and death, especially when they involve the ear-canal and the nerve. The cool sort is different; it causes no fever unless it suppurates.

II. The Treatments

A. Evacuations: At the inception of a warm aposthem use clysters and phlebotomy from a saphenous vein or from a vein on the web between the index finger and thumb. Later, bleed from a cephalic vein on the arm. Wherever feasible, use ipsilateral veins for the bleeds. If phlebotomies are interdicted, apply cups over scarifications between the shoulders or on the neck. When the lesion seems to be bilious, purge for bile, as we have taught. The evacuations for the cool aposthems should be clysters of suitable substances and by laxatives used for phlegm.
B. The Diet is what we prescribed for scurf.
C. We use three kinds of Topicals. For the warm aposthems we introduce warm oil of roses into the ear, or oil of water-lilies or willow, with vinegar, juice of lettuce or morel or the decoctions of those plants. Anoint the outer ear with goose-, duck-or hen-fat, and oil of roses etc. You may combine these substances to make ointments and plasters, to suit the case.

If the local pain does not relent, use narcotics to relieve it, such as opium with castoreum, saffron and warm milk from a nursing mother. When the narcotics take effect, instill some castoreum boiled in oil. If even that fails, apply some very dry maturatives, as we have cited them. Then open the abscess and treat it as we do others.

The topicals for cool aposthems are injections of holy oil, oils of laurel or castoreum or costus, rue, juniper and brandy. Anoint the outer surfaces with the same oils and use ointments of dialthea, tarragon, martiatum and lard from foxes. If those fail to reduce the mass, try to promote suppuration with maturatives for cold lesions, or with others mixed with yeast. When pus appears, open the abscess and follow with the mildly warm medicines used for treating cool aposthems.

III. Three Explanations

A. These treatments differ from the usual for aposthems in two ways. 1. Sometimes we use repercussives, and that is contrary to what we taught in our General Chapter. 2. We avoid maturatives, emollients and suppuratives because they generate pus, to be avoided, if possible, in aposthems at the ears.

B. From the standpoint of any aposthem it seems proper to use maturatives and other humid tipicals. But from the point of view of preserving the natural complexion of the ear, which is dry, we must use topicals with that property, and we must mix them with and modify the humid topicals.

C. The Aposthems and the attendant suffering can destroy very young patients. That is less a peril for older children and rare for oldsters.

CHAPTER 13.

TREATMENTS OF APOSTHEMS AT THE EMUNCTORIES OF THE BRAIN WHICH ARE FOUR FINGERBREADTHS BENEATH THE EARS, BETWEEN THE MANDIBLES AND THE THROAT, IN THE REGIONS ABOVE THE JUGULAR VEINS

I. Descriptions: Two Subjects

A. The three General Causes : 1. The frailty of the vital forces of a critically ill person. 2. The density of the matter. 3. Its abundance. The two latter causes can be suppressed when the patient's vitality is energetic and he is healthy.

B. The Pathogenesis: Each aposthem forms in one of five ways: 1. During a crisis. 2. By derivation. 3. When matter is rejected from the brain into the emunctories. 4. According to Galen (*Megatechni,* Bk. XIII, at the end of Chapter 1) and Avicenna (Bk. II, Fol. 3, Tr. 2) these aposthems appear in patients who are wounded or have ulcers, other aposthems, scabies or bruised arms, legs or heads. 5. They may have external causes, as a blow at the emunctory itself, resulting in an aposthem in two ways: a. when the body is replete but frail and the humors are not malignant, or b. In a strong person the causative humors are plentiful and malignant.

II. Treatments

A. The *General Treatments* here, as for all aposthems, is twofold. A good dietary regimen with correct abstentions and topicals applied during the entire course of the illness, which avoid repercussives except in healthy persons who have suffered contusions.

B. The *Specific Treatments* for particular aposthems caused by external force in healthy persons may include some weak repercussives and resolutives that are only weakly attractive, such as rosat, oil of camomile, egg-yolk and wool-fleece with its natural lanolin. The treatment does not use purges. Bleed someone who is barely replete and whose similar aposthem derives from non-malignant humors from an hepatic vein between the fourth and fingers of the ipsilateral hand. The following day bleed him from the internal saphenous vein and use the topicals described above, which are weakly attractive. When the aposthems in this region have internal causes and occur in fully replete persons, and when the humors are very malignant, the treatment is the same, excepting when the matter is warm, a frequent occurrence. In that case begin your treatment with a general diversive bleed Use a purge against phlegm if the matter is cool.

The topicals to use at the onset already have been described as partly repercussive and partle resolutive, with mitigatives and comfortives. After the purges and the proper topicals have been used for a time and you observe no shrinkage of the mass, or if it enlarges, then apply attractives, beginning with the mild ones, to prevent the matter from hardening and to distract the vapors from the noble organs. As you see the need for them, use increasingly potent attractives as prescribed by Avicenna. Then use the maturatives, and do not wait for a complete suppuration. If the patient's vitality seems threatened do not delay your open drainage. However, when the matter is not toxic, wait it out.

After the abscess is open, treat the noxious matter with mondificatives which also can hasten the maturation of what has not suppurated. If the pus is smooth and not toxic, use your ordinary detergents and follow them with the regeneratives and consolidatives. You will find the proper medicines in the chapter on Aposthems in General, and in the Antidotary.

III. Four Explanations

A. Opening an aposthem in this region and wherever there are many nerves is fraught with risks for hemorrhage. You must make a direct in-and-out stab into the abscess and not turn you knife to one or another side. In that way you will avoid the nerves, arteries and veins.

B. Treat all painful maladies of the ears and their region by instilling the oil of almonds. Also lay on wool fleece with its own lanolin, and other good medicines. The fleece is better if taken from the region between the jaw s and shoulders of the sheep, because that is where the animal brushes against plants and herbs and absorbs some of their medicinal properties. The fleece at the rear end of the sheep does not have that added quality.

C. Do not use repercussives for aposthems in this region when the cause was external. There are four reasons for that: 1. An

external cause is single. 2. As it forms it diverts from a noble
organ and preserves it. 3. It is near a noble organ. 4. It is in an
emunctory. Other aposthems that form in this region should
not be repercussed for the last two reasons.

D.   Elangy (Rhazes' *Totum Continens*) recommended that we should
never promote suppuration before the resolutives have failed.
The best resolutive is the oil of camomille, because it
penetrates, points, mitigates and expels the harmful matter.

CHAPTER 14.

TREATMENT OF ORDINARY APOSTHEMS IN THE NECK OTHER THAN QUINSY. ENCYSTED LESIONS SUCH AS GLANDS

Some of the considerations of causes, symptoms, pathogenesis
and treatments of the ordinary aposthems may be found in the General
Chapter. Here I will add nine more items about treatments.

1.   Aposthems in the neck often are warm. If by chance they attract
some cool matter, they become scrofules and glands or some
other encysted masses with which we will deal in separate
chapters.

2.   When we diagnose these lesions and if the patient's conditions
allow it, perform a phlebotomy from a vein on the web between
the ipsilateral thumb and index finger. If that not permissible,
scarify and apply a cup between the shoulder blades.

3.   Early in the course apply wool fleece with its own lanolin, wet
with oils of roses or camomille.

4.   The diet until the abscess is drained is that used for fevers.

5.   After it has been opened and some or all of the pus has drained,
fill the ulcer with oil of roses and sugar-alum.

6.   If the opened abscess continues to bleed, add egg-whites to
#5.

7.   After the drainage begins, the patient may eat easily digestible
meats and chicken-eggs and drink wine diluted by one third
with water.

8. If the mondificatives in #5 and #7 are inadequate, use the Green Ointment.

9. Avoid the nerves, veins and arteries, especially the two large vessels that accompany the trachea on each side. If you cut them, the patient will die.

Chapter 15.

Aposthems In The Axillae

Be reminded that these aposthems are identical with those in the emunctories as to causes and all the rest. Here I will simply add six items about them which will apply as well to aposthems in all the emunctories and will explain several topics more clearly.

1. Axillary and inguinal aposthems are called buboes, because they resemble owls in two ways; they lodge in hidden places and they grow large heads.

2. Some famous practitioners claim that axillary aposthems are acompanied by burning heat. Galen (*Megatechni,* Bk. VIII) said that we should treat hot aposthems with phlebotomy if the patient's vitality is strong. If he is frail, we should use prolonged massage and exercise of the opposite arm. Avicenna agreed (Bk. I, Fol. 4, Ch. 62), saying, "You should drain some pus from these abscesses with the knife, not bothering to distinguish the warm from the cool." That precept is not immediately clear. I cite Galen (*Megatechni,* Bk. XIII, Ch. 6). "Early in the formation of glands in soft flesh prescribe abstinence from food." It follows then, that patients with aposthems in emunctories should be bled and should abstain as we have prescribed.

3. Avicenna (Bk. IV. Fol. 3, Tr. 3, on Treatments for Aposthems In Glandular Regions) wrote, "We should be chary with repercussives which will dispel the matter to the viscera, and with emollients which will attract more matter to the aposthem. We want to avoid those kinds of actions. Having avoided them

(ie by open drainage), we then may repercuss and use emollients."

4. The Handbooks often give the authors' recommendations to use weak resolutives that have few or no attractive properties soon after the purges have taken effect. Avicenna disagreed (Bk. I, Fol. 4, Ch. 25); he used cups (ie after the purges) to attract the matter. He repeated the advice in Bk. IV, Fol. 3, Tr. 2 On Glandular Aposthems. I must admit that the Handbooks and their authors favor the resolutives only at the onset, and only until they see if they work, whereas Avicenna advised the vigorous use of attractives when the resolutives were insufficient and the aposthem is not affected, neither arrested nor augmented. He wanted to impede the drift of the matter to the principal organs.

5. You may ask if buboes can be accompanied by an ephemeral fever. It seems that is not the case. Galen (*Megatechni*, Bk. X., Ch. 4) said that a fever that accompanies an aposthem always is putrid. Whereas Hippocrates (ie before him) had written to the contrary (*Aphorisms*, Part 4), that all such fevers were prolonged and bad except for the ephemeral, which is the only fever that is short-lived.

    A bubo sometimes derives from a great repletion, which also may be the source of putrid and wasting fevers. If there is a putrid fever a potent evacuation and early open drainage are needed, because a prolonged fever is evil. The same is true for the wasting fever, where the fever begs to be fed and the treatment of the aposthem requires abstention from food. Sometimes the aposthem appears when there is little or no repletion. Then gassiness alone can cause the swelling and an ephemeral fever, but not a putrid or wasting fever. The ephemeral fever consumes the gas and favors the treatment of the aposthem; and that is why ephemeral fevers in cases of buboes are not entirely entirely harmful. As to Galen's opinion, I must point out that a fever in the presence of a large aposthem in a fully replete patient is always putrid.

6. The practitioners agree with the Handbooks that buboes will

mature on their own, and there is no need for anything else. We should never incise to drain them until they are completely mature, unless we are forced to do it. That conservative policy recognizes that a premature incision will leave behind some undrained matter which will be very difficult to clean away and may go on to form a melancholic aposthem or a fistula. In the other hand, if you do not promptly empty the corrupt liquid pus, its vapor will flow to the principal organs and will prolong the fever, and the pain will weaken the entire patient.

Therefore, the surgeon who awaits the complete maturation of what he has provoked must do all he can to ease the fever and the pain. The best topical for that is the plaster of bran and mauves described in our Tr. II, Doct. II, Ch. 1, where we deal with the treatment of warm apothems and dyscrasias in wounds. My practice employs this plaster for every aposthem to hasten its maturation. I apply it all over the lesion itself excepting the small site where I expect to open it. On that place I lay an incisive maturative containing dialthea and yeast, et al.

CHAPTER 16.

APOSTHEMS OF THE UPPER EXTREMITY

I. The Apothems. Three Parts

A. General Considerations: The causes, symptoms, et al. For all apothems of the arm are those already discussed in the General Chapter.
B. The Special Considerations of the treatments as to their warmth and coolness also are in that chapter.
C. The Very Special Considerations deal with the requirements for the particular complexions, compositions and locations in the arm. These deal not only with apothems but with other acute and chronic lesions, and I believe that the treatments already described can suffice. I repeat, any surgeon who will follow the lessons precisely and apply everything as he should,

would need to know nothing or very little more. Therefore, the five very special considerations for treating warm aposthems which I add here may be repetitious. 1. Make a diversive bleed from an hepatic vein of the opposite arm or from a saphenous vein of the ipsilateral foot, observing all the rules and precautions given in the chapter on phlebotomies. I cited Galen in saying that we should offer relief by bleeding sufferers of warm aposthems. 2. You may repercuss in such cases if the conditions are fulfilled, and if you follow the rules. 3. After the phlebotomy you must not use repercussives. Go at once to the resolutives. 4. When the brachial aposthem is at a joint, open it before it completely matures, thus making an exception[227]. Do not make a large opening by cutting through muscles, because when they heal the scar contracts and partly or wholly limits the natural range of motions. Therefore, mondify the thick pus by introducing liquids with injectors, as common sense will dictate, to wash out between the muscles any pus that may remain in the abscess. 5. When the aposthem is at the elbow, do not drain at the summit (ie the olecranon) because that will permanently impair flexion. Instead, drain from the side.

Another special consideration is the need to purge the matter several times, as we taught in the chapter on scurf (ie dartre). Then, follow the protocol of treatments, as above.

II Explanations

Whoever has an aposthem at the elbow or who has suffered a transverse wound in the arm which divides muscles, should keep the arm fully extended from the onset, and hold it against his side, closing the axilla, until the scar is fully formed.[228] The patient should sit and recline with his arm bound at his side and keep it dependent when he is standing to prevent himself from flexing the arm, even though

---

[227] The rule not to drain until the abscess is completely suppurated (LDR)

[228] Until the triceps tendon has healed (LDR).

the dependent position and the weight of the arm will lead to swelling of the entire limb. If he does not maintain the position as stated, he will flex the forearm and disrupt the scar and the healing will progress very slowly, if at all. Movement impedes cicatrization, and delayed cicatrization will impede motion later on.

CHAPTER 17.

TREATMENT OF APOSTHEMS OF THE ANTERIOR CHEST WALL AND OF THE FISTULA THAT MAY ENSUE

I. The Aposthems

A. Descriptions: The causes, the symptoms, the pathogenesis (why the matter comes to the chest wall rather than to the interior), all can be deduced from what was stated in the General Chapter. However, when there is an aposthem [229] whose matter comes from within, and Nature pushes it to the surface through the space between the ribs. The more common lesion is cool and there is little pain and the patient does not suffer. But, in time, the matter is corrupted, and, being thick and resistant, it cannot get through the outer skin. The matter reenters the body cavity following the path of egress. Much later, the skin opens of itself or by other means and a fistula forms from the surface to deep within the chest.

B. The Treatment: Aposthems that are not connected with the interior are treated as we do for lesions in the emunctories or are near noble internal organs, such as the liver or the brain.

We have three methods when the matter comes from within:

1.   Evacuations: these include phlebotomies when the aposthem is warm. If it is cool, digest the matter and evacuate it with medicines described in the chapter on scurf. Sometimes, when the matter is partly warm or cool, we use both types of evacuations

---

[229]   Empyema that necessitates. At first it may be tubercular, then secondarily infected as part of a chronic empyema (LDR).

simultaneously. The bleed alone may suffice for some aposthems, and so may laxatives, and sometimes either one or the other. When we use phlebotomy and laxatives, which should come first? The surgeon may find answers in Tr. II, Doct. I, Ch. 1 in Part 6, titled Evacuations and Potions for Wounded Patients, and in the chapter on phlebotomy in Doct. I of this Treatise.

2. Surgeons with different persuasions have their own preferred dietary regimens for patients with thoracic aposthems. The wound-surgeons favor cool foods for the aposthem with warm matter and the opposite if it is cool, without setting limits on the degrees. The old-time surgeons, as we described them in the first of the fifteen Notables dealing with the General Treatment of Wounds, manage all wounded patients as if they have continuous fevers. But we moderns manage our patients with aposthems much as we do the wounded, as I described in Tr. II, Doct. I, Ch. 1, at least those who do not have putrid fevers, in a way that acts contrary to the causative humor.[230]

3. The Topicals must be maturatives and very potent attractives, such as yeast et al. Avoid repercussives and resolutives, and open the aposthems at the first sign of pus. Then deterge with strong mondificatives placed on but not in the open lesion. Use such as a decoction of myrrh if the aposthem is warm or the wine of myrrh if it is cool, or other substances which we have mentioned.

II. Fistulas Afterwards[231]

The definition is well known but the treatment is very hard to learn. The savants debate the issue and it remains a very lucrative course of treatment. The fistula is caused by itself and it endures for a long time without cure, while the patient becomes a wasted invalid. There are many and variously favored procedures.

---

[230] Includes nutritious food and good wine (LDR).

[231] Chronic draining empyema thoracis (LDR)

1. Until the era of Theodoric (ie of Bologna) there were four items: evacuation, diet, potions and other comnpounds and local measures. The evacuations and diets have been described; the potions and compounds consisted of detergents for the chest, decoctions of figs dry raisins, licorice, hyssop etc. Electuaries were made with dia-iris, diapenidon, diatragacanth, diaprassum and penidium. The local measures consisted of dilating the external opening of the fistula with some stents of elder-tree pulp, gentian roots, sponge etc. until it will admit the tip of a clyster-device. Use it to inject and irrigate the recesses in the chest with detergent washes made from water or wine mixed with honey, myrrh, sage and hyssop. After the injections you roll the patient side-to-side and up amd down, and you repeat the injections until the returns are clear. Then you introduce regeneratives and cicatrizers to close the fistula. I have seen those procedures used many times and I have seen no cures. On the contrary, the patients waste away, shriveled and dry as wood, emitting two Parisian quarts of pus every day. If there was any hope for rescue it must come very early and easily.

Now I shall offer two reasons why the treatments failed. 1. By keeping the fistula open with tents etc.[232] the patient's vitality was continuously sapped by the drainage, while the cold ambient air was allowed inside; both actions are destructive of life. 2. The patient is weakened by the taxing and painful maneuvers, and we know that pain saps one's vitality. 3. The ablutions that were injected into the chest reached the heart and lungs, and all of it cannot be recovered. That impeded the actions of those organs by oppressing and compressing them, and by infecting them and corrupting their vital spirit.

2. A second procedure defines our modern concepts, as taught by the monks under Master Theodoric, of whom we claim to be

---

[232] A solid or hollow cylinder of cloth or plant fiber or bark that serves both as a drain and as a dilator in the opening in which it is set. In that sense it acts as a stent. The similarity of 'tent' and 'stent' may be more than in the spellings(LDR).

intellectual sons and heirs. That is in Tr. II, Doct. I, Ch. 8, where we describe the treatment of penetrating wounds in the chest, and also the treatment of old (ie neglected) wounds. We describe the topicals and their applications, how to position the patient, the potions and the desiccative diet. I can recommend the following medicine which serves to dry the superfluous abnormal fluids in the chest and elsewhere, and prevent suppuration. Once I questioned a man who had recovered after a long and failing course, who suffered a serious fistula in his chest; he had been desperately ill, pouring forth very much pus every day. He had been treated as a hopeless case by several celebrated physicians. Inasmuch as I had seen such curable cases fail to recover, I approached the man with curiosity and I paid him a fee to tell me that he was using a compound taken from Averroes' *Colliget* in the section on simple medications. He claimed that 'The CARDO', that is the thistle grown in fuller's earth, deterged all the rot and stench from his entire body when it was against his principles to induce suppuration. He scrubbed the roots until they were free of contaminants, crushed them and dosed about a walnut shell-full morning and night. That is an easy and inexpensive method and available everywhere. You may use other potions in such cases, such as a good spiced wine, brandy diluted to suit the frail patient, at least before the fistula is lined with dense scar. The procedure may have some benefits.

3. The following method is used when the fistula has a tough lining, when 1 and 2 are not effective. We use a hot cautery on the lining of the fistula and then promote the separation of the slough. Others say to use caustic medicines, but I think it is very dangerous to use the corrosives where they can eat into the lung.

Here I will comment on Palliative Treatments for these fistulas, to be used when a timid patient does not want or cannot tolerate the other procedures; either he has no faith in the surgeon's abilities or he lacks the will or the strength to bear the procedures. I must say that you cannot provoke suppuration in a cavity lined with dense scar. The pus cannot penetrate the barrier and it simply accumulates in the softer surrounding tissues. Therefore, the fistula cannot evacuate any pus that is not inside the cavity. The rest of what had accumulated within the chest before the fistula opened remains undrained.

Therefore, the patient will need a sparadrap to seal the orifice to prevent loss of vital heat and the entry of ambient cool air.[233]

4. The fourth procedure is designed to treat patients who have too much pus in the chest and cannot mondify it or expel it through the fistula, and the other procedures have failed or displeased our Lord. In older times the surgeons used them for all fistulas, either because they did not have anything better or because it was more lucrative. Indeed, the more energetic measures (as above) were lucrative sources for the common surgeon.

Inasmuch as pus tends to accumulate at the bottom of a cavity our modern procedure consists of making an opening at the lowest level of the collection in the chest, using the space between the fourth and fifth ribs, above the false ribs, counting down from above. That is where the diaphragm adheres to the chest wall as it passes toward the back bones (see the Anatomy). That is the lowest level of the chest cavity, and as water does in a lake, the pus and superfluous fluids in the chest collect there. An opening there will evacuate the pus. Then you can dry the cavity and close the fistula.

III. Six Explanations

A. Abscesses in the chest become fistulas usually for two reasons. 1. An ignorant or indolent surgeon does not realize that pus is forming inside; the patient had few significant complaints and there were no signs at the surface, no reddening, etc. 2. The patient himself is negligent, because he notices only a foul breath, and he sees nothing at the surface.

B. In the Rubrics of some of our chapters and in other places, we speak of common aposthems. Today we give that term to the lesions that appear anywhere, such as phlegmon, erysipelas, water-aposthems et al. It is a name in common. Other aposthems generally form in particular regions and not in

---

[233] The fistula fails to drain walled-off pockets of pus. After eliminating the scarred lining, the sparadrap will seal the opening in the intervals between the treatments and prevent the closure of the fistula until the empyema has been cured (LDR).

others, such as scrofules and glands at the emunctories. Others form only at certain places and never at others, such as turtles in the scalp or a stye in the eyelid. We do not call them special types of common aposthems, although that name would be applicable for all.

C. In re the methods used in the old-fashioned treatments of a penetrating thoracic fistula, please note that the tent introduced into the fistula or penetrating ulcer must be fastened to a leash to prevent it from falling into the chest, and requiring a troublesome hunt to find and remove it.

D. If the tent material is fragile, such as the pulp of elder tress, you should encircle it with cord from one end to the other.

E. When you use a tent continuously to treat an ulcer (ie or fistula) in the chest the patient will die as a doddering, edematous or wasted sufferer.

F. When you introduce certain corrosive medicines in penetrating ulcers, they will do great harm, as Avicenna said (Bk. IV, Fol. 4, Ch. 3) because the vital organs in the chest will not tolerate their violent actions.

CHAPTER 18.

COMMON APOSTHEMS IN THE BREASTS

I. Descriptions

Refer to the chapter on tumors of the anterior chest wall for data on the causes and symptoms of mammary aposthems in men.

Aposthems in women's breasts are more common, for five reasons. 1. Women are more moist. 2. Their breasts are larger and masses appear at an earlier age. 3. Their breast tissues are spongier. 4. Their breasts receive menstrual blood to form milk, 5, Their breasts are more protuberant and thereby more exposed to external trauma.

The causes and symptoms of the common aposthems in breasts are those described in the previous chapter, but the differential diagnosis between warm and cool lesions and the distinct characteristics of milk-cake and engorgement with blood are given below. The differentiation of cool from warm lesions are the usual signs of a warm aposthem: redness, local heat, acute pain and tenderness and fever. The cool aposthems exhibit no heat or redness; they are not tender but they may be very painful, and they may remain for a long time before they cause any fever. The warm aposthems are unlike the other diseases of the breasts.

The cool masses and the caked breasts are similar in appearance, but they require different treatments, and the surgeon must recognize the differences lest he fall in error more than once. I count five symptoms that are different. 1. Caked breast involves the entire breast, unlike the limited extent of a cool aposthem. 2. The milk-cake causes the overlying skin to glisten. 3. The milk-cake darkens when it suppurates, whereas the cystic aposthem pales. 4. The milk-cake does not occur during pregnancy or in puerperal women, because there is no milk to accumulate. The aposthems can form at any time in any woman. 5. The cool aposthem is much warmer than the milk-cake.

II. Treatments

A. For Warm Aposthems: 1. Evacuations: Bleed from a cephalic vein in the contralateral hand. If that is not permitted, apply cups on the shoulders, buttocks and back. If the cause is amenorrhea, provoke the flow by bleeding from a saphenous vein. 2. The diet is what we prescribe for fevers until pus begins to drain, or until there are signs that the patient is failing (ie from abstinence). You may then provide the meats of hens and verjus as a beverage, or some red wine or pomegranate wine. 3. The topicals: 3 parts of oil of roses and 1 part of vinegar. Anoint the mass with the warm mixture, and that will restore the complexion of the breast by repercussing the bad humors in the breast and repelling what is flowing toward it. If that does not succeed, add a linen compress wet with warm juice pressed from morel. That herb, according to Avicenna (M*edicaments forAposthems)* matures and resolves warm aposthems that are in remote places. If the aposthem persists, try to mature it. If that, too, is unsuccessful, incise it promptly lest its vapors penetrate to the heart. Observe these five rules and all the precautions that we put in the General Chapter. a. Do not allow an explosive emission of all the pus at once. Drain a small amount at a time. b. Open the abscess at a dependent site. The breast is spongy and we can attract the pus to wherever we wish. You cannot do that in a dense structure, and you cannot always open abscesses in their most dependent parts. Try to incise where the skin is the thinnest. This repeats what we wrote about open drainage in the General Chapter on aposthems. c. Do not insert long drains. d. After the abscess is open, irrigate it frequently with watered honey. e. Then deterge with a wine of myrrh which mondifies, regenerates and cicatrizes. If even then the repercussion, the resolution and the rest have failed, then you must suspect something very odd (ie crazy). As Hippocrates wrote (*Aphorisms*, Part 5), "In such cases, shave the heads [234] and anoint them with oil of roses and vinegar, or a similar ointment of the sort that you use for fevers.

B. Treatments for Cool Aposthems in the breasts: The evacuations and the diets are those we use for maladies caused by cool humors, as we have described them in the chapter øn scurf. Then apply very warm topicals such as the oils of camomille, dill, lily etc. all with some vinegar. If the resolution fails, add maturatives and proceed as we have described for cool aposthems in several chapters. If yet you fail and the mass hardens and darkens, discontinue the applications of warm topicals before they can attract even more matter or resolve only the liquid matter and leave behind the denser substance which will be more damaging, or cancerous, as we have explained.

Two Explanations

III. Menstrual blood has five elements.

1.The most pure matter augments the two sperms in the formation and growth of the spermatic structures (ie of the fetus in the womb). 2. The matter from which simple flesh is formed. 3. The substance that forms fat. 4. That which goes to the breasts to form milk. 5. The rest is pure superfluidity. Left over from the previous cycle to keep moist the uterine tissues until the next pregnancy B. Caseation,

---

[234]   The seat of madness. Take note that the patient's are women, and the misguided male exhibits the same ignorance that led him to locate the source of 'hysteria' in the uterus (LDR).

congestion, coagulation and conglobation of milk are all synonyms for a condition (ie the caked breast) produced by an agent that warms and dries the milk, which then takes on the color of saffron, while the entire breast becomes warm. At times the condensing agent is cool, and then the milk becomes watery and pale, and the entire breast is cool. That is called congelation. We will deal at length with all theses mammary maladies in Doctrine III of this Treatise.

CHAPTER 19.

APOSTHEMS OF THE ABDOMINAL WALL OVER THE STOMACH, LIVER AND SPLEEN

I. The Aposthem; Two Parts

A. Description: The causes, symptoms, pathogenesis etc. have been discussed in the General and other Chapters.

B. Treatments: The Evacuations are by phlebotomy or cupping as the situation will dictate. Purges should be mild, such as prunes, violets, senna, whey et al. to avoid attraction of the matter in the abdominal wall into the underlying internal organs.[235] The Diet consists of light and easily digested foods and beverages are what we use in the other aposthems.

The Topicals were listed by Galen (*Megatechni*, Bk. XIII, Chs. 6 and 7, and *De Ingenio*, Ch. 6) These aposthems are a closely related group of neighbors, and we will deal with them as one. Since all of

---

[235] Again we observe the effects of the response to a potent laxative. The 'excited' gut attracts the bad humors from the abdominal wall through the belly where they can contaminate the vital organs. This effect is the reverse of what happens when one fails to purge the gut before using phlebotomy or cauteries. They will attract the bad humors from the feces-filled gut towards the sites of those operations and expose the injured region to the bad stuff as they drift past. The attraction must be properly directed by the surgeon to where he wants the humors to be evacuated (LDR).

them are in the abdominal wall and overlie the organ with which they are in close association, you must not use any medications on the aposthem which may offend an organ beneath it. Now you can understand the sense of Galen's remarks in the cited treatises. When an aposthem appears over one of the organs you should apply equal amounts of resolutives and emollients with added astringents and comfortives, as aromatic as the others. We add the astringents for two purposes. First we wish to resist the humors that are waiting to flow to the aposthem, and to dispel what already is on the way. Second, we want to support the vitality of the organs whose functions are so important for the entire body. However, if the matter already has accumulated in the aposthem you should apply different medicines in equal amounts and reduce the amount of the astringents, because they cannot repercuss what is there, and they may interfere with the actions of the resolutives and the emollients. The rationale is the same when we add resolutives to astringents, when we are concerned that the astringents alone will harden the matter in the aposthem and convert it to a cancer.

The best topicals for our purposes are the oils of mastic, spikenard, mint, absinthe, cloves and lilies, each alone or in mixtures. Use plasters made from the solid residues after extracting their oils and those from quinces, sedge, iris, roses etc.

If those fail to empty the lesion, fortify the resolutives, and if that, too, fails after a prolonged effort, use maturatives without eliminating the astringent. If the liquid matter is dissipated and leaves a more solid residue, use the remedy described in the chapter on aposthems caused by natural melancholy, and try to prevent the development of a cancer or dropsy. If that should happen, refer to the appropriate chapters.

After the maturation and the open drainage, use detergents, etc, as we do for common aposthems.

II. Two Explanations

A. Always treat these aposthems of the abdominal wall with astringents, as Galen insisted in the cited chapters. He added, "astringents have the same effects on the stomach as oak-galls have on leather undergoing tanning. So, too, we see scribes use vernis to dry, toughen and shrink their parchments.

B. The liver is the most susceptible of these three organs, because it is formed from a loose and soft substance resembling blood-clot. That is why a direct application of hepatic emollients will produce a dysentery, and the patient will suffer cold sweats and pass pieces of liver in his feces. The stomach comes next. Applications of astringents will produce anorexia. The spleen is third. It reacts like the liver but its substance is tougher.

CHAPTER 20.

APOSTHEMES AT THE GROINS, WHICH ARE THE EMUNCTORIES OF THE LIVER

In addition to all that I have written about aposthems at the emunctories of the brain and at the axillae, I will offer five items here.

A. Bubo is the common name for apostems in the armpits and groins, usually called 'verble' in France, whether or not it is in a replete body. But, when it does occur in a replete person or if it forms during a crisis, it is called a 'stricto', or, in the language of the common folk, a 'clapoire'. It is more difficult to treat than the abscesses that are secondary to ulcers of the legs or penis or contusions caused by energetic coitus, where the causes are clearly evident. [236]

B. Use curved incisions to drain axillary and inguinal buboes, as we instructed in the chapter on incisions. The drainage incision must go deep enough to empty all the pus from a large cavity and from the interstices and pockets between the normal glands. Furthermore, there always is a large flow of humors (ie lymph) from these regions, more so from the groins, because they are seated lower in the body, near the ducts and openings of the immodest organs.

---

[236] Here the cause is internal, perhaps from the liver. More likely it is from gonorrheal urethritis We see here the origin of today's term 'clap'. The verble is an ooze of pus resembling a maggot (LDR).

C.   Emetics are more effective evacuants here than for the axillae, because the diversive emesis is at a greater distance from the lesion. It rightfully deserves the name 'diversive'.

D.   The surgeon must be especially careful when he drains the inguinal abscess not to enter and injure the spermatic cord or a hernia and damage the intestine. I know of some who have done that.

E.   Wherever the accumulation of humors in a region is as great as it is here, a life-saving diet must be followed to spare the vital forces. Abstinence (fasting) to one's limits must be the rule. Hippocrates wrote (*Aphorisms*, Part 2), "the more you feed the body the more you harm it."

CHAPTER 21.

APOSTHEMS OF THE TESTICLES, THE PENIS AND THEIR REGION

I. The Maladies

A. Descriptions: The causes, symptoms, pathogenesis and the general treatments have been described in the General Chapter and in those which follow it.

B. The Special Treatments for Warm Aposthems of both organs have three elements. Before describing them, I must mention seven conditions which explain why treating affections of these organs is so difficult. 1. The patient is embarrassed and does not reveal his malady until it becomes chronic and more serious. 2. The region is hairy. 3. The affected region is low in the body, and the humors drift downwards to it. 4. The affected organs are sensitive and painful and therein are even more attractive of the bad humors. 5. The channel for disposing of the body's excess moisture (ie the urethra) is in this region. 6. The aposthems are near those channels. 7. The regions are hidden, and that impedes the expulsion of malignant vapors.

The Evacuants include phlebotomies from ipsilateral hepatic

veins if the aposthem is in one testicle, and from both arms if it involves both testes. Bleed from the right arm if the aposthem is at the penis, when the circumstances allow bleeds. If not, prescribe emetics, clysters, and place cups over scarifications at the kidneys, and use a suppository of 1 oz. of honey with 1 dr. each of salt, niter and colocynth. The following day, bleed from an ipsilateral saphenous vein.

The Dietary Regimen forbids wine, all sweet foods and other foods which engender warm humors.

The Topicals. After the evacuations, apply the following repercussives: Grind the peels of pomegranate fruits and dry rose hips, and heat them for a long time. Then add oil of roses with a small amount of vinegar. Another: Mix well equal amounts of vermicularis and oil of roses and add some vinegar. Soak cloths in either of the mixtures and wrap the penis and scrotum. If that causes the aposthem to diminish, add some barley-meal and bean-flour mixed with oil of roses, egg-yolks and juices of morel or similar plants, and if the aposthem really shrinks, use resolutives of bean-flour, camomille, melilot, fenugreek etc. Add some lard from a castrated ram and some honey.

If the aposthem resists the repercussives and resolutives and seems ready to suppurate, use maturatives. Then open the abscess, as we do for other aposthems.

Special Topicals for warm scrotal aposthems, to supplement the preceding general measures: Boil crushed beans with vinegar, then mash the pulp and apply it. Another: Add oil of roses to the preceding. Another: Make a plaster of cimolean mud and barley-meal and vinegar.

Special Topicals for warm aposthems of the penis (ie balanitis). Peel garlic after boiling it in water. Mash it with oil of roses and apply the paste. Another: Leaves of hyoscyamus boiled in water, pureed, and ground with fresh pork lard; it is an excellent mitigant. Another: The same leaves of hyoscyamus wrapped with ashes in cabbage leaves. Grind them and apply. Another: Warm, bread-crumbs moistened with oil of roses; then add bits of opium and saffron. This will soon calm the heat and the intense pain.

How to use the foregoing prescriptions: Insert one or more slivers of medicated sponge, or similar, as tents into the prepuce (ie not the urethra). Replace them as often as necessary until the aposthem heals, thus keeping open the orifice and preventing stricture. If pus is retained, it will putrefy and corrupt, eroding (perforating) the prepuce which will slough away or dangle and require a surgical amputation. During the entire course, the patient cannot indulge in coitus. If you are careless about separating the adherent prepuce from the glans you can cause brisk bleeding and the glans may ulcerate, all or in part. The ulceration may erode the shaft and eat into the urethra, causing the urine and the sperm to be discharged from the shaft rather than from the end of the penis.

C. Cool aposthems of the scrotum (the penis rarely develops cool aposthems), usually are watery swellings (edema) or priapism. The treatment consists of purging the cool humors, as we taught in the chapter on scurf. After the purge, apply ointments made from oils of spikenard, absinthe and artemisia. Add bran, sulfur, elder, dwarf-elder, etc. Use any or all of them mixed with fats, oils or butter, which will tend to provoke suppuration. If the lesion hardens, use this plaster, which I have tried with success: Make an oxymel with 2 parts of vinegar and 1 part of honey. In 1 lb. of the mixture macerate 1 oz. of ammoniac overnight. Filter it in the morning and grind the ammoniac with wheaten bran. Mix the solid matter with the oxymel and apply it warm. Persist with it until the matter of the aposthem either resolves or begins to suppurate. If the latter, use maturatives and treat it as any common aposthem.

II. Seven Explanations

A. Aposthems of the penis and scrotum often are followed by fevers. The penile aposthems are locally very tender and painful; those of the testicles affect the entire body, because they are related to the heart, in health as well as in disease.

B. When purgation at the onset is indicated in these illnesses, use clysters or suppositories rather than laxatives taken by mouth. The oral medicines produce feces from the upper body and add to the burdens of the illness.

C.   The cool aposthems in the scrotum always occur in debilitated persons, and in the dropsical ones. Their treatments are more difficult. They continue to accumulate watery and malignant fluids in the aposthems.

D.   Testicular aposthems seldom suppurate because they yield to repercussives and they resolve quite easily. The matter in them tends to be thin and the tissues are sloose. But if suppuration occurs, the testicle is doomed.

E.   A special and nicely surgical (ie artificial) bandage should be used as a suspensory for the scrotal contents.

F.   Avicenna (Bk. III, Fol. 20, Tr. 1) said that scrotal aposthems may not involve the testicles, and that is easily recognized. At other times a testical and not the scrotum is diseased, and that is not so easy to distinguish, because the two are involved together. In the diagnosis the testicle alone may respond to a cough and the fluid can be traced to the chest.[237]

A practitioner told me about a patient who suffered pains in his hip, which always extended to a swollen testis. Whenever the pain occurred in the hip, he prescribed a potent clyster which cured the pain and prevented the testis from swelling.

G.   You may question why we use repercussives to treat aposthems of the penis, whereas in our General Chapter on Aposthems we condemned their use in treating aposthems in the emunctories or when they are near principal organs. I reply that in the cases of penile aposthems you should ignore the rule even though the penis is an emunctory for the testicles, and is near them.

---

[237]   The cough from the chest is detected as an impulse in the scrotum if there is a wide open scrotal hernia (LDR).

To explain: Observe that some emunctories are more sensitive than the organs which they serve, as is the case here. The matter does less harm in being repercussed toward the testes than by remaining in the penis. Furthermore, some emunctories are indispensable for the organs the serve, and we must not repercuss them. Others, however, serve as enhancers of a good life and not as vital needs, as is the case here, and we can use repercussives. because life itself is our first concern before living well. Know then: the rule is not all inclusive because it does not refer only to emunctories of principal organs, which are bound up with life itself. Their emunctories are less delicate than the principal organs, therefore, etc . . .

CHAPTER 22.

APOSTHEMS OF THE PERINEUM, ANUS AND NEARBY REGION

First let me say that the lesions here are not different from the other aposthems. When the influx of matter is meager it forms a fig (ie polyp) or a mulberry or hemorrhoids or condyloma, and other lesions that we shall mention.

The treatments are what we use for other aposthems: evacuations and diets directed against the causative agents. I shall add some items here.

1. Six reasons why these aposthems are not easy to treat: a. They are low in the body. b. The passageway for feces is there, as well as other channels. c. The region has many nerves and it is very sensitive. d. The region has a cool complexion and is more liable to accumulate superfluidities. e. It is hidden and not easily examined. f. The aposthems are a source of embarrassment and they are not revealed early in there development. g. Because these lesions are not often exposed, few operations are performed, and only a few practitioners know how to perform them well.

2. Four reasons why the internal anal lesions are more difficult to treat than the external ones: a. One cannot see them. b. Feces soil them. c. Medicines are not retained in the canal (ie they slip out). d. The rectum lies above the anal opening and medicines placed in the canal slip up into the rectum.

3. We discussed the measures for treating perineal and anal disorders in Tr. II, Doct. I, Chs. on nerves and on wounds in general. Also, in the preceding chapter we set forth some principles for avoiding the use of naturally cool medications in this region, and yet we warm them before applying them.

4. Do nothing that causes discomfort, such as inserting long drains or corrosive materials.

5. Use suppuratives only when they are necessary.

6. Avoid violent (ie radical) interventions when treating chronic painful lesions.

7. In such cases, those who have suffered much for a long time, begin your treatments with sedatives. When the pain subsides and is tolerable, begin with a purge.

CHAPTER 23.

APOSTHEMS OF THE HIPS, THIGHS AND LOWER LEGS

I. Aposthems that seldom suppurate.

Aposthems or swellings at joints rarely suppurate. Treatments for them will be found in Treatise III., Doctrine III.

II. Aposthems that suppurate.

These do not appear near joints. Usually they develop deep in fleshy locations in the limb where they suppurate and where they are not easily discovered until late in their courses. The less fleshy the region, the slower is the process of maturation and the easier it is to notice it when it happens/

The treatments are similar to what we use in aposthems in the arms. I shall add two items. 1. Here we use scarifications (ie and cups) on the buttocks with good results. 2. Never incise anteriorly below the patella a life-threatening act. I have seen the lethal effects of incisions at that site in several cases.

Here Ends The Second Doctrine
Of The Third Treatise

DOCTRINE III OF TREATISE III

The meager contents of the Third Doctrine of Treatise III consist only of Mondeville's Introduction and The Rubrics which represent his unfulfilled plans. He placed the rubrics at the beginning of Treatise III (see his comment below) rather than following his Introduction, as in this English edition. I believe that the rearrangement will be more convenient for the reader. Mondeville offered the following apology at the conclusion of Doctrine II.

I must comment why I do not place here the list of chapters for Doctrine III of Treatise III. Instead, I put it at the beginning of the Treatise. I fear that death will come before I can attain my goal, and I can in this way exhibit my general intent to posterity, had I been able to complete the entire Treatise.

Mondeville also had planned to write Treatise IV on Fractures and Dislocations. His final illness defeated that plan. He was unable to submit Rubrics for that section of his Grand Surgery. We are fortunate to have his Antidotary in Treatise V which follows the abbreviated Doctrine III of Treatise III.

Here begins the third Doctrine of Treatise III, dealing with the treatment of certain specific diseases of the limbs, with blindness, with tinea of the scalp, with blockage of milk in the breasts and with quinsy of the throat.

INTRODUCTION
TO THE THIRD DOCTRINE
OF TREATISE III

Galen said that an animal's body and its parts are the tools of its innate spirit, by which the spirit executes its will (*De Juvementis Membrorum,* Bk. I, Ch. 1, Prop. 1). The spirit, therefore, is the prime mover as well as the body. Its parts are the organs and the organic instruments. The hand which holds the axe can hew the tree, whereas the hand alone or the axe alone cannot do it. So it is that the spirit deprived of the body or the body without its driving force cannot perform out-of-body actions. The body and the soul are closely bound as a single unit; when one is sick the other is disabled. That fact is clear to any observer. Whichever is struck with illness, the body or the spirit (as in the case of a mad man) the victim cannot perform his proper functions—be it his work, his studies or his teaching. Galen also said (*De Ingenio,* Bk.I,Ch.5) that you should first treat the sick body and then the spirit. Constantine wrote a reverse opinion in his little book (*Incantations, Sorcery, Curses, Medicines Strung From the Neck, etc.*) saying that you must first treat the spirit and then the body. You can find a lesson on that in my book, Ch.4, at the end of the second Doctrine of Treatise II, titled the Treatment of Ulcerated Cancers, and in the Notables.

Since a sick body and its sick limbs cannot be operated on all at once, nor even be studied as a whole, I felt it necessary that I should write a Doctrine (Treatise III) for treating diseases of each part separately.

The writing of that Doctrine was often so demanding and so depleting of this author in so many ways, that the Lord had to come to his assistance. Jean Mesûe stated (*Pratique* Part III of Sect. 1 on

Diseases of the Organs of Digestion in the chapter on Indigestion), that without God, no man alone could write an all-inclusive compendium of pharmaceuticals, and that was the object of my Doctrine. Without his help a surgeon who is also occupied with treating human patients would fail to achieve that goal.

That is not surprising! God created Nature, which governs our bodies and overpowers them and rules them. If God suspends His controls over Nature, everything would come to a stop, including Nature herself. If a surgeon keeps the Lord's presence in his mind's eyes, The Lord will show the way in the time of need, and he will be able to operate without distraction wherever he may find himself. He must not glorify himself alone for his successes, nor exalt himself. Whoever believes that he knows everything will founder in his dishonesty. And when he treats patients who are easily cured and he falsely suggests that they were in the difficult or incurable categories, and heuses medicines of little value or nostrums taken from ignorant hags, he will be ill-remembered into eternity for his hubris. Only he who is not falsely puffed up, who is fearful of his God, since the fear of God is the first stage of wisdom, will lack nothing. He trusts in the Lord's bounty, mercy and power.

Thanks to those bounties, I have lived, even while languishing, to claim the miracle of that special Grace. I have lasted these past three years, despite the unanimous predictions to the contrary of the Physicians.[238] Now I pray to the Creator that it please him to prolong my life, as he did the life of King Hezekiah[239], until I can complete my work, for the common good, and that the instructions be condensed as rain (ie from vapor) and that my words be distributed everywhere as is the dew.

---

[238] Mondeville espressed his fears for his health for the first time in the Introduction to Treatise III. The first Doctrine of that Treatise was written after 1316, after the death of Louis X (The Quarreler). At the time of this (ie above) writing of the Introduction to the third Doctrine, Mondeville had been very ill for more than 3 years. By his telling us that we can place the time at around 1319 or 1320. His disease progressed and he could not complete the third Doctrine of Treatise III. He went immediately to the Antidotary, to satisfy the pleas of his students (EN).

[239] For 15 years! See OT 2nd Kings (LDR).

THE THIRD DOCTRINE—FORTY THREE CHAPTERS.

(The following was Mondeville's "wish-list". After completing the second Doctrine, overwhelmed by his pulmonary disease, he gave up on the third Doctrine and he used whatever energies he could bring to the task to complete the Antidotary in Treatise V.)

Chapter 1. Care of scalp and body hair. Five sections.
> a. Remedies for too soft and straight hair. b. Remedy for gray hair.
> c. Remedies for undesirable colors. d. Preventing regrowth of depilated hair, or of unwanted growth. e. Regenerating fallen hair.

Chapter 2. Diseases of scalp and body hair. Five sections.
> a. Too short. b. Too sparse. c. Erosions. d. Fissures. e. Hair-loss.

Chapter 3. Disease of the Cranium. Five sections.
> a. Water accumulating in the scalp of infants (ie caput succedaneum).
> b. Turtles, or other scalp-lumps (ie sebaceous cysts). c. Tinea.
> d. Impetigo (ie safati). e. Dandruff (ie furfur).

Chapter 4. Diseases of the Auditory Organs. Sixteen sections.
> a. Partial or Total deafness. b. Slight hardness-of-hearing. c. Purulence (ie corruption) of the ear. d. Tinnitus or hum. e. Earache without obvious cause. f. Earache due to another ailment. g. An internal cause for pain whose source is another disorder, perhaps an ulcer. h. Foreign bodies in the ear-canal. i. congenital obstruction of the canal. j. Wax. k. A worm or the like in the ear-canal. l. Bleeding from the ear. m. Itching. n. A recent ulcer. o. Fistula. p. Disturbance caused by loud noises or shouts.

Chapter 5. Diseases of the face excepting the four organs. Seventeen sections.

> a. A node between the eye and the nose (ie lacrimal cyst) b. Hypopion c. Lacrimal fistula d. Salivary Fistula from the jaw e. Noli me tangere (acne or sebaceous cysts) f. Butigigua (ie probably acne rosacea), and puffiness of the face g. Paralysis h. An Ugly Color of the face i. Pannus j. Red Spots k. Wrinkles l. Sunburn m. Blotches n. Excessive Redness without scales. o. White Pimples (acne) p. Small dry pimple q. Loss of eyebrows

Chapter 6. Headache in the temples and hemicranial headache accompaning disorders of the eyes.

Chapter 7. Disease of the eyes. Thirty-three sections.

> a. Ophthalmia b. Ulcers c. Albugo. (whitening of the pupil, probably. early cataract) d. Sebel (pterygium, pannus) e. Corneal blemish f. Point artapach (scar of a puncture, punctus artefactus) g. Ungula (pterygium, see above) h. Wounds i. Dilated Pupil j. Contracted Pupil k. Cataract l. Corneal Vesicle m. Ruptured Cornea n. Discolored Cornea o. Pannus over the entire eye (trachoma) p. Conjunctival Ulcer q. Conjunctival Granulations r. Lacrimation s. Darkened albugo t. Pus on the Choroid (uvea) u. Button on the cornea (, bothor, pustule) v. Corneal Ulcer w. Corneal Cancer x. Discolored Cornea y. Watery pus z. Defective Vision aa. Clouded Vision bb. Veiled vision cc. Maudite (? frightful visions) dd. dark Wart (ie mûre) ee. Venomous bites ff. Snow blindness gg. Residuum after jaundice

Chapter 8. Diseases of the eyelids alone. Twenty-seven sections.

> a. Itching b. Scabs c. Hypertrophy d. Tumor e. Ectropion f. Sty g. Granulations h. Nodule i. excess hair (from scalp) j. Adherent lids k. Split eyelids l. Warts m. Polyps n. Lice o. Mites p. Seeds q. Drooping upper lid r. Loss of eyelashes s. Unnatural whiteness of lashes t. Ulcer and corrosion u. Glands v. Dead blood w. Superfluous granulation tissue x. Soft granulations tissue y. Greasy granulation

tissue z. Petrification aa. Ectropion of lower lid due to contracture
of a burn-scar

9. Diseases involving both the eyes and the eyelids. Ten sections.
>   a. Redness b. Tears c. Foreign particles d. Blinking e. Hypertrophy
>   f. Strabismus g. Small eye (ie enophthalmos) h. Moist and Bleary Eyes
>   i. Burning and Stinging j. Adherence of the lid to the eye.

10. General considerations in re the health of the eyes.

11. Diseases of olfaction and of the nose. Seventeen sections.
>   a. Nosebleed b. Foreign bodies in the nose c. Narrowness of the nares
>   d. Granulation tissue e. Pustules f. Ulcer g. Cancer h. Hemorrhoids.
>   (ie hemangiomas) i. Polyp j. Warts k. Itching i. Coryza j. Foul odor
>   k. Dryness l. Variable odors m. Sneezing n. Snoring

12. Diseases of the oral cavity excepting the five portions in it. Fourteen
sections.
>   a. Acute inflammation 2. Pustules c. Excoriations d. Ulcers e. Cancer
>   f. Ranula g. Tonsils h. Halitosis i. Garlic odor, et al. j. Tongue-tie
>   k. Ulcers due to applications of mercury on the body l. Nodes m. Fistula
>   n. Bleeding (gums)

13. Diseases of the Uvula: Four Sections:
>   a. Swelling or inflammation b. elongation c. Ulcer d. Bleeding after
>   surgical incision.

14. Diseases of the tongue. Twelve sections.
>   a. Wounds b.Fissures c. Ulcer d. Cancer e. Burn f. Warts g. Bite h. Spasm
>   i. Restricted motion j. Paralysis k. Altered Taste l. Loss of speech.

15. Diseases of the gums. Eight sections.
>   a. Acute gingivitis b. Pyorrhea after burns c. Ulcer or corrosion
>   d. cancer e. Fistula f. Overgrowth of Tissue g. Soft gums and loose teeth
>   h. Bleeding gums

16. Diseases of the teeth. Thirteen sections.

> a. Difficult loss (in child) b. toothache c. Corrosion d. Blackening, discoloration e. Elongation f. Insensitivity g. Worms h. Cavities i. Loosening j. Cold teeth k. Adherent Crusts l. Broken tooth m. How to Remove a Tooth

17. Diseases of the lips. Seven sections.

> a. Fissures b. Pustules c. Hemorrhoids (hemangiomas) d. Quivering e. Paralysis f. Inflamed Cancer (? herpes) g. Cancerous ulcer

18. Diseases of the emunctories of the brain. Three sections.

> a. Tumor in an emunctory b. Tumor in two c. Cancerous inflammation

19. Diseases of the throat and neck. Four sections.

> a. Quinsy b. Goiter c. Scrofula d. Torticollis

20. Diseases of the pharynx treatable through the mouth. Four sections.

> a. Thorn caught in the throat b. A leech c. A wad of food d. A cancer

21. Axillary stench.

22. Stffness at the elbow with hardening of the tissues.

23. Disorders of the hands. Two sections

> a. Cramped fingers (ie Dupuytren's contracture) b. Chapped skin

24. Disorders of the fingers. Six sections.

> a. Supernumerary digit b. Webbed fingers c. Nodules at the joints without drainage from joints d. Fistula without Nodule e. Arthritic nodules f. Trigger finger

25. Disorders of the fingernails. Seven sections.

> a. Paronychia b. Fissures c. Blemishes d. Cracked nails e. Subungual hematoma f. Leprosy of the nails

26. Diseases of the breasts. Fifteen sections.

a. Gynecomastia b. Too large female breasts c. Engorgement with milk d. Caseation e. Clotted milk f. Anomalous lactation g. Mammary induration h. Ulceration i. Cancer j. Fistula k. Painful lactation l. Scrofula or other masses m. Unwanted hair at the nipple n. Retracted nipple o. Faulty development (in a girl)

27. Disorders of the back and the vertebrae. Two sections.

a. Gibbus b. Nepta (large lipoma) dangling from the back

28. Diseases of the anterior abdominal wall. Four sections.

a. Contusion b. Umbilical hernia c. Omphalitis with pus d. Ascites

29. Diseases of the peritoneum in the upper abdomen. Two sections.

a. Loose peritoneum b. Hernia

30. Disease of the didymes in the groins (ie and scrotum). Three sections.

a. Enlargement and hardness (ie epididymitis) b. Loosening with inguinal hernia b. eruption or fissure (ie scrotal hernia)

31. Diseases of the kidneys (ie lumbar region). Six sections.

a. Inflammation b. Ulcer c. Gassiness d. Obstruction e. Stone or gravel with renal colic f. Pain unrelated to stone or gravel

32. Diseases of the hips. Two sections.

a. Steady pain in the hip b. Radiating pain, so-called sciatica

33. Diseases of the pubis and groin. Two sections

a. Ulcer b. Cancer

34. Diseases of the bladder. Nine sections.

a. Stone or gravel b. Hematuria c. Gas in the Urine d. Retention of Urine e. Incontinence f. Dysuria g. Ulcer in the Bladder or Bladder-neck h. Blood-clot in the Bladder 1. Bed-wetting

35. Diseases of the penis. Twenty-five sections.

a. Acute Inflammation b. Excoriation at the Glans c. Ulcerated Glans d. Red or White Small Pustules e. Ulcerated Skin of the Shaft f. Warts, etc. g. Cancer h. Adherent Prepuce i. Urethral Ulcer j. Stricture at the Urethral Orifice k. Purulent Smegma and Balanitis l. Retracted Prepuce m. Infection after coitus with a menstruating or leprous woman n. Sensitivity o. Priapism p. Itching q. Bifid penis r. Impotence due to a curse s. Urinary obstruction in the newborn t. Urinary obstruction by a wart u. Urinary obstruction by a stone in the meatus v. Chapped, fissured, skin w. Nodes x. Anomalous meatus for passing urine y. Painful swelling

36. Diseases of the testicles. Seven sections.

a. Painless swelling b. Painful swelling c. Enlargement d. Shrinkage e. Undescended testicle in the groin f. In the abdomen g. Castration

37. Diseases of the scrotum. Fourteen sections.

a. Fleshy hernia b. Varicocoele c. Hydrocoele d. 'Wind' hernia e. Composite hernia f. Hernia due to an old contusion g. Lax (elongated) scrotum h. Ulceration exposing the testis i. Ulcer and Cancer j. Fissures k. Itching l. Induration m. An Undermining Ulcer (scrofulous) n. Total emasculation

38. Diseases of the uterus and vagina. Twenty three sections.

a. Imperforate vagina b. Obstruction at the cervical os by a growth c. elongated clitoris d. Draining proud flesh e. Swelling at the urethral meatus f. Fistula g. Ulcer h. Cancer i. Itching j. Fissures k. Acute inflammation and excoriation at the cervix l. Proud flesh at the os m. Tissue resembling a penis alongside the vulva n. Enlargement of the Vulva o. Pustules p. Nodes q. Gas, coolness and other things inside the vagina r. Hemorrhoids s. Aposthem or Abscess t. Sensitivity of the vulva u. Nymphomania v. Prolapse of the uterus w. Enlargement of the uterus that may impair the function of the thoracic organs (ie probably large leiomyomas or ovarian cysts))

39. Diseases of the perineum. Six sections.

a. Ulcer following cutting for stone b. Bladder fistula c. Itching
d. Apostheme e. Complete Rupture of the perineum in a woman
(cystocoele and rectocoele). f. Partal Rupture

40. Diseases of the anus and rectum. Thirteen sections.

a. Imperforate anus in the newborn b. hemorrhoids c. Polyp (ie or wart)
d. Fissure e. Condylomas f. Vegetations g. Itching h. Rectal Prolapse
i. Tenesmus j. Intertrigo k. Paralysis l. Fistula m. Rhagades

41. Diseases of the thighs and legs. Nine sections.

a. Hard tumor at the knee without change of color of the skin.
b. Edema of the legs in pregnant women and debilitated people c.
Esthiomene (ie erysipelas, gangrene) d. Cancer e. Varices f. Gangrenous
ulcer (ie mal-mort) g. Elephantiasis h. Salt phlegm (ie flaky skin) i.
Telangiectasia

42. Diseases of the feet. Sixteen sections.

a. Podagra b. Fatigue after marching c. Rhagades d. Chilblains
e. Avoidance of or treatment of frostbite f. Smelly feet g. Blisters
h. Contusion of the nails i. Corns j. Interdigital fissures k. Subungual
cancers l. Ingrown toenail m. Perforation or corn of large toe
n. Perforation on the sole (callus) o. Arthralgia, usually beginning at
the great toe.

43. Painful joints, in general.

TREATISE V THE ANTIDOTARY

INTRODUCTORY NOTE

Chapter 1. Contains the Notables for Mondeville's Surgical Antidotary. He was careful to define it as such and to claim that it is the first Surgical Pharmacopeia.

Chapters 2 through 8 of contain medicinal substances which he favored for use in his topicals. He described them according to their therapeutic indications. Here we place the lists taken from those chapters, adorned by the recipes for his compounds and his clinical comments.

Although redundancy may be burdensome and seem to be unnecessary for some readers, we will find the same Simples appearing in Mondeville's Treatise— and in most of the Medieval surgical texts—as single items in prescriptions and as one of two or many substances in a Compound to be applied as a topical for more than one purpose. The same—or a close kin—ferula resin, for example, may appear in the list of repercussives and of maturatives, or even as a corrosive. We may fail to recognize which of the many Simples included in a recipe for a Compound is its primary substance, therefore, we put them in the same order in which they occur in the text. See footnote 1 in Chapter 2. The lists of Simples are taken verbatim as they appear early in each chapter. Later in each chapter Mondeville added the names of other Simples which he used in his Compounds. I have no explanation for that practice, although I follow it here.

Chapter 9 is followed by a long catalog of the hundreds of medicinals mentioned in Mondeville's text. It is accompanied by a glossary of synonyms, many of them in Arabic and Greek, of 222 of them. The Chapter ends with an index of all the items. Some of the materials in the chapter were the product of Prof. Saint-Lager, Nicaise's botanical consultant in his French editions of Mondeville and of Guy de Chauliac. Much of the material in those pages appears in the Compendium Pharmacopeia which is our

Appendix I. That section is the proper index for all the medicinal substances used by all the great surgeons during the 150 years that preceded and included Mondeville's.

LDR

HERE BEGINS THE ANTIDOTARY OF THE *SURGERY* OF MASTER HENRI DE MONDEVILLE

THE SINGLE DOCTRINE CONTAINS TEN CHAPTERS

CHAPTER 1.

GENERAL CONSIDERATIONS AS AN INTRODUCTION TO THIS TREATISE

To improve the reader's understanding of the material in this Fifth Treatise, 'The Antidotary', we offer these nineteen Notables.

I. 'Antidotary' is a word that means 'given against' and all the topicals for each part is derived from a Greek preposition 'anti', which means against, and from 'datum' (ie the participle of 'dare', meaning to give). Master Simon de Gênes in his *Synonyms* offers antidote, remedy, easement, relief, to explain the same concept. From 'antidote' we coin 'antidotary' to indicate a collection of antidotes, which may include formulated medicaments which may serve many more than just one purpose. We will describe them in an easily accessible arrangement in the chapters to follow, which are limited to the antidotes we use. We list the simple medicines and then show how those are compounded.

II. An Antidotary is a necessary adjunct for treating diseases. We cite Galen (*De Ingenio*, Bk. VII, Ch. 3), "The recovery from a disordered complexion and the cure of a disease are obtained by satisfying two conditions: 1. Know the suitable remedies, and 2. Know how to use them. Those two concepts are fulfilled in books devoted to simple

and compounded medicines, the latter being called antidotes."
Therefore,

III. Three requirements should be met when one writes an
antidotary: 1. It should be brief and easy to consult. One effects greater
brevity and ease of use by providing separate chapters with succinct
descriptions of each particular medicament. That is better than bits
of description in a hundred or a thousand different places. 2. Many
medicines can achieve the same results. Placing all of them in the
same chapter (ie classed according to therapeutic goals) would result
in excessive length and intolerable confusion, all to no avail. Even
the smallest handbook would exceed in length Avicenna's *Canon*. 3.
Inasmuch as any illness which we can cure runs through four phases,
for each of which many medicines may be needed—each of which
may be used for other purposes—we must have an antidotary that
avoids confusion and prolixity, and eliminates what is useless,
cumbersome and costly.

IV. An Antidotary is more important in Surgery than it is in
Medicine. Indeed, before Galen, who was the most eminent physician
after Hippocrates, no one, to the best of my knowledge, had composed
an antidotary. Furthermore, Galen wrote his, not for physicians but
for surgeons and called it the *Catageni*, which is a proof of my opening
statement.[240]

V. I give three reasons why the word 'antidotary' is a title for a
compilation of formulated rather than simple medications. First: all
compounds are of higher worth than each of the parts and since a
compounded topical is related to all of several components, it seems
proper to assign the title of antidotary to a description of how they are
made. Second: formulated topicals act in two ways: first according to
the properties of each of the contained simples, and second by the
effects of their being mixed (fermentation). Thus, they deserve a
higher rating than the simples, each of which acts only in its own way.
Third: I quote authorities. Galen, Avicenna and others used

---

[240] 'peri syndesews twn katageni'. Concerning the composition of
medications..(EN).

'Antidotary' as a title for books dealing with compound medicaments and not for simples. That is why, ...

VI. In the preface to his own antidotary, Galen provides five explanations relevant to formulating topicals. I add two more, making a total of seven. 1. We may not be able to treat an illness with a simple remedy. If, for example, it is possible to keep an entire body or part of it warm by using the correct amount of a medicament of the necessary degree and if we cannot find a simple one, say of the third degree, then we may make one by mixing one of the fourth degree with another of the first degree. 2. Frequently we find it necessary to use a remedy differently from what it was designed for, as when we add ointments which liquefy (ie at body temperature) to solid materials. 3. Often we have to modify the properties of simple medicines, as by lessening the liquidity of an oil by adding wax. 4. We formulate topicals to treat complicated illnesses where they cannot be cured except with compounds. 5. On occasion we may need on short notice a specially compounded topical medicament for some acute severe disease. Therefore, we should formulate that surgical preparation in advance of the need and hold it in readiness. 6. We may need to fortify or dilute the effects of some simple topical when we come to use it, as when we strengthen a corrosive with sublimated arsenic or add some cool medicines to weaken the arsenic. 7. When treating fistulas and deep ulcers we sometimes need or wish to use a simple medication. But, it may be too hard or the piece may be too large, such that we cannot introduce it. Then we may put it into some decoction or the like and mix some of it with the liquid.[241]

VII. I have seven reasons for composing this Antidotary. 1. Every day new cases show up which call for new topicals and every new situation requires new solutions. 2. Even if nothing new crops up, the routine and familiar cases offer opportunities to try new treatments, and that creates a need for new topicals. 3. Suppose that a novel case appears for which a new topical has not been devised, a careful search

---

[241] Conceive an attempt to introduce a chunk of a solid medicinal resin into a tortuous narrow fistula (LDR).

of the older remedies may lead one to discover anew some properties in them which had been overlooked and had been passed-by unmentioned. That is why *contrary to the opinions of all the medical authorities, we have demonstrated in our own era that all wounds are curable if they are promptly, simply and securely treated with a single topical.* Applying Dialthea will destroy salty phlegm, etc. 4. Every Antidotary does not list all of the compounds used by surgeons. Some may be found here or there in any of several works, none of which is complete. That is why I felt the need for a new Antidotary which will gather up all the scattered formulas for local medicaments and put them together. 5. Since many of those topicals which were popular in former epochs in their antidotaries have been abandoned by the surgeons of today and have been omitted from modern books, there may be little purpose to include them with those described here. But, do not be surprised! Horace, a long time ago, wrote "Much that seems dead and buried suddenly is revived and has value."[242] 6. It would be absurd, even heretical, to believe that our glorious and sublime Lord accorded Galen a sublime spirit such that no human after him could discover anything new. What then! God would thus have given away some of His powers. Has He not given to each of us, according to his merits, a natural spirit such as he gave to Galen? That spirit would be pitifully lacking if we were allowed to know only what had been discovered in the past. We moderns relate to the past as a dwarf perched on a giant's shoulders. He can see as much as the giant can, but even farther. So it is that we can learn what was not known in Galen's epoch, and it is our duty to record it in ours. 7. In that sense, less should become more, as we see in the mechanical arts—as in architecture. If the temples and the palaces built in Galen's time were to be rebuilt, they would be inadequate for an architect of today. We see the temples and the palaces which lie in ruins as examples for us to rebuild in better ways. And so my strongest reason (ie of the seven) is this: the concepts of the ancient liberal sciences can be improved. We must add to each of them and describe what is new.

---

[242]   From the *Ars Poetica,*, a rhyming phrase (EN).

VIII. I was strongly compelled to compose this fifth Treatise, The Antidotary, for three reasons all of which appeared just as I got as far as Doctrine III of Treatise III, which was to deal with special surgery of particular regions of the body and other matters including taenia capitis, deafness, cataracts etc. And I set aside Treatise IV which was to deal with diseases of bones, fractures, etc. The first compulsion was the realization that this Treatise V would be more useful than the other two. The second was the plea from my students who begged me to hasten its completion. The third was two-pronged: my prime motive was the love for my Art of Surgery and the second was my fear that I was not destined for a long life, unless there would be a special dispensation from God that would extend my years.

I am asthmatic, bronchitic, phthisical and consumptive, and consequently I thought it a better that I get on with this part of my toil.[243]

IX. You may find it worthwhile to recall what I wrote in Tr. II. Doct.II, Ch.1, 'Treatment of Virulent Ulcers', where I described why a surgeon needs to change to a stronger or a weaker remedy. In the same chapter I discussed the purposes and the manner of changing. Occasionally the change will be to another equally strong topical but with different properties. There are two reasons for that. 1. Nature becomes less responsive as it adapts to a medicament, which in the end becomes ineffectual as Nature develops her scornful tolerance. The acquired resistance will not be obvious as a problem until the medication has been used for several days. Then the surgeon will see no improvement with continued applications. 2. Rarely, one may say never, will a single remedy be equally effective in all cases of the same disease. That which has been good for most may not serve the one special case. That is why you should have available for use several remedies. When one of them is not effective, the immediate application of a new topical may succeed. In fact, in cases of the same disease the same medication may help Peter but not Paul; it gives

---

[243] Nicaise inserted a footnote here in which he compared Pagel's edition of 1892 with a Ms of 1437, to confirm what Mondeville wrote in the Introduction to Doctrine 3 of his third Treatise, in regard to his own ill health (LDR).

immediate relief to one, not to the other. The diversity of responses may depend on four factors: a. Every individual, no matter in what group he belongs, or whether he has a warm or a cool complexion, he has his own personal make-up, different from all others. b. Similarly, every remedy has its own individuality beyond that of its general class. All things being equal, if those complexions (ie patient's and remedy's) match, the remedy will cure; if no match, no cure. Let me add: the same medicine will produce different effects as it deteriorates with age and when it has spoiled. c. Changes in the constellations which are ever on the move (ie astrological) can account for the diversity of the actions of medications. d. Another factor is the specificity of each individual dose of a remedy, aside from its complexion in common (warm or cool) with other doses. For example, every piece of a magnet (ie loadstone) has four properties which account for its four modes of action. First, when a chunk of it is laid on the body, the weight of the material is felt by the body, and its hardness affects the wound. Its general actions on things are to cool and to dry them, and when it is applied to human bodies it cools and dries. A third property is specific for magnet and is exclusive in it, and characterizes all magnets and nothing else, just as laughter characterizes humans. It belongs to a magnet as it appears in any matter or in any shape, but it is a property distinct from matter and form; what may be shaped like it may not share magnet's specificity and vice-versa. In summary, the prime quality of magnet is not its complexion but its specificity, that which some call its species-specific perfection by virtue of which magnets applied on the surface of a body will attract iron buried within.[244] In addition to those properties, every magnet has a final specificity (ie its potency) by which it can attract a bit of iron more or less rapidly than another magnet of equal size. Given that the fragment meets no resistance in its path through the tissues, two magnets may have different powers of attraction on the same bit of iron.

---

[244]   The use of the magnetic property of Loadstone to attract buried bits of darts, arrows, spears etc. was mentioned by ancient writers, including Pliny. Mondeville found it more effective as a metaphor than for use in the therapy of wounds (LDR).

X. I shall not repeat in this antidotary the recipes for topicals in common use as they appear in the Antidotary of Nicolas, including the plasters of Diachylon with wax, Apostolicon, the Oxycroceum, the brown Popoleum ointment and others.[245] This Antidotary, both in French and in Latin, already is large and expensive enough without a repetition.

XI. Never apply a topical on a plethoric patient before you purge him as much as is necessary. That was Galen's rule (*Techni, Treatise on Causes*, Ch. 24, 'sufficiunt autem manifeste', and in the canon 'si ergo fuerit totium corpus'). That rule was developed at greater length in the chapter on treatment of infections. In offering that general rule Galen failed to include the exceptions; as a result some less discerning physicians (ie the clerical scholars) accept the rule completely as given. They try to force surgeons to delay their more practical applications until they (ie the physicians) have purged exactly according to the written rule. Therefore, no sick person leaves their care unless he has been classified as plethoric and has been purged.[246] For their part, some surgeons have actively protested against that Galenic rule, claiming that it is ill-formed and without any basis in at least fourteen types of cases. Because it has so many exceptions, they want to use their topicals indifferent to the rule, to the great exasperation of the physicians— and, often, to that of the patients—notwithstanding the precept. They claim that their patients never are plethoric and that they have no need for the advice of a physician (ie and the need to share fees).

The fourteen exceptions are:

1. Plethora does not forbid the use of defensive medicines. In fact, the more marked the plethora, the greater the urgent need for defensive topicals. 2. When the pain is intolerable and continuous the patient may die before having been purged, unless you apply a calming medicament. 3. When the offensive matter is hard[247], emollients are useful even before purgation as they make the purges

---

[245] Nicolas Myrepros (fl 1250, Nicaea) wrote the 'standard' pharmacopeia of that epoch, to be found in every apothecary's shop. The compounds listed here all had many ingredients, with many minor changes made by various surgeons, often more than one recipe in a single work. (LDR).

[246] And charged fees, accordingly (LDR).

[247] Sloughing infarcted tissue, sequestra of bone, etc. (LDR).

more effective. 4. When infections supervene as a result of congestion, it will be better to apply remedies that restore a normal complexion and improve the patient's digestion and assimilation. 5. When the diseased matter lies deep, as in a hip, apply extractive remedies, such as Levain, before purging.[248] 6. You should not use purges during pregnancy [249] or, 7. During a crisis (ie as a malarial chill), or, 8. Above all, in children, or, 9. In old people. In the four preceding cases you may use topicals before purgation, because you must not let someone die before using them 10. Do not purge in infections that complicate a recovery (ie after a wound), or, 11. In an infection involving emunctories (ie auditory canals or regional lymph nodes). Indeed, in Nos. 10 and 11 use extractives whether or not you have purged. 12. Do not use purges when the infection is near vital structures, or, 13. Where the diseased matter is excessive, or, 14. Poisonous. In the last three instances even when no purges are needed you may, nay, you must, apply the topicals which counteract and direct the material away from the internal organs.

All in all, Galen's rule is Rational and consistent. One concludes, however, that no rule regarding a medicine is necessarily and absolutely true. Let it be accepted for the majority of cases when that seems rational, but with exceptions noted. The rules are useful for patients as well as physicians, and they were so in former times for conscientious and rational surgeons. But that is no longer true today, in our time of mendacious, perfidious and perverse contemporaries. We are forced to be on guard against malicious people and to conform with their practices. It is preferable to falsify the falsifiers rather than be the victims of their fraudulence. Yet sometimes we are constrained to misrepresent and to vitiate our art and to take precautions of all sorts. It is like when Jupiter changed an actor into a bumble-bee because he buzzed in front of his statue. In view of the dangerously unsettled conditions and with fears of what may result after we are

---

[248]　Levain, leaven, baker's yeast(LDR).

[249]　In nearly all his statements about patients Mondeville used the masculine forms of nouns, pronouns and articles. Here he is constrained to use 'La grossesse' (LDR).

called to a patient, we immediately apply a topical, paying no heed to the contingents and the particular conditions. The result of that practice is that the topicals now in use must be harmless even if they accomplish nothing. Earlier, at the beginning of Treatise II, in the preliminary gerneralities (ie The Notables), I spoke of ruses and subtleties, related to the malicious mouthings which some surgeons direct against their fellow surgeons and physicians and patients and the public at large.

XII. First, I arrange and classify the simples which are compounded into the topicals. Then I group the surgical topicals into nine categories: oils, ointments, plasters, stupes (epithems), sitz baths (encathisms), poultices (cataplasms), porridges (bouilles), liniments and lotions (embrocations) and mustard plasters (sinapisms).[250]

Such a classification has two advantages for topicals which may be composed of the same materials and have similar properties but which appear in different forms. One advantage is real and the other is apparent. The real advantage has three facets. 1. The same medicine does not suit all patients nor act with equal efficacy. 2. Even if the effects are the same in all cases they may not work all the time. 3. Diverse forms of the same compounds may have provide benefits differently. The apparent advantage is that different forms of the applications may satisfy the needs of a patient differently: one may like an ointment, another may prefer a plaster and a third may not be happy with any of the topicals in this book. In truth, if you prefer one medicament more than another and if all goes well, use it for long periods. In response to annoying complaints from patients some surgeons have adopted the use of the same ointment made with different colors; one is green with plantain or rue leaves, another is yellow with saffron, another is blackened with ink (ie india ink) or reddened with a pill containing different red substances.

An *Ointment* is a slightly pasty topical, made smooth by grinding the mixture until all gritty particles are gone. We have five reasons for making ointments. 1. The contents penetrate better than those of a powder. 2. They adhere better. 3. They are gentle. 4. Rubbing on

---

[250] See the Glossary which follows this section of Notables (LDR).

some ointment before you lay on a plaster opens the pores and improves the deep penetration. 5. Ointments have been in good repute with the masses as well as with surgeons for a long time, and they believe that you cannot obtain healing unless you use them. Therefore, some surgeons use some ointment in all cases and their patients tell them that they notice the benefits. And you may find surgeons who use the same ointment in all cases, varying its colors as I described above. Occasionally a surgeon will use two, one for patients who pay well and another for the misers and the deadbeats. The first ointment has the weakest effects and is used for the payers, because the longer the course of treatment the larger will be the fees. The second ointment is stronger because they do not want to be burdened for long with stingy patients.

Some surgeons use three ointments: a cold one, a warm one and one that is a mixture of the the other two. Some others use four ointments, adding a fourth ointment made from the leftovers after making the other three. This last is the best and is milder. It is like the bread made for the local bakers and millers, which is said to be better because it is made from a variety of grains and doughs.

A *Plaster* is a solid topical, melted by heat, not as oily as an ointment, which we can keep in storage in the form of madeleines.[251] Topicals applied as plasters provide three benefits. 1. Their effects last longer. Avicenna (*Canon,* Bk. IV, Fol. 4, Treatise 3, 'General Comments re Ulcers') said, "we know the importance of having available viscous preparations in the form of plasters which adhere and keep the humors from escaping." 2. They are more acceptable and less disagreeable than other topicals. 3. Fewer changes are necessary. Sometimes they can remain in place for a year or longer and still retain potency.

An *Epithem* is composed of fine powders dissolved in liquids in which cloths or *stupes* are soaked and then applied to the affected part. It has the advantage of working rapidly when needed, more quickly than a thick plaster or ointment in which wax and oil act as obstructions (ie over the pores).

---

[251] Madeleine: a name for a confection and also for cylinders of a medicament, shaped while they are still soft (EN).

An *Encathism* (ie a Sitz-bath). It is a bath for an immersion of limited depth, perhaps up to the level of the umbilicus, where the water is not agitated nor douched from above. It has two advantages as a topical. 1. When we may be embarrassed to use an embrocation, as at the anus or vulva. 2. When we wish to bathe only the affected part and not the whole body, as in persons with diseases of kidneys when we want avoid disturbing the humors of the rest of the body and increasing the discomforts of the patient.

The bouilles (porridges) are topicals made by diluting powders in water or mixing resinous saps with oil or honey and then boiling them. They have two advantages. 1. They are easy to handle and easier to spread than the others. 2. They adhere well enough so you may not need to cover them with a wrapping. Examples are mondificants that do not contain fats.

A *Cataplasm* (poultice) resembles a bouille, but contains more ground herbs than the bouille. We commonly use them as maturatives. Its three advantages are: 1. Each medicine in the mixture maintains its own function. 2. It is easy to prepare, requiring no special skills. 3. It is less costly.

An *Embrocation* is a lotion or liniment poured on a dressing composed of warm water and additives. It has two advantages. 1. It dissolves what has been desiccated inside the dressing. 2. The sensitive patients love it (ie removing a less adherent wet dressing causes less pain).

A *Sinapism* (ie usually a mustard plaster) consists of dusting a powder on the affected part after applying an ointment. The cooperative effects seem to favor recovery. This alone of all the methods of applying topicals has a succession of two actions; softening by inunction plus the powder's effects. That seems to satisfy the patient, because it appears to him that the surgeon is trying to complete a cure in one day.

XIII. Just as we do in the chapters on regeneratives, incarnatives, cicatrizers and in other places in this Antidotary, we should mention some medicaments that are incinerated and washed. Here we find it of value to discuss the methods of combustion and lavage, according to the instructions contained in the hard-to-find book by Serapion,

The Servitor, where he described the methods of preparation of medicines. Let us go on to combustion, its advantages and its process, ignoring the rest for now.

Among the uses of Combustion favored by physicians there are four of value for surgeons. 1. When a simple medicament has two inherent healing properties, we may use combustion to increase one and diminish the other. Thus we may increase the drying actions of alum, india ink and couperose (zinc sulfate) and decrease their corrosive effects. 2. Combustion rids medications of impurities, as when we heat litharge (lead oxide) and others to purify them.

3. Combustion renders the actions of some medicaments more intense, as happens when we heat oyster shells and egg shells to make fresh lime. 4. Combustion makes some medicaments easier to grind, as it does for lead, silver, iron slag and other metals.

Surgeons use six methods for combustion. 1. For bones, shells of molluscs and of eggs, cream of tartar etc. apply heat until they crumble into small pieces. Then place them in a clean pot and leave it in an oven at least overnight. The process is complete when the materials are snow-white. Continue the heating if needed to obtain that result. 2. For bol d'arménie (ie iron oxide in clay) or mineral stones do this. Morcellate the material into bean-size pieces, then heat them on a tile or a griddle plate. 3. For india ink, alum, flowers of bronze, couperose etc. first crush them into small pieces and place them on a tile or a shell and set it on a charcoal fire, placing some of the coals atop the material. Intensify the heat with bellows until the pieces melt. When they stop foaming and bubbling or when they begin to turn green the process ends. 4. For metal slag, chips and flakes of iron (ie as from blacksmiths' forges) first grind them and mix them with wine or vinegar or other liquids. Put the mixture into a glass jar which you leave in an oven for one or more nights until the material has been reduced to bits of cinder, each bit retaining its own keen effect. 5. For tuthie (ie zinc oxide), heat it until it is red, using a charcoal fire and bellows. Then add liquid (wine or vinegar). When cool, reheat and remoisten as before. Repeat the process until you get what you want. 6. For bronze flower (ie verdarain), take 7 parts of red bronze filings and one part of sulfur, mix and place it on a cloth over which you had thoroughly smeared with Lut de Sapiens (ie philosopher's clay) and allowed to dry.[252] Then set the cloth for two days over a slow fire fueled with animal dung. By then the material will be perfectly incinerated. Most of what I have described usually is carried out in pharmacies.

In respect of Lavage (ie washing) we will discuss two issues: the advantages and the methods.

The usefulness of lavage derives from four motives. 1. When you wish to dilute the potency of a medicine, as when you wash lime, green flower of bronze etc. 2. When you want to eliminate impurities, as when you cleanse oils. 3. When you want to soften something for easier grinding, as you do for Tuthie after the repeated incinerations. 4. To eliminate a malignant quality, as you do to terebinthe (ie a resin from pistachio plants) to completely eliminate its irritating properties.

There are four processes of lavage which are necessary for surgeons to know.

1. Washing Bronze Flower: After it has been ground, soak it in fresh water at least four times until it floats and the wash water remains clear. The same procedure is followed in washing the incinerated slag and chips of bronze. 2. To wash oils, use a jug with a narrow neck which you can cover with your thimb and which has a pin-hole size vent at the bottom, which you plug. Fill the jug half or two-thirds full with a mixture of oil and hot water Shake vigorously, with your thumb over the top opening. Set it to rest for an hour, then open the bottom vent and allow the water to flow out. Repeat with fresh water as often as is necessary to obtain clean oil. 3. Lavage of lime and similar stuff is the same as for bronze flower except you need not grind it beforehand. 4. For red terebinthe use a jar containing terebinthe and fresh water which you shake for a long time, as you would beat eggs for an omelet.[253]

---

[252] Luteum sapientas: Nicaise notes that the formula is fully explained in Notable XV. The clay of wisdom was composed of wheat flour, egg white, fresh lime and clay. It was used by the savant alchemists, hence 'clay of wisdom' (LDR).

[253] At 'La Mer Poulet" a restaurant at Mont Saint Michel in Brittany (a few miles from Mondeville's birthplace), I observed omelets being beaten for at least 6 minutes before being set to cook. A local tradition? (LDR).

Decant the water and renew it and agitate again until the decanted water is clear and the terebinthe is white, having lost its bitter and irritating qualities.

In all the above I provide what the average surgeon should know about preparing topicals. However, it does not include what must be known by those few who treat eyes and other delicate structures.

Certain preparations, as sublimated arsenic and others like it which are seldom used by surgeons, are left to the alchemists.

XIV. At the conclusion of Averroes' commentary on the *Song of Songs,* in which Avicenna divided Medicine into Theory and Practice, he said that Theoretical Medicine has no purpose except as a science without the practical aspects; one learns it by discussion etc., whereas, one learns Practical Medicine through experience. That is why, after his study of theory, a surgical physician must turn to intensive practical training. Averroes added that lectures and dissertations were only a small part of the education in Surgery and Anatomy. In fact there are only a few matters in those fields which may be learned by exchange of words alone. Haly agreed (*Regalis Dispositio,* Part II, Lesson 9 of the *Treatise on Medicine*), stating, "A surgeon must seek formal operative training spending a long time at centers where experienced surgeons perform operations. They must observe attentively and engrave what they learn in their memories. Then they should perform operations under the supervision of Masters until they are ready to be on their own. Those who obey the precepts will become well-trained and able surgeons.

When it comes to operating, in following the teaching of such authors, Surgery leans little on theory and much on practice. A surgeon need not fulfill conditions (required of the clerical physicians) that he have a good basic knowledge of all the properties and nuances of the simple and compound medicines. Elangy[254] agreed with us (*Totum Continens*) when he wrote about preparing compounds. He said, "A physician need not know every remedy.

---

[254] Nicaise explained that El Hangi was another name for Rhazes' *Continens,* not the name of an author, similar to the error that confused Albucasis with a name of another book by Rhazes' (Nicaise, p 630) (LDR).

Certainly, knowledge is more noble and praiseworthy than ignorance, but some inexperienced pup may say that such total knowledge belongs in Medicine (ie and not to pharmacy) in addition to other adjunctive functions, such as making surgical instruments, etc." Rhazes added, "It is not proper to call someone a physician simply because he knows (ie what is written) about medicaments. He who understands their real actions and knows what results to expect when they are used according to the standards of medical practice, he is the proper physician rather than an a priori philosopher.

XV. Preparing each of all the topicals used for every particular case may be difficult, easy or somewhere between. Preparing some of the oils may be difficult, painstaking and even dangerous, as in making Holy Oil, oil of Terebinth, oils based on human feces and others like them. You will find it a better policy to leave the preparation of those to the alchemists (ie the apothecaries) who usually are more experienced at it. I prefer to buy those oils at the apothecary's shop where they are made rather than to prepare them myself.

Other oils are not so hard to prepare, those which are so very necessary in our daily work. They include oils of roses and the like. We will describe the most useful and the easier to make, as well as the best ways, leaving out alternative methods of no use to us.

There are many different methods to prepare those oils, all of them in the not-very-hard class. 1. All of them, as a general rule, use one or one and a half parts of the medication for two parts of oil, but that does not hold for simple oils (ie not compounded with medicines) such as oils of olives, nuts, egg-whites and wheat-flour. 2. If stronger effects are desired, increase the ratio of medicament to oil. Do the opposite to reduce the potency. 3. If you find that the potency of the remedy is too feeble, you may find that the medicines are too old, as occurs naturally with violets more than rue and with cinnamon more than euphorbia. In such cases add more volume (ie if a liquid) or weight (ie if solid) of the medicine. 4. In very complex oils, as the oil compounded from mastic, mandragore and lily, you must not use (apply) them unless you are certain that strict controls over the contents were in effect. 5. Olive oil is the commonest of the basic vehicles used in preparing compound oils.

These are the methods of preparation of the oils. 1. Crush the fruits, grains, tender rootlets and other materials to be used. Measure out the prescribed amounts and allow them to simmer in the oil. Allow to cool until tepid, filter out the solids and store the oil for later use. 2. Instead, you may add the crushed materials to the oil and allow them to macerate for approximately six days, unheated. Pour them into a cloth sack and twist the sack forcibly to filter out the oily product. 3. Sometimes you heat the mix and filter it while hot, not waiting for maceration. 4. Certain oils (as roses, violets etc.) are made during the summer by exposing the mix to sunlight for forty days before filtering it. 5. The initial mix for preparing the oils of flowers (roses) contains four pounds of oil and one pound of roses, all placed in a glass jar for solar exposure. During July (ie hot summer months) stir the mix every day for a succession of eight days. On the eighth day squeeze it through the filter and add anew one pound of roses to the oil and repeat the whole process. After this second round, add three pounds of roses and set the jug in the sun for forty days before filtering and storing the oil. 6. During the winter when the sun is not warm enough you avail yourself of the heat stored in the soil. Cap the jar and bury it in a pit for two months before filtering the contents. 7. You may bury the glass or glazed earthenware jug containing the mixture for three months after sealing the mouth with clay (ie not renewing the roses) and then filter it. 8. An oil may be made by distillation. Use an alembic to distil the water from flowers, rootlets, grains or other dry-milled materials after which they are macerated in oil, and then the oil is prepared as above. 9. We prepare some oils by using the *descensio*[255] process, as we do for wheat flour, cinnamon and other sorts of grains. After they have been softened in oil, put the mixture into a glazed clay jug which has three or four tiny perforations in the bottom. Plug the mouth and seal it with philosphers' clay. Set the jug in another dish that just fits over the bottom of the jug and seal the two together (ie see below for the sealant). Bury the lower dish in the ground and surround the exposed upper jug with a fire made from blazing cow dung, or, if that is not available, with wood charcoal.[256] 10. A special

---

[255] Descensio: the extraction of the oily stuff dripping from the materials (LDR).

[256] The material that drips into the lower dish is the medicinal oil (LDR).

procedure is used to make an oil of wheat flour and egg yolks. The flour and the hard-boiled yolks are mixed and placed between two heated iron plates. The liquid which is expressed is the oil. 11. Another special method is used to make oil of egg-yolks which have been hard-boiled and separated from the whites. Gently roast them on a glazed clay dish until they are blackened. Take care when turning them gently lest they crumble during the cooking. After all are blackened, remove the dish from the oven and tilt it to spill off the oil that has been cooked out of the yolks. 12. Some oils, such as from tartar et al.,[257] are made as follows: Take some kneaded tartar and grind it with some vinegar. Wrap it in leaves and lay them under some coals (ie charcoal). After they have cooked a while put them in a glazed clay jug tilted to one side, kept in the open. A reddish oily liquid will ooze forth, to be called oil of tartar.

13. That same oil may made in another way. Calcine the tartar and place it in a linen cloth sac which you suspend in a damp place, such as a humid cellar. The oil of tartar will collect, drop by drop into a dish which you place beneath the sac. 14. Another way: Place the calcined tartar in a bladder (ie probably leather) which is immersed in water until the inward seepage dissolves the tartar and produces a greasy material, also called oil of tartar.

We make the "philosophers' mastic" which we use to seal the jugs and jars in which we prepare our oils and the sublimates (including those of arsenic and mercury) in this way. Take ordinary clay, potter's clay or any other, and grind it until it is a fine tan powder. Add ashes of grape-vine wood, burnt red clay, goose-droppings or horse-manure and mix with the ashes of human hair. I myself obtain a seal which works well for me by adding horse-manure or cow-dung or goose-droppings to ordinary clay.

The white wax which we use in ointments and plasters and for waxy applications is whitened as follows. Obtain as white a wax as you can find and melt some of it at the bottom of a bowl with a wide mouth. Keep a jug filled with cold water close-by. Pour over the melted wax a large cup (about as large as a urinal) half-filled with cold water, and pour out what

---

[257]   Tartar was the dried lees after making wine rather than the white 'bloom' which accumulates on the grapes during ripening (LDR).

is left of the melted wax after some of it adheres at the bottom of the jug. Then plunge the heated bowl into the cold water while agitating it until the wax (ie the film that adhered) comes loose and falls into the water. Repeat the maneuver until you have accumulated enough of the thin lamellae of wax. Put them on a table exposed to the sun and sprinkle cold water. When that evaporates, sprinkle on more. The wax will become very white within about one day.

You may bleach wax in another way. Spread the films in sunlight atop plants in a garden and turn them frequently until they are white. That will occur after about three days of hot summer (July) sunlight. You may leave them to themselves, not turning them, during days or nights, exposed to sunlight as well as dew, until they whiten. That is a common practice from May until September.

This is how to make mucilages which we usually put in our plasters. Crush the roots, grasses etc. and put them in a pot with enough water to cover the materials. Bring all to a boil and set the pot aside to allow maceration during three entire days and nights. Boil again until half the liquid is gone. With time, the thick liquid is scraped out and filtered to become what we call mucilage.

The difficulties in the preparation of certain plasters, including diachylon, ceroneum (ie mostly wax), apostolicon, ysis et al, have limited the correct methods to a few apothecaries. One can do justice by writing only about the methods of preparing the few that contain litharge, such as diachylon. You mix two thirds of the prescribed amount of oil with some mucilage and warm it over a low fire, stirring until it is about to boil. Remove it from the fire and add the rest of the oil which had been mixed with all the litharge. Boil it over a low fire while stirring continually with a wooden rod. The indicators that the decoction is complete are three: When a bit is tossed into cold water, it thickens and slowly 'swims' toward the bottom, seeming to stop halfway down before it proceeds to reach the bottom. It is less sticky on your fingers than you may expect. According to some practitioners, the bubbles will tend to separate. Then you take the pot from the fire while continuing to stir vigorously for about an half-hour, using the same rod. Then you pour it all on to a marble plaque that is wet with oil and you form it into madeleines. However, you cannot develop the necessary skills from watching (ie or reading) or after making only one such a plaster. You must assist at many performances and try many times on your own, often without success.

When a plaster contains plant saps (ie resins), use as described in Treatise I, Doct. I, Ch.2, concerning head wounds with fractures, etc.

I will not waste time and effort to describe how to make the extremely easy-to-prepare epithems, cataplasms, embrocations, bouilles, encathisms and sinapisms. When you concoct a bouille with honey be sure to add the flour to the water as the first step and then add the juices, and never in reverse order. When the bouille contains fatty materials, first combine the flour with a little water before adding the remaining water.

You do not need much skill to make any of the moderately difficult ointments. During summer months use four ounces of oil to one ounce each of wax and powders. In winter-times reduce the amount of wax by half. After a good mixing, heat it until a smooth fusion is obtained. Remove it from the fire for a few moments until it cools to avoid burning the powdered contents. Then reheat gently while stirring with a spatula or in a mortar, as you deem best.

When an ointment contains fatty material use less oil and add your wax and powders to a correct amount. If the fat is not melted or if the ointment contains soluble gums which are already in solution or which can be blended in the ointment without being in solution (such as Ammoniacum, Serapinum et al., which when added do not render the ointment too thin or too thick after the filtration), then the amount of wax should be adjusted as needed. If the gums can be ground, as are frankincense, myrrh and mastic, they may be considered as powders when you mix them with oils and waxes in amounts suited to your needs, corresponding to the proportions indicated above. When the liniment contains juices or other watery materials add a premelted mixture of oil and wax and pour it all into a mortar and stir it with a warm rod to get as smooth a product as possible.

XVI. When the surgeon wishes to use a repercussive cooling oil or a local fortifier (ie a tonic) as oil of roses or myrtle or water lilies et al., use oil of unripe olives. To make a resolutive oil which also is a maturative or a suppurative and emollient, use oil of ripe olives. And if you want a resolutive as well as a fortifier, use oil from half-ripened olives.

XVII. The surgeon whose goal is to become a skilful and a learned practitioner should reread all the twenty-six Notables and the fifty-two Contingents applicable to all Surgery and the fifteen Notables which are devoted to the treatment of wounds. He will discover many useful general principles which appear in no other surgical texts. He will find instructions about the precautions and the subtle and tricky ruses which surgeons employ to protect themselves against the deceits of other physicians and surgeons and deceitful patients. The Notables also teach you how to obtain suitable and honorable fees from the last named. Finally, you will read, clearly stated, the Declarations and the Considerations (the General Principles) of Surgery which should guide surgeons in the canons (ie of surgery) and in the techniques of performing operations. Certainly, it would be supremely unjust for a surgeon to master all the surgical books and to become adept at operating, to work every day, from morning into night, ceasely going about through the streets and the squares of the city to attend his patients, to toss sleepless at night while he reviews his day's experiences and makes his plans how to expend himself in the service of others during the entire day ahead—I repeat, it would be an injustice for that man, in return for the admirable services which he alone has provided those people while restoring their health, if he does not receive a decent payment for his labors. A law says that one does not serve his term of service in the army at his own expense. The common folk say, "All work deserves payment in return." Cato wrote, "When anyone works unpaid, the sum of human misery increases."

XVIII. In this Antidotary you will find nearly all the topicals required in the various surgical situations. We will avoid the confusion resulting from frequent repetitions about the same medicines in many different chapters. In preceding pages we have refered to topicals but did not give exact references to the chapters in which the surgeon could find the best one for his purposes in a given case. What we did not do then, we now will do here. For example, if we simply refer a surgeon to the chapter on repercussives or another, he will not know whether to choose an ointment or a plaster. Or if he is refered to an ointment he still will not know which to choose unless we have designated one by name, as ointment of roses or popoleum or Brown ointment. But all ointments do not have special names; therefore, I decided to give each its own identifiable number in its own class. For example, I put all the repercussive oils together in one place, and I do the same for the ointments and for the other classes of topicals in the various chapters of this Antidotary. And in each category the oils, the ointments etc. bear numbers 1, 2, 3 or 4. The surgeon will have a sure guide.

XIX. An operating surgeon is forewarned about preparing regeneratives, incarnatives, cicatrizers and medicines for use in the eyes, not to use a wooden rod for grinding or a mortar which previously had been used to prepare harsh medicines, like peppers, garlic, or corrosives like arsenic and realgar. Unless one has scrubbed them thoroughly, the utensils will harbor bits of the dangerous materials which may contaminate the newer preparation and not only impair its own function, but may render it corrosive. Therefore, always use a fresh pounding rod and mortar. If you lack those tools, have the medicines prepared by an apothecary. Also, do not put ophthalmic medicines in the same storage cabinets that hold corrosives et al.

harbor bits of dangerous materials which may contaminate the newer preparation and not only impair its function, but may render it corrosive. Therefore, always use a fresh pounding rod and mortar. If you lack those tools, have the medicines prepared by an apothecary. Also, do not put ophthalmic medicines in the same storage cabinets that hold corrosives.

Chapter 2.

The Repercussives[258] [259]

This chapter has three parts: Introduction; The Medications; The Explanations.

I. The Introduction: In his *Canon*, Bk.II, Tr. I, Ch. 4, "The Mode of Action of Medicaments", Avicenna said that repercussives and

---

[258] In this and in the following seven chapters, I have given the Simples their more common English names where that is suitable. They are further identified by a number set in brackets which will lead the reader to a number in the Compendium Pharmacopeia which follows Chapter 9. There one may trace the synonyms and botanical terms. When a Simple has been named more than once in any chapter, the identifying number may not be repeated, in the hope that a one-time label will be sufficient for the curious reader (LDR).

attractants had opposing actions. They cool and thicken where they are applied, lessening the heat of attractants, and they close the pores. They condense and thicken the humors which flow to the affected region and deter their reception there. Solathrum (431) is an example. Serapion said the same in his *Agregation, Discourse 4,* 'On the twofold actions of medicines' "The virtues of repercussives should be cool, their substance be thick and styptic if they are to act with strong repulsion." He added that there are several kinds of repercussives. 1. Some are cool and moist but are not astringent. 2. Some are cool and astringent. 3. Some are comforting. 4. Some cause thickening. 5. Some are contractive. 6. Some are expressive (ie dispersive). 7. Some obstruct the entry of humors.

In those chapters Avicenna and Serapion gave examples. Serapion said that cool, humid but not astringent repercussives act contrary to rarefacients by closing some of the large and small channels of entry but not all of them, as do the styptics. The latter actually shut off the veins. They are potent thickeners and are cold; although styptic they are not erosive because they are dense and adhere on the surface, and they do not penetrate via the small pores to penetrate deeply. Their coldness contracts, condenses and strengthens the body. Serapion used the term 'impulsive' to describe their actions within the body to dispel the humors they encounter. The comfortive repercussives fortify and temper the complexions of the region[260], to maintain its normal state and enable it to prevent the superfluous accumulation of harmful humors. His example was terra sigillata (460) and theriac (462) which are innately comfortive. In others, such as oil of roses (388) that property is a result of the complexion at the time of application.

Thickeners increase the density of the humors, even to solidify them. The contractives cause the affected part to block the entry of

---

[260] The terms 'temper' and 'temperament' had special meanings. Sarton (*Galen of Pergamon,* Univ. Kansas Press, 1954. P. 51 ff) explains the Latin equivalence with the Greek crasis, meaning a mixing (ie a mixture). Therefore, idiosyncrasy refers to a particular person's body's (or a part of his body) temperament, or the state of his 'complexion' at the time. Or, in other words, the admixture of his humors and his vital spirit (ie pneuma). (LDR).

the humors. The expressives eject the humors in the sense of someone pressing wine from grapes. The opilatives block the channels and pores by drying and condensing the matter in them.

Although I said that repercussives oppose attractants, I have yet to explain the nature of the property of attraction. By virtue of its warmth and subtlety an attractive medicament may draw humors from elsewhere in the body to the very region beneath where you apply it. That property is useful, say, when you treat sciatica, after you have purged the patient, or after you have extracted darts, arrows or other foreign materials from a wound. Beaver musk (96) and leaven (502) are examples.

You may conclude from the foregoing that true repercussives are cool. That quality is the basis for their action. They can act only to oppose warm things. Whereas a warm topical can dissipate or consume a repercussive, at the same time the reverse happens, the repercussive eliminates the warm. Therefore, although it is commonly held that some warm medicines are repercussives, that really is not the case. Again, we say that only cool substance can be repercussives, in the strict sense of that term. It holds for warm substances only when they mix with the cool ones as they flow together from one region to another.

Cool repercussives are of two types: one has all the necessary qualities and virtues; the others are less well endowed and are less potent.

The first type can dispel what already has accumulated,which by its abundance, its intensity and its heat is a source of grave concern. Such it is when the source is a sanguinous humor. An experienced surgeon can feel and see the local signs, and observe the patient's responses.[261] In such cases we always use styptic repercussives, some dry, some moist.

All the second types are moist and they lack the power to disperse an accumulation of bad humors from a lesion into other regions of the body. However, they may be able to block the inflow of more of it by

---

[261]   Celsus (@ 20 AD) described the four classic signs of inflammation. Mondeville did not have his book. However, Celsus' statements were transmitted by Paul of Aegina (7thC) and then via Arabic sources (LDR).

cooling and thickening the matter. Here we are more concerned for the heat and the intensity than for the amount of the superfluous stuff. Such is the case when it is entirely bilious.

II The Repercussives may be Simples or Compounds.

The Simples

Derived from Plants

Morel (431), orpiment (329), large and small purslane (368), shepherd's purse (487),plantain—pucier (353), hyoscyamus (232), ivy (120), oxalis (328), chicory (113), water-lily (491), large and small plantain, both wild and cultivated, leaves and twigs (428), leaves and twigs and unripe fruits of trees and bushes, including oak (309a), medlar (274), pear (339), quince (372), ash (432), dogwood (63b), plum (355), grape-vine (205), eglantine (389), willow (495), aspen (364a), rushes (378a), various marsh reeds (378a), barley (323), wheat (187), oats (310), darnel grass (255), sumac (449), hawthorn (215a), blueberry (58b), green raisins (473), oak galls (195), pomegranate-peel (369), pomegranate flowers (47), roses (389, rose anthers (389), acacia (2), ache (4).

Mineral and Metal

Bol d'armenie (59), sandalwood (407), cachymia (121), litharge (253), iron filings (289), coral (131, antimony (28), cimolia (289), ceruse (105), terra sigillata 460), lut (119a, clay (32).

Partly Repercussive Substances—cool to some degree

Orach (42), Mercuriale (282), malva (273), violets (486), cold water, vinegar (484), rave (378), zucchini(137), cucumber (144), melon (280), seeds of malva (273), the four cold seeds (123a), mandragore (263), verbena (477), hepatica (268), polygonum (361), poppies (367), umbellicus venus (136), venus-hair (476), joubarbes (229), hypocistus (224), moss (291a), bean-broth (178), hyssop (226)

Some warm simples commonly, but improperly, called repercussives spikenard (437), absinthe (1), cabbage leaves (71), cypress nuts (149), both marrubiums (271), fumeterre (191), mountain germander (200), both stoechas (446a), myrrh (299), frankincense (188), mastic (272), lupin (259, alum (20), salt (404), sulfur (450), oil of roses (388), schoenanthe (416a), artemisia (35), thistle (100), two aristolochias (33). And you may include any bitter simples which do not exceed the second degree of heat, whether used as decoctions, oils, juices, powders, or flours, and whether used alone or in compounds.

The Preparation and The Application of Compound Repercussives: Three general rules:

1. A surgeon should obtain in advance enough juices or leaves of the special herbs etc. to fulfill his needs, and enough of the following to make a base with three parts of oil of roses, one part of vinegar, one-half part of bol or other clay, altogether to have the consistency of honey.

2. The surgeon must assess the humors he is opposing, sanguinous or bilious, as to their heat or coolness, and he should compose his topicals to match in degree what he is opposing. For example, he will use more of his cool and moist medicines if the bad matter is bilious. In the same manner he will modify his medicine to match the predominating sanguinous, phlegmonous or other humors. When the morbid substances are an admixture of equal amounts of humors, the surgeon should mix his medicines to be suited for each of the humors and with the same proportions.

3. Before applying a topical a surgeon should ascertain if the matter he treats is cool or warm and if it is sanguinous or bilious. Against the former he will use only simple repercussives which are styptics. If bilious he will use moist medicines which are not styptic. If the matter is cool he will use warm simples or

compounds which are comfortives. In all cases use only one type of remedy without interruption, not mixing one sort with another or changing from one to another; persist until you attain your goal or you fail. Do not wait too long in such cases. You may prescribe other medicaments, one or more in succession, each known to have virtues and potencies equal to the first.

Those are our rules.

Compound Repercussives

In addition to the simples we have twelve ointments (ie inunctions or plasters).

1. 1 oz. bol d'armenie, 3 dr. oil of roses, 3 oz. vinegar. If erysipelas or herpes are feared, add terra sigillata, 1/2 dr. This ointment will prevent those complications if applied around wounds. If they already exist, the ointment applied around those aposthems will prevent them from expanding by blocking the entry of more humors.

2. 1/2 dr. each of juices of morel and joubarbe (229); 1 oz. bol d'armenie, 1 oz. oil of roses, 1/2 oz.vinegar.

3. 2 dr. each of white sandalwood, spodium (439) and acacia (2); 1 dr. camphor (77), 1/2 dr. opium (320), some violets, vinegar and a suitable herb (ie aromatic).

4. 1 oz. red sandalwood, 2 oz. camphor, 1/2 clump each of morel and joubarb; oil of roses, 2oz.; rose-water, 1 d.

   After having evacuated the patient by phlebotomy or another, as you deemed fit, place the repercussive topicals around the diseased region and not on it. The exception is a plaster of No.4 which may be placed directly over the heart to protect it from venomous matter.

5. The above warmed with strong vinegar.

6. Galen's Wax: white wax and oil of roses. This warms what is cool and cools what is warm on the surface. It should not be

used at night. By repeatedly immersing the paste in cold water or pouring it on snow it acquires repercussive properties against bilious affections and against priapism.

The following are especially good for painful joints:

7. Equal parts of wheat, or other grains, and camphor ground together with a little rose water.
8. 10z. white bread crumbs, 1/2 oz opium, 1/2. Moisten with cow's milk (286).
9. 1oz. oil of roses, 1/2 oz. wax (493) Moisten with rose-water and add 1 dr. saffron (398), and 1/2 dr. opium.
10. Lanolin sprinkled with tepid vinegar.
11. Vinegar containing a decoction of red rose petals.
12. Wheaten flour mixed with morel and a small amount of vinegar.

III. Five Explanatory Notes

1. Do not look askance at my citing so many remedies with the same effects. Be aware that many medicines affect patients differently and may affect the same patient differently at different times. Furthermore, any of these medicaments may not be available for use at all times and some of them may be too expensive for poor patients 2. When a surgeon prolongs the use of repercussive topicals because he senses an over-abundance of morbid matter and the affected region turns dark, he must discontinue the applications of cold medicines and mix in some resolutives, as the juices of morel and coriander (132) and cabbage with barley flour, bean flour and the likes.

3. If the abscess, as above, resists the repercussives alone and the matter is not dispelled and the darkened surface does not pale or the mass remains unchanged, neither enlarging nor shrinking, and the local heat and burning pains persist, you must continue with the mixture of resolutives and repercussives until the signs clear before you go on to resolutives alone. However, even then continue to put repercussives around the inflamed region and especially over the part that is closest to the primary source of the bad humors.

4. Many surgeons insist that we should not use repercussives and resolutives in inflammations (ie before they suppurate) and that we should encourage the formation of pus and then drain it. They claim that is the better way to rid the body of harmful humors. However, I say that the surgeon who promotes the accumulation of pus does harm. On the other hand, if one has common sense and uses repercussives to dispel the bad matter and aborts the development of an abscess, which will be an added burden for an already sick patient, that surgeon may be subjected to calumny from the same critical persons who accused him of prolonging the treatments for his own personal gain. What should one do? If he does what he thinks is right he may not escape blame. If he cannot find a middle way, willy-nilly, what then? I say we should allow the patient to choose; then treat him accordingly. If the patient cannot make up his own mind and tells the surgeon to choose the treatment which he thinks is best, which, I say, will avoid suppuration, then he will do as I instructed in Notable 2 at the beginning of Ch.3 of Doct. I, Treatise II, which is titled "The Treatment of Head-Wounds With Fractures of The Cranium". Do as I wrote in the paragraph where I pose the question "What treatment is better for those and similar wounds, when there is a choice: to avoid suppuration or to promote it?"

5. Inasmuch as we use opium and other stupefacients in some of our repercussives, take note that they are cool by virtue of their stringency, whether they are taken internally or are used as topicals. When they are taken internally in small amounts and are well diluted by other simples, they are sedatives (ie calmants) and narcotics and they dull the patient's senses. If you deliver too large a dose of the undiluted oral medicines along with them in the repercussives, you may kill the patient.

The same effects are seen when the (ie opium containing) repercussives are used as topicals. A small amount will numb the senses but a large dose will permanently destroy the complexion of the limb, in the same way that one may say that a dead man can feel no pain.

The narcotic medicines are opium, mandragore (263), belladonna (217), all varieties of hyoscyamus (232) except the white and all species of poppies (337) except the white. The following medicines are effective when they are dry, especially the bark and the roots of mandragore and the seeds of the white poppy. The seeds of the black poppy are less potent. One should be cautious with the strong narcotics and use only small doses of the attenuated substances and never give them to a terminally ill patient.

CHAPTER 3.

THE RESOLUTIVES

I. Introduction

A. Where cited in Ch. 2, Avicenna and Serapion also agreed that resolutives rarefy and evaporate the accumulated humors from the site from which they escape, bit-by-bit until nothing is left. They heat, they dilute and they dilate but they do not desiccate. By their heat they evaporate the diluted humors; as diluents they reduce them to vapor; as dilators they open the pores to allow the vapors to escape. They are not desiccants lest they foil the actions of wet substances which soften the matter to be dissolved. The definitions offered by those authors are easily understood.

B. There are two proper occasions to use resolutives: 1. When we have not used repercussives for any of the reasons we have stated in the Nineteen Notables in Ch. 1 of this Tr. V. 2. When we have been unsuccessful with our repercussives because the body has been too full of humors to be able to accept more of them. We failed when we tried to dispel them from the affected region, or the humors resisted dispersement, or because the inflammation was caused by an internal derangement rather than by a wound, etc. We must not use resolutives before we have brought about an adequate purgation. We taught that in our Rule 4, in the chapter titled "The Treatment of Aposthems In General".

The resolutives resemble maturatives in two ways and they differ in one at least, as I will show here. They resemble each other in their analgesic and soothing effects, and, at times, they may substitute for one another. For example, the resolutive may provoke maturation (ie bring to a head) when the accumulation of bad humors is too large to be accomodated by the 'escape-pores' of the overlying skin. Or a maturative may cause small amounts of thin pus to be absorbed rather than come to a head.[262] That explains the double actions of compounds such as the diachylon ointment.

The resolutives differ from maturatives in that they are warm diluents and dilators, whereas the maturatives, although warm, act to thicken (ie make viscous) the matter which then will block the escape-channels (ie opilate).

From the foregoing you may conclude 1. that resolutives may have the same effects or reciprocal effects as maturatives; they may resolve the thin portions of certain accumulations and thicken the residue; or, 2. They may resolve the thin portions and cause what is left to suppurate.

II. The Resolutive Simples:

Camomille (76). This is the only herb that deserves being called noble, because it never attracts more than it resolves; melilot (277), pellitory (334), wild or white mauves (273), fumitory (191), bindweed (416), dill leaf and seeds (26), cabbage leaves and seeds (71), nettles (305), spikenard (437), bugloss (69), borage (63a), elder (165), dwarf elder (406), valerian (475), parsley (344), wild celery (29), fennel (180), fava beans (178), chick-peas (112), vetch (483), crumbs of dense breads (65), fats of geese, ducks, hens, pigs (206), bone-marrow of various animals (275), mastic (272), frankincense (188), myrrh (299), hore-hound gum (23), galbanum (194), sagapenum (399), opoponax (321), gums of ladanum (234), hyssop (226), and terebinth (459), wax and lees of wax (492), butter (70), and others.

The Compounds

All of the simples may appear as compounds: oils, ointments, plasters, cataplasmas, poultices and fomentations (ie as defined in Chapter 1.)

---

[262]   The maturative may act as a resolutive (LDR).

Six Resolutive Oils.

1. Oil of dill crushed in oil of ripe olives.
2. Oil of camomille flowers prepared with oil of ripe olives.
3. Oil of costmary (135): Take 1 oz. of costus; 1/3 oz. each of pepper, pyrethrum and euphorbium; 12 oz. of beaver musk (96). Grind them and filter. Mix the product with the oil of lily or spikenard. This oil dispels cool humors and restores nerves that have been chilled.
4. Oil of large camomilles: Take 2 oz. each of dried fresh flowers, fenugreek (181) and flax seeds (252). Place them in 20 oz. of oil of ripe olives (317) in a glass jar exposed to sunlight or placed over a low flame or in a dry well or buried in the ground. This oil warms and relieves pains caused by cold humors.
5. Oil of lilies: Take 2 oz. of oil, 30 lily blossoms (248) from which you have removed the yellow stamens, 1 oz. each of cassia bark (94), mastic, carpobalsamon (49) and saffron (398). Grind all (excepting the lilies) and place them in a glass jar in the shade. The lilies are isolated in a cloth sac immersed in the jar. Remove them after one month lest they rot and pollute the oil. This prescription warms and relieves the pain caused by the cold humors which it dispels, especially in the kidneys and the uterus. It does not incite inflammation.
6. Oil of mastic: Use 3 pounds of oil of ripe olives containing 6 dr. of mastic. Heat it in a double boiler until the mastic dissolves. This oil energizes and is effective against inflammation near the stomach, liver and spleen, and when there is pain and disordered function caused by cold humors.

Five Resolutive Ointments

These may be prepared with one of the oils or with powders (see below) by adding wax as directed.

1. Add 1/2 oz. of wax, 1/2 oz. of seeds of mauve and guimauve (273), 1/2 oz. each of fenugreek flour(181) and flaxseed (252) to 3 oz. of oil of camomille and bring to a boil. This

ointment will resolve and mature without attracting humors to the site.

2. Add 1/2 oz. of wax and 1/2 oz. each of mauve and guimauve seeds to 3 oz. of oil of lilies. This ointment resolves and matures warm matter.

3. Add 1/2 oz. each of bdellium (53) and serapinum (429), 2 oz. of terebinth and dissolve them in vinegar as you add the last of the terebinth. Then add all to 3 oz. of oil of lilies. This ointment will resolve cool inflammations.

4. Take 6 dr. of oil of camomille or dill and add 2 dr. of wax, 2 oz. each of duck and chicken fat, 2 dr. each of dill seeds and camomille flowers. Use as above.

5. The oil of hysop (lanolin) really is not an ointment or a plaster, but it is something between. Put sheep fat from the midriff of ewes (206) in a pan and spread rain-water to cover it completely. Set it out during a full day and night. Then simmer it over a low flame and then cool it. Filter it though a cloth and then gently bring it to a boil in a pewter dish, stirring it with a spatula until it thickens. This will resolve like an ointment and can be used as an oil-base in preparing other ointments of all classes.

Three Resolutive Plasters

1. Rhazes' diachylon: take 5 oz. of oil from ripe olives, 5 oz. of finely ground litharge (253), 1 oz. of a mucilage made from the seeds of greek fennel and flax, 2 oz. of a mucilage of guimauve. Prepare as above. If you treat scrofules, add 1 oz. of powdered dry orris root (228). It will not always resolve cystic scrofules, but when used without the orris-root it will bring carbuncles to a head.

2. Mesûe's diachylon: Take 12 oz. of powdered and sifted litharge, 8 oz. each of oils of camomille, dill and iris (orris), 12 oz. each of mucilage of althea (18), 12 oz. each of fenugreek, flax-seeds, plump dry grapes (208), seedless raisins (473), orris-juice (ie of roots), squills (443), moist hyssop, gelatin made from cow-hide (124), 3 oz. of terebinth, 2 oz. each of white resin (379

and yellow wax. Heat all until it is midway thick between plaster and ointment. This topical perfectly resolves cool matter and softens the firm stuff (ie inspissated pus).

3. The ordinary diachylon described by Nicolas: Take 4 oz. of old olive oil, 36 oz. of flowers of silver (186, 430b), 12 oz. each of wild mauve and guimauve, fenugreek and flax-seed. Make 1 lb. of mucilage from the foregoing and use it to make a plaster as I have described in the preface to this antidotary. Use it to resolve furuncles and small hot aposthems if their matter is at a suitably early stage. If not, it will mature it and still later (ie after drainage) it will act to digest, cleanse and remove the matter. It will regenerate (ie granulation tissue) and cicatrize (ie close the wound). Alone it is a sufficient treatment from beginning to end. You can use it to treat large hot abscesses after you have opened them and they need additional deterging to remove what remains of the diseased matter.

This plaster also may be used to treat cool abscesses, in which cases you must reapply it twice a day. After draining and deterging, freshen the wound—surface by scraping with your thumb-nail before you apply the plaster.You will find it useful in treating painful chronic swollen joints and painful intestines. Mixed with oil of mastic it helps to cure incisions where nerves have been cut and they contract (ie convulse, twitch); the plaster favors regrowth. And, finally, it is good in treating wounds in decrepit patients.

Four Resolutive Cataplasms

1. Take 2 oz. each of camomille flowers and dill seeds; 3 oz. each of flours ground from fenugreek, flax-seed and barley; 1 oz. each of oils of dill and camomille. Boil all in water and mash them. Before applying the cataplasm you should evacuate the patient and use repercussives. The cataplasm resolves inflammation and gets rid of hot humors, and it especially readies the lesions at their firm stage before you use the maturatives.

2.   Take 1 small clump each of wild mauve, cabbage leaves and camomille flowers. Boil them in water and mash them in the broth. Add 1 part of dill-seeds and cabbage seeds and 2 parts of powdered sulfur.

3.   Take 2 oz. each of seeds of fennel, anise and dill, 1 oz. of flour of lupin, 3 oz. each of fenugreek and flax-seeds, 1 oz. of oil of lilies. Boil all in water and mash all of them in the broth and add some more of the oil and a bit of vinegar.

4.   I have described my own cataplasm and plaster of mauve in part 7 of Ch. 1 of Doct. I of Treatise II, where I dealt with the treatment of warm aposthems. In Notable 5 preceding that section I described its many advantages.

Resolutive Poultices

These were used by the Ancients to act as resolutives in inflamed wounds. However, I think they are too gentle and that they promote suppuration. Indeed, that is precisely why the Ancients preferred them Their recipe: Put enough wheaten flour in 4 parts of water to which you have added 1 part of oil. Boil them until you make a soft paste which you apply warm. (ie *see item 5 in Chapter 4 below*)

Resolutive Fomentations

Resolutive fomentations are made as broths of the simples we have named. They are used just before you apply any of the other topicals.

III. Four Explanations:

1.   When you are preparing a resolutive topical and you make a broth of a simple medicament, set aside some of it to use as a fomentation and use it warm until the affected part reddens and swells. Then apply the complete prescription.

2.   After the application has dissipated the liquid portion of the bad matter and what remains is thick and firm, you should

alternate with emollients and resolutives until you achieve your goal; the emollient will finish the job.

3. When you fail to resolve all the bad matter because it is too dense or too viscous or there is too much of it and it shows signs of suppurating, go directly to the maturatives (see below). The surgeon should always follow Nature's lead.

4. The resolutives, maturatives and mondificatives (ie detergents) should not be too harsh lest they injure the affected regions and by their violence cause pain and attract more of the humors from the rest of thebody into the affected part.

CHAPTER 4.

THE MATURATIVES

As in the preceding chapters, here are three sections.

I. Introduction

Avicenna (ibid.) wrote that maturatives bring about the digestion of and the uniform heating of the humors, and keep them in situ until they are completely broken down, not separating the wet from the dry or the thin from the thick. Galen (*Simple Medicaments*) wrote that the aperitive qualities,[263] of their own natural heat, penetrates the matter without desiccating it or changing its basic substance. Serapion (ibid.) said the same, and he added that when treating a naturally hot region you should apply medications which are even hotter, thereby to provide heat in excess of what the body as a whole or the local region can generate. Therefore, use the maturatives suited to the case. And, since the complexion of the humors is naturally warm and moist, our maturatives should be more intense than a warm hand or another warm part placed over the diseased region, even if it were possible to hold it there for a long time.

---

[263] The term describes the action of opening pores and channels as well as laxative actions (LDR).

II. The Maturative Substances

When a surgeon has not used repercussives and resolutives, or he has not succeeded with them, and the aposthem shows signs of maturing (turns red and does not shrink, and the pain increases and the fever climbs) then he should apply maturative topicals, simples or compounds.

The Simples: Garden-grown mauve and guimauve (273), acanthus (brancus ursinus)(64), bryony Roots (67), patiens (336), bread crumbs (65), seeds of flax (252) and fenugreek (181),dry figs (185), wheaten flour (187), dry raisins (473), egg-yolks (163), onions (319), garlic (198), barley meal (323), yeast (502), fat of pigs and young hens (206), butter (70), olive-oil (317), bone-marrow (275), lily-roots (248), honey (220), terebinth (459), and all the resolutives mixed with viscous maturatives to attain a viscous consistency. Take note that some of the entirely cool substances by chance may act as maturatives. They push the already-liquified matter on the surface of the indurated mass into its center where the heat will bring about the maturation of the remainder, whereas if it remains on the surface, it would lose its heat.

Maturative compounds: Six Warm and Thin Ointments:

1. 1/2 lb. Of Guimauve-roots with 3 oz. of lard or butter.
2. 1/2 lb. Of guimauve-roots; 1/2/oz. each of flours of flax-seeds and fenugreek-seeds, rose-honey (385); terebinth.
3. 3 onions and two eggs, both cooked over charcoal. Peel the onions and shell the eggs. Mash them and mix in about 1/2 that volume of butter or lard.
4. The surgeon's poultice: Mix 5 parts of water and 1 part of oil with enough wheaten flour (494). Heat it until it has the consistency of a paste. This poultice will mature inflamed wounds and other abscesses caused by insidious humors, such as bile. In the cases where the matter is already thick (ie an indurated inflammation) you should apply strong penetrating medicines. They also serve to relieve the pain in the flanks (ie kidney regions) and over the pubes caused by stones. Their sedative actions will settle the imbalances of the complexions

of the suffering regions, and the warmth of the ointment serves well both when the complexion is warm or cool.

5. Equal amounts of honey and butter, thickened with a maturative flour, such as ground from seeds of fenugreek and flax.

6. 2 oz. each of leaves of mauve, pellitory (334), acanthus, scammony (416) and henbane (217). Heat them in some water and then squeeze out the water and mash the cooked leaves and add back 2 oz. of the decoction. Add 1 oz. of lard and some barley-meal.

Five Maturatives for treating hot, thick and violent aposthems.

1. 6 plump dry figs, 1 oz of dry raisins, 1 clean whole clove of garlic, 12 seeds of peppers (34), 2 dr. of salt (404). Grind and add 2 oz. each of old oil and vinegar and an amount of active yeast [264] equal to half the total weight of all the rest.

2. For use similar to 1, when the inflammation is not so indurated: honey and oil thickened with barley-meal.

3. Equal parts of yeast, woman's breast milk (286), and hard-boiled egg-yolks.

4. The poultice in 3 with added equal amounts of gum of horehound (23).

5. Add to number 4 about 1/4 part of pigeon droppings (179) and a little borax (63).This fortified poultice will rapidly bring an abscess to a head.

Six maturatives for cold aposthems.

1. 1/2 lb. each of grilled and peeled onions and garlic, 5 hard-boiled egg-yolks, 6 oz. of clean roots of guimauve, 4 oz. of pork lard. Mix all.

---

[264] Yeast was not a pure culture. The apothecary, the surgeon and the baker had leavened dough which they kept alive by renewing it frequently. The medieval prescriptions which contain yeast measured out a sizeable clump of day-old yeasty dough (LDR).

2. 1/2 lb. each of peeled garlic and of terebinth, 2 oz. of oil of spikenard (437), thickened with flour of fenugreek.

3. 1/2 lb each of honey and resin (379) thickened with 1/2 oz. each of ground olibanum (316) and fenugreek. Add wheaten flour or mill-dust (177).

4. The Greater Basilicon Ointment: equal parts of wax (493), resin, pitch (352), beef-fat or butter and oil. This will soften indurated lesions and heal malignant ulcers.

5. The Lesser Basilicon Ointment: lacks the fatty materials.

6. Cooked and peeled garlic cloves and leaves of absinth (1). Boil in a small amount of water. Decant and mash the solids with pork-lard. This compound is a penetrating maturative.

III. Seven Explanations

A. Here we have given all the maturative plasters according to their various forms (ointments etc.) for ease of classification.

B. If the surgeon feels that his medication will dry faster than he anticipated and he cannot remain in attendance with his patient (ie to renew it), he should spread his topicals on some leaves and apply them. In that way the medicines will work less rapidly than when they are applied on cloth. The thick stiff leaves (ie probably cabbage) will block the exhalation of the vapors which come from the plaster as well as the wound. To the contrary, they will hold them where they can continually reinforce the moisture of the maturatives.

C. When you treat with simple or compounded medicines, first foment the wound with the decoctions obtained during their preparations. Foment the region until it reddens before laying on the maturatives.

D. Apply all maturatives as hot as can be tolerated by the patient.

E. When you use the roots of althea (18) or guimauve prepare them as follows: Wash them until they lose their roughness. Mince them fine and boil them. Strain (ie clarify) and grind them with care and make madeleines (ie small patties) which you may store for use when desired.

F. Never apply a maturative containing fenugreek on an angry inflammation. Experience has shown how it excites and inflames.

G. When you wish to provoke the suppuration in an aposthem caused by hot and thick humors or when the matter is a mixture of hot and cold stuff, always mix your maturatives 2 parts to 1 with yeast which is only moderately warm. The more moist the yeast, beyond the second degree, the less nitrous or salty it is. When it is less than the second degree it is more salty.

Heed the words of Serapion et al.: leaven has both warm and cool properties (ie virtues). The coolness derives from its sourness; its warmth from the flour and salt (ie when in a dough). That is why it dilutes and resolves by attracting the moisture from the depths and it does not interfere with the treatment.

CHAPTER 5.

MONDIFICATIVES (IE DETERGENTS)

I. Introduction in Four Parts

Avicenna (ibid.) and Serapion (*Agregations*) agree in claiming that there are twenty types of mondificatives, and they both list them: abstersives, lavatives, attractives, etc. In their opinion, a medicament which removes any superfluous or abnormal material from the body or a part of it, either from its surface or from within, sometimes one and then the other, that medicament is a mondificative. Those which attract from within the body are clysters, electuaries, syrups etc. Those which remove from the surface vary much more as to their natures and their virtues and their modes of action. Those which have both internal and external actions are honey, barley-water and the like, and all the compounds which contain them. Setting aside the others, we will devote ourselves to the surface-acting mondificatives. There are four types. 1. The weak ones which cleanse little or incompletely. They are honey and its compounds, celery-juice (4), absinthe (1), wheaten flour (187).

2. The second types contain strong detergents which are active. The items in 1. are fortified and improved by the addition of sharply acting substances as sarcocolla (410), myrrh (299), aloes (19), and lupin (259). We use the terms 'weak' and 'strong' in this chapter to denote the substances in the two preceding sentences. 3. The third group includes the very strong detergents which both mondify and corrode, such as asphodel (40), verdigris—verdarain (481), orpiment (327), all kinds of salt (404), alum (20), couperose (106) and niter (402) and all the varieties of ashes (117) and corrosive plants such as the tithimals (173). 4. The fourth category includes the exceedingly potent detergents such as sublimated arsenic (34), quick-lime (251), realgar (380), soap (430a), roots of arum (35a), clematis (120), aconite (4a), ranunculus (376), euphorbias (173), spurges (442), cantharides (80), the two hellebores (164), laureole (239), etc. They eat away the skin and are called eruptors. In this chapter we will use the terms 'very strong' and 'extremely' strong for items in the third and fourth groups.

II The Methods of Use

I need not repeat the instructions in Ch. 10, Doct. I, Treat. II (The Use of Certain Remedies for Treating Wounds). The use of these remedies is explained in the Seventh Notable preceding that Treatise, at least their employment in wounds. Elsewhere we deal with the uses of the very strong corrosives. You also will find there a discussion of the similarities and differences among these medications and others used in treating wounds—where and how to apply them and the frequencies of applications. There, also, we deal with detergents in special cases, and I deem it not necessary to repeat it here. Furthermore, in the Eighth Notable we give the names of the authorities, the titles of their books and the chapters where the medications are described.

C. More on Nomenclature

No author or practitioner and no surgical handbooks (ie practica) uses only the one general term, *Mondificatives*, to designate all surgical detergents, because they all are not of the same sort. The may be detergents (abstersives), lavatives, purifiers, etc. It is as if we could use the word animal as the sole name for man or bull or a donkey, or use a general term to designate a particuilar individual. It is as Porphyry said, "Happy are the species, happy are the genera because they are abstractions and have no reality beyond our spirit, which can conceive them."[265]

Surgeons rarely use simples for their detergents, preferring compounds. I consider it of no value to list all of them We will limit ourselves to the commonly used detergent compounds we will emphasize that the surgeon who uses detergents must consider the region where they will be used and the nature of the impure matter that he must cleanse. When the matter is thick he should use detergents which will thin it. When the matter is thin his detergents should thicken it. When the affected region happens to be too humid, the desiccants should predominate. When the matter is viscous or penetrating, the surgeon should depend on simples. If it needs to be broken down, he should use maturatives. If it is sticky, use lavatives or lubricants or other simple detergents. Etc. Etc. For every sort of impurity one should use a mondificative with opposing properties. Contraries should be treated with contraries.

In the chapters we have cited, Avicenna listed simple mondificatives in the categories state above. Galen did the same in only one chapter in Vol. I, proposition 3 of his *Simple Medicines*.

---

[265] Nicaise commented that Porphyry, a 3ʳᵈ C Syrian philosopher, was famous for his *Introduction to Aristotle's Categories,* and for developing a philosophy dealing with genera and species. Albertus Magnus and Thomas Aquinas and all the medieval philosophers who wrote commentaries on Aristotle cited Porphyry's dissertation which appeared in Chapter V of his book.

Nevertheless, we add that Porphyry is not well served in Mondeville's statement about the language of images to which he lays less than a faithful claim. Indeed, Porphyry did not completely adopt Aristotle's views. He was indefinite in his commitments to Plato versus Aristotle, and he definitely was more inclined to Plato, who considered the particular to be a consequence of the general (LDR).

D. Miscelany

In all wounds in vital regions where there has been some loss of tissue

Two superfluous collections interfere with and delay healing. One is a solid kind of granulation tissue which corrosives can obliterate. The other is a liquid of two kinds: One is fatty (ie probably liquefied subcutaneous fat) and the abstersive detergents can void it. The other is thin (ie probably edema) and is dried by desiccative detergents. The liquid superfluidities may also appear within the body. The thin kind is carried off in sweat while the fatty stuff appears at the surface in the forms of scabies or tumors.[266]

II A List of Mondificative Compounds

A. Ablutions (washes): The weak ones are not really abstersive because they consist of water, wine (496), whey (494a), barley-water (323), decoctions, juices etc. for the same purpose.[267] They may be fortified with added strong cleansing agents such as fava beans (178), lupin (259), myrrh (299) etc. You may even use some diluted potent abstersives as alum, aloes, sarcocolla, lye (254) and urine (472a).

Here are four abstersive ablutions: 1. Warm water with sugar alum for treating infections of ulcerated emunctories.[268] 2. For aposthems which are cool, use urine instead of water. 3. A decoction of sedge (434) to treat ulcers which produce little drainage. 4. Vinegar with some ashes of willow-tree-bark (495) to treat dry indurated ulcers.

B. Here are ten poultices or quasi-poultices:

---

[266] Probably seborrheic scales or scurf and sebaceous cysts (LDR).

[267] These were used simply as douches, ablutions or baths (LDR).

[268] Openings which surgeons believed served for drainage, especially from the brain, as the nares, otic canals, etc. The term also is used to describe lymphnode bearing regions, as the neck, the axillae and the groins because lymph fistulas were thought to drain particular organs, as the groin nodes were special for the liver. (LDR).

1. 3 oz. of thin rosat (385), 1 oz. each of flour of wheat (187), barley, rye (395a) or oats (310). Use pure water and bring all to a boil while stirring. This is good for fresh wounds which already have pus.

2. 1/2 lb. of thin rosat (385), 2 oz. each of flours of fenugreek (181) and barley. This is good for the cases in 1, and when wounds are inflamed.

3. No.1 with some washed terebinth (459), for use when wounds of nerves contain pus.

4. 1/2 lb. each of resin (379), terebinth and honey; 1 oz each of myrrh, sarcocolla, fenugreek and flax-seeds. Roast the items which can be melted and add the powders to the melt. This will cleanse nerves and abscesses (ie after they have drained).

5. The recipe for the Surgeon's Poultice was given in the chapter on maturatives.

6. Partly cooked egg-yolks with wheaten flour applied without preliminary fomentations are useful in the same cases as in 4.

7. 1/2 lb. of fresh honey mixed with 3 oz. of wheaten flour and 1 oz. of celery-juice. Thicken them over a low flame and stir for a long time as they cool. This is good in abscesses before suppuration (ie maturation) is complete. Also, it is useful after some abscesses have been opened, as for anthrax, carbuncles and some venomous pustules. The poultice will mature the remaining solid material because itself it is warm and viscous.

8. When one sees that a fistula or cancer may develop from an ulcer, replace the celery-juice in 7 with juice of absinthe.

9. If you fear that the matter will overheat and become venomous, replace the above with juice of plantain (353). Although the absinthe cleanses better than the celery, it is more irritating (incisive), whereas the plantain cools and rectifies the poisons.

10. 1/2 lb.of honey, 4 oz. of wheaten flour, 1 dr. each of the flours of fenugreek, vetch (489), lupin and myrrh, 1/2 lb of juice of absinth. Grind the solids and make a powder, added to the honey. Heat it slowly; when it thickens add 4 oz of washed terebinth. Mix thoroughly. This poultice cleanses putrid ulcers and produces a thick and viscous pus.

C. Five Mondificatives Between Douches and Poultices in Consistency

1. 2 parts each of rosat mixed with 1 part of oil of roses (388). This will cleanse the dura-mater when it turns dark in a wound.

2. 2 parts each of juices of absinth and celandine (98) mixed with 1 part each of wine and vinegar. Thicken it with myrrh and aloes. Two to four drops applied on an ulcer will prevent it from becoming a fistula.

3. For the same purpose: mix equal parts of honey with feather-alum.

4. The Common Mondificative: honey thickened with powdered sarcocolla, aloes and myrrh.

5. Honey thickened with a little wheaten flour. No heating is needed.

III. Three Explanations

A. Instructions for using the detergents: The ablutions (douches) will be poured on the ulcers, or when necessary injected with a syringe (ie injectorio). The poultices are spread on a cloth or on leaves and applied after fomenting the region. Furthermore, in some cases, the detergents may be taken internally or be used both as topicals and potions.

B. To prepare rosat: Take 6 lb. of honey and 1 lb. of rose petals. Heat them in a double-boiler until the honey thickens. Add more honey to make it fluid.

To wash terebinth: shake it with cold water as you do when you prepare egg-whites. Let it settle and then decant the oil and add more water, and shake. Repeat, a third time. When it is truly washed, the terebinth is white and it is free of irritating contaminants.

C. How to wash oils: Use the same method as for the terebinth. An alternate technique: Use a jar that is shaped like a gourd. Drill a hole in the bottom with the point of a stylet. Then cover the hole and fill the jar with equal amounts of oil and water. Shake. Set it on end until the oil rises. Open the hole and let out the water. Refill the jar with fresh water and repeat the maneuver as often as necessary to purify the oil.

You will do well to deterge all your rosat, terebinth and oils.

Chapter 6.

Regeneratives, Incarnatives and Cicatrizers[269]

I. Two Preliminaries

Four Generalities for all the three categories:

1. Galen (*Megatechni*, Vol. III, Ch.1 and *De Ingenio*, Vol. III,Ch. 1, and *SimpleMedicaments*, Vol. I and Avicenna (*Canon*, Bk. IV, fol. 4 doct. 2) and Serapion (*Agregations*, ibid) explained at length the nature and the complexions (ie see Ch. 8, fn. 31) of these medicaments, and their modes of action. We conclude that all of them are drying agents, different only in how much they do. The Regeneratives which promote new tissue are desiccants in the first degree.[270] Their weak drying ability is sufficient to convert blood into granulation tissue, which simply is clotted and slightly condensed blood. A greater degree of 'siccicity' will completely dispel all the wet adhesive properties of blood.

---

[269] The uneven and occasionally confusing presentations in Chapters 6, 7 and 8 may be signs of Mondeville's haste to complete his work while being wasted by his terminal illness. He offers explanations elsewhere in the text that we may take as apologies (LDR).

[270] Nicaise's explanation: The drying action of a medicament was graded in the same way as was its heating or cooling or moistening effects. For the sake of brevity we speak of heat of the 2nd degree or cold of the 3rd degree, etc, as we here say dry of the 1st degree (N).

Incarnative medicines also are called conglutinatives and aggregatives; they are dry to the 2nd degree. Their drying property is not sufficient for regeneration of proud flesh for three reasons: a. Weak drying capacity will not penetrate deeply enough into a wound which already is united and is dry. b. The primary substance in incarnatives will not permeate beyond the wound. c. It will not dry fresh blood coming into a wound and it lacks the power to incarnify the surface of a wound.

The cicatrizers, also called sealants, consolidatives and generatives of skin and callosities, are dryer than the others, approaching the 3rd degree. They transform the moist surfaces of a wound beyond the capacity of the medications in the other two categories. They are styptic but not abstersive, and their action should not be too vigorous.

2. Although we have described here and in other places the amounts of oils, waxes, powders etc. used in these medicines, you should understand that the amounts are not invariant. Take, for example, cases where wounds or ulcers are too humid (ie edematous) or when they are situated in a region or in a person already dehyrated. In the first you will increase the amounts of powders and decrease the oils and waxes. But, if the wound or ulcer is too dry and the region is indurated or edematous you will decrease the powders and increase the oils and waxes.

3. Cooperative Actions: We discuss the three categories of medicaments together in this chapter because we may use one or more of them at the same episode of treatment. It may happen at the outset or be seen later on, that, by chance, a mondificative will do more than cleanse, and that it will promote the formation of granulation tissue. Similarly, a consolidative will act beyond its primary function and will also mondify and regenerate. Sometimes two medicaments applied together will produce a single effect. An example: A conglutinative and a light corrosive mixed together may have a consolidative action. You must have a master's abilities to create such combinations, because you must know the particular qualities of the medicines and you must have tried them and you must be familiar with the complexions of the person and of the part to be treated.

4. The question of new skin: Here we will investigate whether new skin will replace that which is lost in a wounding. According to Avicenna (Bk. II, Ch. 4, On the 'Functions of Single Medicines'), where he described consolidatives and sealants, a kind of 'bark' can form on the surface of a wound which will protect it from further damage until new natural skin grows over it. Galen (*Simple Medicaments*, Bk.V, Doct. 4, Ch. 2) wrote that skin was a kind of dense tissue, and since soft tissues can become hard, skin can be regenerated. Serapion (*Agregations*) agreed. All the other authorities, in many places in their books, disagree. They state that once destroyed, skin cannot be regenerated; it will be replaced by a hard and hairless callosity, as seen in scars, even in wounds in hair-bearing regions, even when there has been no loss of skin. I dealt with this question in Notable VIII preceding the chapters on the treatment of wounds. I said there that the term 'Natural Skin' has two meanings. It may be the skin with which a person was born, or it may be that which Nature causes to regenerate without a surgeon's intervention. The classical authors refer to the first sense of the term whereas the other authorities use the second, skin repaired by Nature, derived from hardened flesh (ie scar).

Each of the three categories of medicaments in this chapter has three species.

I. Regeneratives:1. As we stated in Notable IV preceding Tr. II, Doct. 1, Ch.10, where we indicated the nature and degrees of potency and the modes of action of the medicines, all of them have detergent properties. We also said that they must be desiccative in the same way, and they have their own cleansing action. In Notable IX we wrote that neither the regeneratives nor any other topical should remain in place long enough to alter the substance of the underlying organs. They should favor the absorption of the blood coming into the wound while cooling the excess heat (ie blood-borne). Finally, we stated that mondificatives should precede the application of regeneratives. 2. Toward the middle of Ch. 10 (ibid.) we discussed what these medicaments

are and how they act, how often to use them and how they should
be used.

II. Incarnatives: 1. Refer to the last citation above. We explained
there that when you treat fresh and uncomplicated wounds in which
there has been no loss of tissue, you need use only incarnatives which
are dry in the 2nd degree. We also described how they act, and we
listed some of them, such as wine, as having been well tested in use by
modern surgeons, and some powders, dryer than wine, as used by the
Ancients. In their method different kinds of pus were formed whereas
in our modern method no wet superfluidities (ie pus!) are formed.
2. I agree that the writers of Medical Handbooks have described the
methods of the ancient surgeons in their treatment of simple wounds
(ie no lost tissue), and this is all I agree with: the incarnatives used in
their methods must be dry in the 2nd degree, must be slightly styptic
and not at all abstersive.[271]

III. Cicatrizers and Consolidatives: Three Preliminaries: 1. Refer to
Notable III (ibid.) There we explained that in fresh wounds without
loss of tissue, regeneratives and cicatrizers are all you need. But if pus
already has arrived you should use detergents. Further, In Notable V we
explained how to use consolidatives which should be dry in the 3rd
degree and left it in place until a firm callosity forms. 2. Refer to Tr. II,
Doct. I, Ch. 1, Part VIII where we discuss scarring in all kinds of wounds:
in those where the wound edges are securely united; in those where
that is lacking; and, finally, where either from within or at the surface,
granulation tissue forms and extrudes between the wound-edges. We
describe how to obtain lovely scars and how to ameliorate the ugly ones.
In a marginal note we described where it is proper to apply medication,
their selection, the importance of precise timing—neither too soon
nor too late. 3. In Bk.V of *Simple Medicaments* Galen said that a consolidative
should be styptic and desiccative. The true consolidatives are balaustia
(pomegranate flowers) (47) and psidium (pomegranate peel) (364),
both of which favor the growth of granulation tissue as well as consolidate.
It follows, therefore, that after a good cleansing of wounds and ulcers,

---

[271] Mondeville defies the Medical Establishment, but grudgingly makes a small
bow. He calls a penny a penny, but he will not sell his soul for that price (LDR).

when the defect is not full of granulation tissues, do not delay in applying consolidatives until the scar forms, before the proud flesh bulges abovethe level of the surrounding skin.

II. The Regenerative Medications

First we will describe two categories: The Modern and The Ancient.

A. The Modern: Refer to Part III of Ch.1, Doct. I, Treatise II, to the preliminary Notables where we show that wine is the best local remedy for all simple wounds, and that it is the only one you need to use. We also discussed whether the wine should enter the wound, and we stated that we should avoid it in fresh wounds that are still bleeding, and we stated that it is not proper for painful suppurating wounds. Furthermore, we answer the question whether the wine should be warm or cool and we why we prefer warm wine for reasons given there. Wine soaked oakum pads are an excellent topical for all uncomplicated wounds. I will cite three passages from Galen: 1. (*De ingenio*, Bk.III, Ch. 4) There he wrote that wine is an excellent medicine for simple wounds when applied on the surface. 2. In *Megatechni*, Bk. IV, Ch. 1, he said it is good to irrigate wounds with wine. 3. In *De Ingenio*, Bk. IV, Ch. 7, he said that wine is harmful for patients with inflamed wounds, even when taken by mouth, but that is not so for patients with uncomplicated wounds.

Therefore, we confirm by our own experience that wine applied on a wound as well as taken by mouth is an excellent remedy for all uncomplicated wounds. I offer further proof of the external application in Notable I of Part III, Ch.1, Doct. I, Treat. II. A proof of its value when taken internally is given in Notable I of Part VI, where we discuss the diet for wounded patients. Besides these attestations we have the opinion expressed by the authors of surgical handbooks and of the practitioners who agree with the views of Galen (*Simple Medicaments*, Bk.V) that regeneratives should be dry in the 1st degree, incarnative in the 2nd degree and cicatrixatives in the 3rd degree. As Isaac stated (*Special Diets*), wine has three levels of dryness: fresh warm wine always is of 1st degree, two-to-four year-old wine should be of 2nd degree and four-to-seven year-old wine is of the 3rd degree. He added that warm wine is drier; the less warm it is the less dry it is. It follows that new wine regenerates, medium-old is incarnative and the very old is a cicatrizer. Use each of them as indicated.

B. The Regeneratives used by the Ancient Surgeons: They are of three types:

A. 1.Weak Simples: olibanum (316), mastic (272), myrrh (299), aloes (19), colophony (126), barley-meal (323), bean-flour (178), flour of fenugreek (181). They used them to treat the naturally humid parts of women and especially of children whose wounds usually are slightly moist. Because they are weakly desiccative, when they are used in wounds which are very wet or on parts of the body which are dry, they will generate very little granulation tissue.

2. Very strong regeneratives are more desiccative: aristolochia (33), cadmia (96a), tragacanth (466), its gum (466), vitriol (490), incinerated iron (228), flours of lupin (259), vetch (483) etc. Use these on the body over organs which naturally are dry and on wounds which are complicated by edema (ie humid). In Tr.II, Doct.I, Ch. 10, ('Certain Remedies') I wrote about the body and its parts, and on the wounds and their complexions in which we should use these medicaments. I explained for each case the deciding factors in the complexions of the patient as well as the medicaments, and I discussed the treatment of complications of wounds.

3. Here are some simple regeneratives which I commend from my own experience for treating deep ulcers: centaury (100), sarcocolla (410), incinerated lead (354), incinerated antimony (28), limails of incinerated iron (289), polium (200), collagen from fish (124).

4. Compound Regeneratives used by the Ancients :

a. Six powders: 1. 1 oz of frankincense (188), 2 dr. each of vernis (379), and flour of fenugreek. Place the mixture in a wound to regenerate flesh, excepting when one of the 17 caveats in wound care forbids its. 2. Equal parts of pounded frankincense, mastic, and fenugreek. This will regenerate flesh and relieve the stench of fetid ulcers. 3. 1 oz of frankincense, 2 oz each of mastic and flour of fenugreek. 4. 1 oz. of camphor (77), 3 oz. ceruse (105), 4 oz. litharge (253), 1 dr. sangdragon (408)—use only 1/2 dr. in warm weather when treating inflamed wounds. 5. 1/2 oz. of colophony, 4 dr. each of sarcocolla (410), frankincense, iris (228), aristolochia (33). This is for use in cool weather in patients who are cool and moist with cool moist wounds. 6. Capital Powder: as in Ch.5, Doct. I, Treat.II (Operations for Fractures of the Cranium).

Five Ointments: These usually are preferred over the powders because they are oily and smooth. The powders take longer to work. 1. The Nutritious Litharge (441) Ointment, as described Galen (*Catageni*) where he describes two similar ointments.: 1 part of ground and sieved ordinary litharge (253), 2 1/2 parts each of oil and vinegar. Simmer over a low flame for an entire day, stirring it until it has the consistency of an ointment. Even better: In summer time place it in the sun for 15 hours. Then stir it in a mortar every day and add a little wine. Let it dry overnight and repeat the process the following day. This ointment resists the influx of humors, it is a strong desiccant and it should be used to treat resistant ulcers as well as recent ulcers that bleed. It is good in fistulas which are not yet scarred. It is not corrosive. 2. Brown Ointment, as found in Nicolas' Antidotary. It gets rid of pus, it cleanses, dries and cures. 3. 2 oz. of litharge nutritum, 1 dr. each of frankincense, sarcocolla, galbanum (194), fenugreek and colophony.

4. 1/2 lb. of oil. 1/2 oz. each of waxes of thin frankincense and fenugreek; 1lb. of resin. Strain through a cloth. 5. Yellow Ointment: 3oz. resin, 1/2 oz. wax, 8 oz. oils, 1/2 oz. each of flour of fenugreek and frankincense. Filter. Use this around but not within wounds. It will regenerate flesh.

The Incarnatives

Simples: Leaves of date palm (150) and plantain (333) and cabbage (71) and pomegranate (364) and cypress-trees including the twigs (149), potentilla (465), honey, leaves of sorrel (363a) and pear—and apple—trees (363), leaves (249) of lily and bryony (67), mill-dust (177), burnt barley, flowers of sorb-trees (mountain ash) (189), equisetum (170), sour milk (286), etc.

Compounds are powders and ointments:

Four Powders: 1. 1 part crushed frankincense, 2 parts sangdragon,

3 parts quick-lime. 2 parts sarcocolla, 1 part each of aloes, sangdragon and balaustia; 1/2 part of olibanum, This powder is hemostatic as well as incarnative. 3. Equal amounts of crushed frankincense, aloes, and sarcocolla. This powder also regenerates. 4. Equal amounts of sangdragon, aloes, sarcocolla, mastic and couperose (106).

Two Ointments: 1. The Ointment of Master Anselm of Gênes: He gave this recipe to our most illustrious Prince and Ruler, Philip the Fair and Pious, of celebrated memory, who was king of France. 1 oz. of white wax, (4 oz. oil of wild roses (389). This is an excellent incarnative, regenerative and sealant. It subdues excessive heat. 2. Add powder of bedegar (rose-hips) (389) to Anselm's ointment. Cut the hips in half and discard the seeds and fibers. Grind and dry the remaining substance and mix it with the ointment until it turns gray and has the proper consistency. The thicker it is the better is its incarnative and sigillative effects.

Cicatrizers

Simples: Pine-tree bark (350), frankincense (188), balaustes(47), cuttle-fish bone (442a), tan bark(309a), bark of barberry bush (55), pomegranate peel (364), cypress nuts (149), curcuma (146), cadmia (96a), silver (430b and gold (43a), four kinds of alum (20), fig-leaves (185), dried turds of dogs that eat bones (179), incinerated and crushed and washed bronze flakes(14), centaury (100), wild tansy (454), juice and pulp of crushed leaves of ash-tree (264), bones (62), artemisia (35), ash-tree wood(189), large wild garance (197), earth-worms (258), both aristolochias (33), eleven types of incinerated vitriols (490). All simples whose innate heat or coolness do not exceed the 2nd degree and which are strong desiccants and weak styptics Compounds:

Seven P:owders: 1. Equal amounts of aloes, balaustes, cadmia, silver, incinerated ground and washed bronze. 2. A very strong powder (ie 3rd deg.): Equal amounts of aloes (19), curcuma, earth-worms, balaustes, myrrh (299), oak-galls (195). Pulverize all. 3. 4 oz. of dry bugloss (69); 2 oz. each of gum adraganth (486), mastic, sangdragon. This powder cossolidates old ulcers and old wounds without corroding them. 4. Equal amounts of pomegranate peel and flowers, powdered rotted oak wood (309a). This powder is good for edematous ulcers and wounds. 5. Equal amounts of sangdragon (408), mastic (272), gum arabic (212) and gum adraganth (466). 6. Equal amounts of litharge (253), pine-tree bark, lead filings (241), myrrh and galls. 7. 2 oz. each of aloes, olibanum (316), sangdragon, 1 oz. each of aristolochia (33), litharge, ceruse (105), pine-bark, and centaury, 1 oz. each of galls and balaustes. This is an excellent powder.

Seven cicatrizing ointments: 1. The Diaphoenicon or Palm Ointment. Galen described its composition and mode of use in *Catageni* where it is given first mention: Take 3 lbs. each of litharge and old oil and of calf-lard; 1 oz. each of couperose (136), and verdegris (482). Use a fine sieve for the litharge before adding the couperose. Remove the fibrous matter from the fat and then mash it, Heat it in a double-boiler. Then grind everything in a mortar before cooking it over a low flame while stirring vigorously with a recently cut (green) twig from a date—palm (150), frequently replacing the twigs as they dry, until the mix is what you want. If you don't cook too long you will have an ointment; if you continue beyond that, you will have a plaster. If green date-palm branchlets are not available, use roseau (386) or laurel (238). This Palm Ointment was praised by Galen and Mesûe for treating ulcers, salt-phlegmatic disorders, herpes, malignant ulcers, fistulas, abscesses of plague, and others, including anthrax, infected wounds, nerves, burns and many more.

2. Rhazes' White Ointment: 4 oz. of oil of roses, wax (1oz. in warm weather in warm regions and 1/2 oz. during cool weather), 1/2 oz of ceruse, 1dr. camphor (77), 2 oz egg-whites. Before you make the ointment, grind four almonds in a mortar, discard them and lightly rinse the bowl with water. Then grind the camphor and set it aside. If you mix it with the others at the outset, it will lose its essence by evaporation by the time you have finished. Add it to your mix a little at a time. Now to return to the recipe: First add the ceruse and the melted and mixed oil and wax. Finally add the egg-whites. Stir vigorously for a long time until you have a finished product. Some people add 3 dr. of litharge to make it more consolidative. I usually

add as much as possible until it becomes a paste. With it I have cured all cases of gangrenous ulcers who have also followed a proper diet and had good bandaging. The recipe as given requires the work of a master. I have used it also in treating cancers. The ointment without the litharge will cicatrize all wounds, warm ulcers during hot weather, burns of various degrees and excoriations of all sorts.

3. Linen Ointment: described by Avicenna and Mesuë. Shred a piece of fine linen cloth, or scrape it to obtain fibers. Take it as a 1/2 part with 2 parts of opoponax (321), 5 parts each of wine, honey and oil of roses or myrtle (300), 1/3 part each of litharge, aloes, sarcocolla (410) and myhrr. Add the liquids drop by drop in the order listed. This ointment regenerates and cures malignant ulcers, resistant wounds and non-callous fistulas. You can improve its cicatrizing effect by adding 1/4 part of vitriol.

4. Mesûe's Gold Ointment. 6 oz. of yellow wax (493), 2-1/2 lbs. oil of myrtle, 2 oz. terebinth (459), 1 1/2 oz. each of resin (379) and colophony (126), 1 oz. each of olibanum, mastic, saffron (398). This treats and consolidates wounds.

5. A topical which really is neither an ointment nor a plaster. Boil some resin in extremely strong vinegar (ie 4th deg.). Add cold water. Fish out the (ie precipitated) resin with your hands after coating them with oil of roses or myrtle. In warm regions and during warm weather add half as much white wax. When spread on a cloth and applied it will consolidate wounds.

6. 1lb. Wax, 1/4 lb each of pitch and resin, 1/2 oz verdegris (482), 1/2/oz terebinth. This is an excellent consolidative.

7. 3. oz. of oil of roses, 2 oz. resin, 1 oz. each of wax, cypress-nuts, mastic and colophony. This also is an excellent consolidative.

In addition, I cite a well known surgeon who claims he has obtained good healing by scar of wounds when some tissue has been lost or the skin-margins have been destroyed. He uses dog-feces (179) cooked with honey. We already have described a method of cicatrizing wounds and ulcers where tissue has been lost and where there is no suppuration or unhealthy drainage. I refer to Ch. 1, Doct. 1, Treat. II. There, too, we gave the recipe for a corrosive ointment.

III. Four Explanations

A. General: Earlier in this chapter I wrote that regeneratives should be desiccative of the 1st deg., incarnative of the 2^{nd}nd and cicatrizers of the 3rd. I questioned whether any single medicine could at once be all three. That would seem to be impossible since it would have to be dry in the 1st deg for all three categories. Nevertheless, some of the authorities accept the possibility of the coexistence of the three properties, depending on how one formulates it. But I will now admit that a single compound can fulfil three functions. Galen agreed (*Megatechni*), Bk. III, Ch.3). He wrote, "Nature is wonderful; one thing can help another that is like it; it supplements what is deficient and it reduces what is excessive, at least when the differences are not too wide." It seems that the differences between the 1st and 2nd degrees cannot be very great, nor between the 2nd and 3rd. So it seems that Nature can emend and compensate for narrow differences. Consequently, it is possible that a single medicament can fulfil three roles. I add that it can not do it equally well everywhere in the same person. Furthermore, it cannot produce all its effects without the help of Nature, even in the same person or in other persons and for the same complexions. Therefore, an incarnative which is of the 2nd deg. will cause proud flesh to grow in a body which is firm and dry. And frankincense which is dry in the 1st deg. produces a similar effect in a soft and moist body. So it goes that a medicament (ie incarnative) that is dry enough to regenerate tissue in a humid subject cannot do it in a dry person, and so on for others. Galen (*Megatechni*, Bk. III, Ch.3) also said that it is necessary that the flesh regenerated by us should be similar to the natural tissue of the body. For that, the humid region should be treared with humid medicines, the dry with the dry, because the new tissue cannot be nourished except by what is normally assimilated in the region. Besides, when we speak of a dry or a humid body we mean its natural dryness or humidity. And when we intend to restore it we mean to restore its normal complexion. We use our regeneratives to repair that which has not been returned to a natural state, otherwise we will obtain a bad sort of granulation tissue. Other authors agree.

B. Frankincense is a weak desiccant and regenerative. It has three forms: 1. A coarse granular powder obtained by shredding the resin obtained from the sacs. It is not ground. We call it the lesser frankincense, and it is weak. 2. The masculine frankincense also is called olibanum. It is very dry (3rd deg.) and viscous and adherent (ie conglutinative)., and it has regenerative powers. 3. The bark of frankincense is extremely desiccative (4th deg.) and astringent. It cicatrizes. We can generalize about frankincense but all types are not of the same degree. We may generalize that it regenerates, conglutinizes and cicatrizes.

C. In regard of consolidatives and cicatrizers we take note that many of them are used with just a few corrosives to obtain an eschar beneath which a pellicle forms by hardening softer tissues. To make such a mixture well you must know by personal experience the complexions of the various compounds, of the body as a whole and of the diseased region.

D. Galen (*Simple Medicaments*, Bk. I, Doct.4, Ch. 2) said, "The function of consolidatives is to dry and toughen the granulation tissues, not to diminish them. That is what incinerated bronze does, especially when it has been washed. It also affects the tissues around the diseased part. Consolidatives are better when they also are styptic and desiccative in the same degree." Galen added that consolidatives which are corrosives work differently than the consolidatives such as alum, oak-galls etc. which consolidate only when large amounts are applied. Those with corrosive actions are effective only when very small doses are used. Furthermore, if they are not combusted they ulcerate and erode the surrounding tissues.

E. Some consolidatives function on their own, as do the non-corrosive and desiccative types. There are other medicaments which have consolidative side-effects, as are the corrosive styptics, like flowers of bronze (186a), which suppress bad proud flesh.

CHAPTER 7.

THE CORROSIVES

I. Introduction in Three Parts

A. According to the authors of our time the term corrosive in common usage has two connotations. In general a corrosive both destroys and dissolves a part of the body which is solid. There are five types, in two categories. a. the *caustics* and the *suppuratives*: 1. Caustics burn and form eschars. 2.Suppurativess break down tissue into pus. b. three *corrosives*: 3. Corrosives, 4. Ulceratives and 5. Excoriatives.

The first group (ie the caustics) includes soap (ie the lye)(254), lime (251), etc. The suppuratives are gum of rue (394), and arsenic (327 and 34), etc. The so-called weak corrosives include both simples and compounds. The excoriatives are costmary (49), aristolochia(33) and others that are effective against morphia.[272] The ulceratives are the milky saps of euphorbia (174), esula (442), tithymals (442), daphne, rue, etc.

In the narrower sense of the term, we include corrosives which do not leave behind any eschar, such as mildly corrosive ointments. Those that are escharotic are usually called caustics or burners.[273] However, at times the mild corrosive ointments produce eschar, according to how potent is the corrosive contents in respect of the non-corrosives. The more they are corrosive the more rapidly they form eschar, and the more rapid is the condensation (ie of the bad

---

[272] Morphia: an ill-defined motley of cutaneous lesions, including vitiligo, leprosy, pigmented nevi, etc. (LDR)

[273] Nicaise used the Latin term 'adurent' (burning), which emphasizes the concept of the virtual cautery in contrast to the actual cautery of red-hot metal (LDR)

humors) Also, the intensity and the duration of the local corrosive action will vary with the consistency of the medication, ranging from powders, pastes and clays to stones. Also, the drier the part on which you apply the medication the more rapid will be the topicals own desiccative and thickening effects. And, the more potent the corrosive the less of it you will need to produce an eschar. However, a dry and hard form of a weak caustic applied in usual amounts will produce an eschar where too little of a soft potent medicine will fail.

The operating surgeon should know why an eschar is or is not desirable, because it may lessen his difficulties afterwards and decrease the need to change his medications. There are two factors: 1. The eschar-producing topicals may be hemostatic, whereas other medicines may actually increase the losses of blood by eroding the veins. 2. Where an eschar has formed the surgeon cannot treat other than simply to apply dressings until he sees the eschar begin to separate and come away. But, when he has used corrosives that do not produce eschars he is not delayed, and he can continue to treat according to his own schedule.

B. The surgeon should know which corrosives cause local pain immediately after they are applied and those that may not cause discomforts until many hours have elapsed. That information has two values. 1. Some patients will tolerate pain at certain times better than at others, such as before a meal rather than after; for another patient the opposite may be the case. Some do not want to be alone when they suffer and some want solitude. Because the surgeon cannot always stay with his patient, he should learn if the patient wants to undergo the painful application only while his surgeon is in attendance. 2. when the surgeon has applied a corrosive with immediate action on a sensitive patient, that sufferer may want the surgeon to take it off at once. But if the corrosive action has a delayed onset, after the surgeon has departed, the patient will not know how to get rid of the stuff and the surgeon will not be able to support the fellow through the period of suffering.

C. One should have concern for the particular properties of each of the five categories of topicals in question, according to their ranks and primary functions, and he should know the information provided by Galen (*Simple Medicaments*, Bk. V, sect.1) and by Avicenna (*Canon*, Bk. 1, Tr. 1, Ch. 4). Thus, the escharotics deterge the outer layers of the altered skin.[274] Otherwise they are similar to nearly all the medicines used in the treatment of morphia, nevi, lentiginous spots etc., which we described in earlier chapters. Most of them are weakly corrosive but are potent detergents.

Ulceratives are stronger than the foregoing but are much weaker than the true corrosives. They get rid of most of the moist portions of the skin and may produce an ulcer when they extract the bad humors. Examples are the milky-sap corrosives: rue (394), squills (443), soap (412), leaves and seeds of nettles (305), etc.

What we usually call corrosives destroy the flesh on which they are applied. They reduce the tissue by their powerful and energetic resolutive actions and by causing ulceration. They are stronger than excoriants but weaker than suppuratives. Because they are weaker they act only on the surface and do not penetrate deeply. They promote the healing by scar of ulcers in which superfluous proud flesh appears. Examples are verdigris (482), ink (490), and all other types of couperose (490).

According to Galen (ibid.,Bk.V.) and other writers, the suppuratives or putrefactives are of two classes: the hot, humid and true putrefactives and the others which resemble it only because both destroy soft tissues painlessly. That is not the primary benefit of the first class whereas it is of the second, which is warm and dry and burns feebly. It is very different from the potent hard medicine (ie the true putrefactive) which forms eschars as it burns. The feeble ones burn painlessly, or simply with a brief prick of pain. They really should not be called putrefactives.

Examples of the weaker types are orpiment (327), two gums of rue (394), lye made from ashes of plants with milky saps, such as figs (185), esula (173) and anabula (173), vinegar lees (176a), worm wood (1), horehound (23), borax (63) etc.

The stronger types are more potent than the common corrosives

---

[274] A strange claim. The outer layers of skin are charred by the medicament and form a thin eschar which must be cast off. The detergents get rid of it (LDR)

but weaker than the intense escharotics. These resolve the more subtle (liquid) part of the bad matter, and leave behind a cinder-like char. They eat away the flesh, burn the skin and dry it, harden it and make eschars. Their innate heat approaches the fourth degree. They are sort of fatty and they work slowly and they make themselves felt for quite a while, as the little pricks continue while the medicines work. These corrosives are anacardus (cashews) (25), ink (490), airain (14), pyrethrum (371), meerschaum (162), staphisagre (445), soap (430d), Lime (251), tragacanth (466) colcothar (490).

II. The Degrees of Potency of the Corrosive Simples

There are four degrees of intensity: weak, strong, very strong and extremely strong. This conforms with the writings of the ancient masters and agrees with current surgical manuals because all of us have better access to the sources, and the current practices of the apothecaries are more uniform.

The weak corrosives are hermodactyl (219), aristolochia (33), bryony (67), gentian (199), all types of burnt vitreol (490), chalcanthum (106), bronze flower (14), ink (490), etc.

The strong are: filings and flakes of bronze (106), ink, thapsia (461), crow's-foot and ravens-foot wort (kite's-foot) (142), various ranunculus (376), bryony bark, etc.

The very strong are: quick-lime, lime stone, oyster-shell and egg-shell lime, realgar (380).

Compounds made from the above simples may be powders, juices, ointments, caustic pastes (eruptoria) and others.

Eight Caustic Powders

1.   The Premier Powder which can be used in several ways is made with powdered ink or vitreol or verdigris, muriate of soda (alcalin) (404), rock salt (404), niter (402), orpiment, two alums (20), various types of tartar (456, 176a), quick-lime and any of the many substances in common use by surgeons.
2.   The powder of asphodels : 6 oz. of the roots of asphodels (40);

3 oz. of quick-lime; 1 oz. of orpiment. After mixing, expose it in bright sun-light until it is very dry. Then make troches and set them to dry in shade. Keep them in a glass jar. When you use them, grind one and spread it on a piece of cloth moistened with saliva. Lay it atop the sick part and it will erode strongly, only at the site: it will not spread.

3. For the same use: Take equal amounts of verdigris and aristolochia. Make a powder and apply as in 2.

4. Human feces (179) mixed with honey, incinerated, powdered and applied on a cloth, as above. It corrodes well and causes little damage beyond the lesion.

5. Equal amounts of red arsenic, alum (20), dry oak galls (195), quick-lime. Grind and mix with vinegar to make troches, etc.

6. 1/2 oz. each of white burnt niter (402), live sulfur (450), verdigris, galls, ink, green vitriol; 3 oz. of orpiment. Grind and dilute with strong vinegar and juice of tithimals to the consistency of honey (173). Make troches, etc.

7. Powder of moss (lichens)(291a) which grow at the trunks of trees. They are mildly corrosive.

8. Take equal parts of live sulfur, orpiment, rock salt and dilute with vinegar, Place all in a covered pot in a hot oven until incinerated. Keep the ashes for use. Before applying it, wash the region with warm vinegar and then apply the remedy twice in the same day. It is a potent corrosive and it destroys all cancerous (ie gangrenous) tissues.

There are several corrosive plant-juices:

1. Oxalis (329), 2. Gallitric (196), 3. Centaury (small) (100), 4. Gentian (199), 5. Bryony, 6. Cyclamen (114), 7. Geranium (209), 8. Moss (296), 9. Radish (374)

The mashed pulp of these plants placed over a fistula not only will cure a recent and still soft lesion, it will not do further damage. On the other hand, if the lining of the fistula is indurated (ie fibrous scar) you must use the pulp-free juices of some plants, for example, the six varieties of tithymals, which contain a milky sap. Only three of them grow in France. They are marsilium (122b), catapulla (173) and anabulla (173). Other corrosive plants are ranunculus (376) and asphodels.

There are eight corrosive ointments:

1.Theodoric's Green Ointment: 4 parts of dialthea (273), and one part of verdigris. You may fortify it or dilute it as we have instructed. 2. A stronger penetrating ointment has 2 parts of honey with 1 part of verdigris. 3. The Egyptian Ointment: 3 parts of bronze flower to 2 parts of vinegar and 8 parts of honey. Mix and boil until thick. This is a potent corrosive, but it will produce no eschar although it penetrates into the depths. 4. Ointment of Skin: 1 oz. each of cannabis-seeds (78), rye-flour (395a), roasted in a flat dish until the seeds blacken and can be powdered. Add 1/2 oz. of verdigris and dilute with honey. This compound is corrosive and escharotic. Treat necrotic ulcers etc. with repeated applications, removing the eschar each time.[275] In my experience with those cases there is no better corrosive than this. Yet, when a German surgeon brought it to Paris, his own tattered clothing barely covered him and he had no underclothes.

The green ointments which follow are called so because they usually retain that color. 5. 1 oz. verdigris. 1/2 oz. rock salt, 1 oz melted old pork-lard. 6. As in 5. except to increase the amount of salt to equal the weight of all the rest. It is a potent corrosive. 7. The Green Ointment of the Twelve Apostles, so named because it has 12 ingredients. Some call it the Apostolikon or the Ointment or the Plaster of Venus: 14 dr. each of white wax, resin, gum ammoniac, 6 dr. each of bdellium (53), long aristolochium and thick olibanum, 4 dr. each of galbanum (194) and myrrh (299), 8 dr. litharge (253), 3dr. each of opoponax, bronze flower and oil—3 lbs in winter, 2 lbs. in summer. Mix well and macerate in vinegar. Melt all over a low flame until you have the consistency of an ointment. It will cleanse ulcers, fistulas, polyps etc. 8. The Surgeons' Green Ointment: A clump of celandine (98), roots of alleluia (16), leaves of lovage (256) and scabious (414). Mash them and mix with 1 lb. of fat ewes and oil. Gently simmer until the leaves settle to the bottom of

---

[275] Precisely as today's chemosurgeons do (LDR).

the pot. Then filter and add 2 oz. each of mastic, olibanum and verdigris. This ointment will clean out old ulcers by eating away the bad flesh and it will promote healthy granulations.

Five Caustics (eruptors)

1. 1 part of cantharides (80) and 2 parts of yeast and a small amount of vinegar.

2. 8 parts of quick-lime, 1 part of soot (502) mixed with French soap (412) and saliva (403) to make a paste. This commonly used caustic is easy to apply and has been well tested.

The following two caustics were described for treating impetigo and dandruff in Ch. 8 of Doct. I in Treatise III: 3. The ulcerative caustic: liquid pitch (352) boiled with cashew-honey (ie the sap of anacardus) until it is thin. You may store it. 4. 2 oz. of powdered quick-lime thoroughly mixed with 4 oz. of soap. Add some vinegar and saliva or urine. 5. Split in half a clove of peeled garlic and lay it on an incision in the skin. It will cauterize and vesicate. It will relieve stabbing pain in the affected part.[276]

None of the corrosive compounds that we have described really are powders. Nor are any of the six that follow.

1. Cashew-honey: If no apothecary is available to prepare it, take 2 oz. of ground cashews (25) and 3 oz. of honey. Mix with an equal amount of vinegar. Simmer over a low flame until the vinegar evaporates. Filter. This remedy is just as good as cashew-honey made with freshly picked nuts (ie probably including the green cortex).

2. Capitellum (83) according to some authors is the fluid of the soap-maker, or their lye. I am not sure about that claim, because the soap-makers have jealously kept their secrets. The capitellum of surgeons is made by mixing 2 parts of quick-lime with four parts of the ashes (38) of bean-vines. Powder them and make a paste by adding some water of yeast. Then

---

[276] A form of local anaesthesia (LDT).

tamp it down in a pot with an opening in its bottom. Then drill a pit in the center of the impacted material and drip water into the pit which will percolate through the stuff and out the bottom, That is the socalled Capitellum. You can increase its potency by repeating the percolation a number of times.

3. There is another capitellum made with fireplace ashes when bean-vines are not available. You heat the capitellum until it just begins to harden, yet not completely dry. Then put it in a glass jar and bury it in a compost pile until it liquefies. This liquid will burn and char excrescences (ie warts etc.) anywhere on the body.

4. Another capitellum is made as in 2 and 3 but uses quick-lime freshly taken from the kiln and extinguished with cold water and immediately covered with powdered fireplace ashes. Allow it to interact before cooling it and making a lye-paste.

5. Next in order is sublimated arsenic. The process of sublimation cannot be described in full detail except by those who have themselves often carried it out. Much of the operation requires great mastery of the details. The capitellums and the sublimated arsenic are powerful caustics which destroy and corrode all excrescences if beforehand you lightly scarify or incise them.

6. You may use certain cool liquids such as rose-water mixed with some of the corrosive simples which we have listed. Place them in a small jar for use. They have weak or strong corrosive effects according to the potency of the simples that you use.

III. Seventeen Explanations

1. The surgeon should know that there are two kinds of granulation tissues if he intends to deal with them properly. He should recognize the good type and when there is an excess of it. The same holds for the bad kind. He should know what happens when one applies medications against the proud flesh before the wound has been prepared with a complete

mondification. The good granulations are more firm than the bad and will need stronger corrosives for control.

2. Some medicines corrode when applied on the surface and will not have an effect when taken internally. Examples are mustard, garlic and onions.

3. The surgeon should not use corrosives unless they are really indicated. When not needed they do harm. The reason: Just as the authorities laud a physician who treats by means of diet-and-regimen, so it is with the surgeon who treats without corrosives. And if he is one who uses them, he who uses the weaker corrosives is held to be more worthy of approval (ie by the Physician-clerics).

4. Here and elsewhere it has been said that some of these medicines, both the simples and the componds, have two assets: they have powerful caustic actions and they cause a minimum of pain. I have never made that claim, and I believe it is not well founded. On the contrary, the more caustic the medication the more painful it is. Because it is so potent, it diminishes the humors and weakens the natural spirit, the complexion of the patient. A strong caustic of necessity causes sharp pains unless the part has not first been numbed by a narcotic. However, I know one way, also described by the authorities, to ascertain which of two equally potent caustics will cause less suffering. An example: unsublimated arsenic causes more pain than the sublimated form because it is coarser, more destructive and less pure. Therefore, its effects are more difficult to control and are prolonged, and, as a result, it attracts more humors and causes more suffering. On the other hand, sublimated arsenic is more delicate and more pure. It is better tempered and rectified and it acts more subtly and with less delay. It attracts less and thus causes fewer changes and less pain. For the same reasons, realgar (380) (ie unsublimated) causes more suffering than (ie unsublimated) arsenic.[277]

---

[277] See Explanation 10, below (LDR).

5. Realgar is a type of poisonous and malignant clay that comes from India. It is more corrosive than arsenic. Its venomous and destructive properties can be attenuated to a degree that allows us to apply it even on delicate and noble parts of the body. Spread powdered realgar on an iron plate that has been heated to redness. Then staunch it in vinegar or juices of cool herbs, keeping it there for eighteen hours. Filter it through a fine cloth and set the retained powder in sunlight to dry. Repeat the wet to dry maneuver four or five times. Small amounts of the mitigated realgar may be applied anywhere except, as one famous authority said, on the penis, the lips and on other parts that have little soft tissue beneath the skin, such as the fingers and the toes. It has a tendency to resolve everything where it is applied.

6. Whenever you apply corrosives and caustics follow immediately with cooling medicines to relieve the burning pain. After the pain and the suffering have been assuaged, apply topicals which favor the separation of the eschar and have maturative (suppurative) properties.

7. The eschar produced by these medicines must not be torn away; that will cause bleeding from openings in veins; that is of special concern in regions bearing large veins and arteries.

8. Never wash (ie irrigate) ulcers immediately after you have applied corrosives or have removed the eschar. That will cause bleeding as in 7.

9. Never apply strong corrosives at night when the patient is delicate or feeble, especially in a sensitive part of the body, because the pain will be better tolerated during the day and the patient's natural sleep will not be disturbed.

10. You may attenuate caustics in three ways: 1. Use small doses. 2. Use powders. 3. Incinerate or dilute them or mix them with coolants.

11. The term ointment indicates its oily content, although it is not unusual to give that label to compounds that resemble them. However, the poultices and other mixtures of honey, plant-juices and flour are not true ointments.

12. Some corrosives are escharotic, others are not. The eschar is a barrier to the surgeon and he should be aware of that in advance, knowing the factors which favor the formation of eschars: a. When the corrosive topical is applied as a glob or a powder or a paste. b. When it is applied on a hard surface or on a dry ulcer or other such. c. Whenever the corrosive is not used as a fluid or an oil.[278]

13. All green ointments, especially the green verdigris, have a bad reputation for causing pain and suffering; the same holds for corrosive powders containing verdigris. Furthermore, verdigris colors bones green, and the surgeon should avoid using it near a bone.

14. In using corrosives you should know these three things about their modes of action: a. The particular qualities of the corrosives. b. The nature of the part to be treated. c. The nature of the patient you are treating.

In respect of a: There are four types of corrosives: Weak, strong, very strong and extremely strong. That holds for the simples in them as well as the compounds. All distinctions are not absolute, and the practicing surgeon will learn from his own experience. He will learn enough about the simples and about the compounds to provide a solid basis for judging weak and strong. Furthermore, he always should use less of a medicament when he is uncertain about its potency, until he has tested it in his own practice.

In regard of b: The lesion to be treated may be small, large or very large. It may be a soft bit of viscous proud flesh or firm mature granulation tissue or a tough nerve or warty callus.

As to c. we will note that a patient's body or the part of it that will be corroded can be very moist, or a tissue can be flabby as in a fat young woman, or it may be less moist in a healthy pubescent girl or a nursing baby, or be dry as in healthy young men and mature women, or very dry as in adults approaching

---

[278] In other words, when the dose is concentrated where it is applied, or it cannot soak into porous tissues or it cannot ooze away from the desired place of application (LDR).

senility. Furthermore, the patient may be feeble or strong, very strong or powerful. Also, we encounter patients who are delicate or sickly or cowardly etc. Further, as we wrote in the chapter on phlebotomy, we should abstain from using corrosives when the patient's fever is at its peak and when the Moon is in the wrong phase.

The surgeon who is conscientious in performing his art should take into account all of these circumstances, and at the same time he must set himself to learn to the best of his ability all of the particulars that bear on what he wants to accomplish. The corrosives are the most dangerous of all the topicals, especially when applied heedlessly. That is why I find it necessary to explain to all young students all of the precepts in this section. For example: if, during hot weather etc. a growth of soft proud flesh appears in an ulcer in a part of the body which is flabby and the patient is a weak young woman, you should use a tiny bit of weak corrosive, be it an ointment or powder of something like hermodactyl applied on some snippets of oakum. A contrasting example: if, during a mild season, a mass of firm granulation tissue forms in a solid fleshy part of a young man or woman or a pubescent girl, all with moderate complexions and hardy enough to tolerate it, you may apply a moderate dose of a strong corrosive, such as crystalline alum or incinerated couperose or the green ointment. Measure the dose to suit the case—the extremely potent corrosives for the extremely strong; the strong for the very strong, etc.—when all the particulars permit your decision. But when the circumstances do not permit a corrosive, as in a nursing infant with a wart or other tumor, you must compound your topical accordingly.

The surgeon must suit the particulars of every case as well as the general rules, by additions, subtractions and modifications as they seem worthwhile.

15. Caustics and corrosives differ in two ways: 1. The so-called corrosives eat away exposed flesh but do not affect healthy intact skin. When applied within an ulcer up to the skin-margins they will corrode just at the edges while eating away at the adjacent ulcerated flesh. The caustics (eruptors) will mortify and destroy the intact skin on which they are placed in

order to expose and attack the subjacent flesh. In so doing they corrupt, suppurate and ulcerate. 2. The corrosives which are set aside for later use are just as effective as when freshly made, whereas the caustics lose their potency unless applied fresh or within eight hours after the composition.

16. Caustics should be left in place for twelve hours in a robust adult and for only six hours in a weak child with a flabby complexion. In the last case the caustics act soon after it is applied; in an adult they take longer to produce their effects.

17. Soon after cantharides is applied, even on the head, the patient will produce large amounts of urine and will fill his bladder as if to burst unless it is voided. The urine will burn and cause griping pain. You can relieve that by sitting the patient in a bath with water up to the navel, the fluid being a decoction of mauve, violets, pellitory, water-cress and others like them.

CHAPTER 8.

EMOLLIENTS

I. Introduction: Four Parts

A. Galen (Bk.V, Doct. 2, Ch.1, *Simple Medicaments*) described at length the many ways to harden and to soften parts of the body. He said there are as many of one as of the other, each being analogous to its opposite. That is an example of his doctrine of opposites.

Induration[279] is induced in four ways: 1. After the subtle (ie liquid or vapor) inflamed matter is slowly absorbed by the actions of the weak emollients, the firm matter remains. 2. Chronic induration persists when dislocations are not promply reduced; that will require strong softeners. 3. Firm swelling persists after mal-unions of fractured bones, as when one fragment over-rides the other. Here you must persist with softeners until the callus is destroyed, using the very strong kind. 4. The induration that involves nerves and tendons is of

---

[279] 'Induration': the term describes edema, fibrosis, acute and chronic inflammatory infiltrates (cellulitis), hematomas, callus, scar, etc. The reader should interpret the term in its context (LDR).

the 4th degree and it persists after fractures, dislocations, distorted (ie twisted) green-stick fractures and painful arthritis. This requires emollients of the 4th degree.

Induration of the first degree, of recent origin, is mild, as seen in babies, girls and young women with soft tissues. In such cases use emollients if the 1st degree, accepting Galen's teachings that emollients exist in four degrees (*Catageni*, Bk. VII).

The second degree of induration is called strong; it has been present for a little while longer than the 1st degree, and it is seen in very phlegmatic persons and dry women. Use strong, that is 2nd degree, emollients to soften it.

The third degree of induration, called very strong, is found in slothful people. When it occurs in parts of the body described below, it will impair mobility. We use 3rd degree emollients to treat it.

The extremely indurated conditions of fourth degree are found in hard, dry bodies and parts, as in the callused hands of farm-laborers, which are so stiff that they are moved clumsily. Treat that with extremely potent medicines (ie 4th deg.)

Avicenna (Bk. II, Tr. 1, Ch.4, *Uses of single medicaments*) said that we make use of their innate heat and wetness to enlarge the tiny channels in all parts of the interior of the body,[280] and that facilitates the expulsion of the trapped humors. Serapion said the same (*Agregations*, ibid.), that emollients produce two effects simultaneously: they resolve what they can and they leave behind the firm, lumpy matter. For that reason, he used them to treat the firm, phlegmatic and thick aposthems which occur most often in nerves and tendons. Inasmuch as these aposthems vary in degree, we must use medications with various degrees of potency. That is how we treat indurations near joints, in nerve-bearing regions and after fractures, as described above. According to Avicenna (ibid),we know that emollients are warm and moist, but we also are aware that we should avoid moist medicines when we treat nerves. In such cases we should seek the less moist emollients. Serapion stated (ibid.) that these medicaments, when warm, as they should be, should be moderately dry, and their warmth should exceed their dryness in degree. It is precisely for that reason that properly labelled emollients are warm and humid (ie barely moist), because they are close relatives of medicaments which

---

[280]   The aperitive action (LDR).

are neither desiccative nor moisturizing. Following Serapion's teaching, we assert that emollients should be warm and dry, with the warmth predominating.

C. Medicines used to soften indurated nerves and nerve-bearing regions,[281] where the induration restricts mobility or when it is a result of an incised or penetrating wound or after contusions or after fractures, dislocations and twists and bends of bones or after treating painful joints, they must fulfil nine requirements: 1. Be warm. 2. Be dry. 3. Be neither too hot nor too dry. 4. Their heat and dryness should not just be simply noted but should be defined in degrees. 5. Be composed of a subtle matter (ie soft or semi-liquid) and be endowed with subtle qualities. 6. They should have mildly aperitive properties. 7. They should be strongly attractive (ie to extract humors). 8. They must be emollients(ie softeners) and 9. Be slightly styptic. Now let us consider each of the nine.

1. Emollients should be warm: The degree of warmth can vary in the entire gamut, ranging from temperate to hot or cold. They should not be moderately warm, when treating nerves which are naturally disposed by their complexion to coolness, and when they are in open wounds they are exposed to cold.[282] The exposed nerves have imbibed cold and moist humors, and, according to the doctrine of contraries, the treatment should be with opposites. A temperate medicament is

---

[281]   Again, I remind the reader: The terms 'nerve', 'tendon' amd 'sinew' were loosely applied labels for structures whose appearances and functions were often confused by the medieval and ancient writers. Here, the 'nervous' region could be a laborer's hand, stiffened by his callused palms (LDR).

[282]   Comments about two words: *Complexion* in its original connotation of folding together of various elements. The ancient and medieval usage of the inclusive term had little to do with the color or texture of skin. The very lack of precision suggests a need for cues and clues and insights into the entire nature of a complex person or thing. That usage existed even in Shakespeare's epoch. Cotgrave's French-English dictionary of 1611 stated, "The complection (is the)making, temper, (ie and the) constitution of the bodie; also the disposition, affection, humors or inclinations of the mind." (see HH Furniss, New Variorum Edition of *Merchant of Venice*, Dover Pub., 1964, p. 125).

   Accident: Any complication occurring in a wound, especially that which was attributed to the over-exposure to air of a recent wound (LDR).

neither hot nor cold, ergo, it is not contrary. In the same sense, the emollients should not be cold, because any cold medication, whether it be dry or humid, is harmful to nerves. I refer you to an *Aphorism* of Hippocrates (Sect. V, "Cold is Harmful to Nerves, Bones", etc.) There are two reasons for that. a. Cold obstructs nerves and thereby causes over-loading and spasms (ie of repletion). b. Nerves naturally are cool, and if you expose them to cold you exacerbate that innate quality. Galen (*Commentaries on Hippocrates,* Sect. 2) said that the elements of a sickness should be reduced to lessen the risks. Since emollients should be neither cold nor temperate they must be warm.[283]

2. Emollients should be dry: That is obvious inasmuch as they must be warm. If they were moist as well they would rot; heat is the sire of putrefaction and moisture is the mother. Nerves are at risk for putrefaction, at least that part of them that derives from menstrual blood which has been coagulated or congealed by cold, as stated by Galen (Bk. V. Ch.4. *Megatechni*).[284] "When the nerves are not rotted they will not be dissolved by heat because anything coagulated or congealed will be melted by heat." (Galen, ibid.).

3. Emollients must not be too hot; they could affect the nerves and the skin as a fire shrivels leather and renders it less elastic. Even when you only singe skin you cause suffering and a great influx of humors.

Emollients should not be extremely dry (ie 4th degree) because that would deprive nerves of nourishment and cause spasms and abnormal contractions (ie cramps and convulsions).

---

[283] The reader must accept such seemingly vapid and unnecessary conclusions as signs of the times. This was the epoch of Aristotelian logic as taught by the scholastics of Paris. Syllogistic reasoning etc.was an indicator of one's education and erudition (LDR).

[284] The medieval embryology held to a double origin of the tissues and structures. In brief:the 'spermatics' derived from the two sperms, male and female, and those tissues, once lost, could not be replaced. The 'nonspermatics' derived from menstrual blood which was retained during the pregnancy and furnished the embryo with replaceable tissues, such as blood and fat (See Ch.1 of Treat. I, The Anatomy) (LDR).

4. Emollients must be graded for heat and dryness. See Galen (*Simple Medicaments*, Bk.V, Doct. 2, Ch. 6) and Serapion (*Agregations*, Disc. 4) where he discusses emollients. He said that medicaments that are used in treating cold aposthems, which arise in nervous and tendinous regions, which interfere with movements, should be warm in the 2nd and 3rd degrees and dry in the 3rd deg., which befit the complexions of the rest of the body under treatment. If you apply 3rd deg. (dryness and temperature) medicaments, they should be used on bilious young men and phlegmatic young women in whom the complexions of the nerves are of the same 3rd deg. If necessary, you should dilute the emollients to match the complexions of the patients by adding warm resolutives. Resolutive topicals used for inflamed nerves in bilious young men should be warmer and dryer than other medicaments, because, in those patients there is a much greater gap between the natural complexions of their nerves and what they acquire by accidental exposure to 3rd deg. cold and moisture. The gap is narrow as observed in young women, between their natural complexions and what happens after the same exposure to cold and moisture.

When the gap is large and you want to restore the natural complexion, you need to use the more potent medications. Galen (*Techni, Treatment of Causes*, Ch. 27) said that you must reduce all that is in excess of normal by using the most energetic medicines. Haly said the same (*Commentaries*), "Everything that has reached a 3rd deg. of cold should be reduced to normal by a medicine of the same degree of heat."

5. Haly (*Techni.*, ibid., in the passage "Nerves and tendons . . .") explained why emollients should consist of a subtle (ie thin liquid or vapor) substance. Nerves usually lie deep within the body near bones, and thus are shielded from external violence by the overlying structures. Furthermore, because medicines of dense consistency cannot penetrate into the depths, emollients must have subtle qualities if they are to be effective in ridding nerves of bad humors, and be able to dry them.

6. Emollients should be aperitive if they are to enlarge the pores in the skin and the nerves. a. That will enable other medicines to

reach the interior, and b. The bad humors in the nerves will be more easily extracted.

7. Two reasons why emollients should be potent extractives and attractives: a. As stated, nerves lie deep and therefore more time is needed to extract the harmful humors. b. Nerves are composed of both solid and viscous matter.

8. Emollients should soften the affected part matter—treat the disease to suit the affected part. Haly (ibid. Ch. 7, beginning "Principium nobis.") wrote, "Nerves are cool and dry, consequently the bad humors that affect them are cool and dry." Avicenna agreed (Bk. III, Fol. 2, Ch. 1) "Nerves are especially affected by cold, which also is a quality of their natural humors. Because nerves are cool and dry and therefore firm, the emollients must be able to extract the humors and dry the nerves." Galen also said (*On Accidents and disease*) "Nerves which are most suitable for providing movements are the driest, if the dryness does not exceed the natural degree."

9. Emollients should be mildly styptic in order to maintain their prime functions. Serapion (*Agregations*. Ch. "On Styptics") said that styptics dry as well as bring local comfort; their stypticism must be mild so not to interfere with other medicines; they should not exceed the degree needed to strengthen the part.

Armed with what we have learned up to this point we can observe two errors made by the ancient authors. 1. Some of them used desiccative and potent corrosive agents to soften indurations of nerves, sometimes to the point where they induce spasms. 2. Others used emollients in such cases and continued to the point where the nerve-fibers themselves were destroyed. Therefore, in the case of one or the other, never use emollients alone. The more you compound them with comforters the better; at least alternate the two kinds frequently. Change them at least once a day or oftener.

D. The common opinion among the writers of the (ie surgical) handbooks has it that the restrictions of the movement of nerves and nerve-bearing regions are caused in one of two ways:1. External humors imbibe the humors of the nerves (ie extract them), or 2. Their natural humidity is consumed (ie within the nerves). For this to be fully apparent, you have to know that at their origin the nerves, at least their more liquid part, were formed from two sperms;, to which later was added a gristly matter which fortify them and give them tensile strength.[285] Then, according to Galen (*Megatechni*, Bk.iv, Ch. 1) what was solidified by cold is melted by heat. Therefore, when suppuration occurs in wounds involving nerves, as well as in fistulas and in inflamed joints, the humid part of nervous structures is dissolved by the heat of the pus and other unhealthy matter. The remaining parts of the nerves are desiccated. The observed results are spasms due to inanition (ie deprivation) and desiccation. It cannot be cured.

II. Emollient Medicaments: Simples

The four kinds are classed as mild, strong, very strong and extremely strong.

1.The feeble emollients will soften 1st deg. Indurations: butter (70), white wax (493), common oil and its lees (175), fresh animal fats (206) of pork, capon, rooster and duck, etc., camomille (76), melilot (277), mauve (273), guimauve (18), mercuriale (282), violets (486), brancus ursinus (64), pellitory (334), and other similar plants. 2nd degree emollients affect indurations of 2nd deg. which restrict movement: Seeds of fenugreek (181) and flax (252), lily petals (248), oils and lees, oil of lilies (248). 3rd degree emollients for use in cases of 3rd deg. induration are very strong: Humid hyssop (226), litharge (253), old unsalted animal fats, mucilages made from all varieties of emollient plants such as fenugreek, flax, roots of guimauve, etc. 4th degree emollients are extremely potent and are used to treat 4th deg. indurations: Liquid styrax (447), opoponax (321), galbanum (194), serapinum (429), bdellium (53), horehound (23), and other gums, bone-marrow (275) especially of deer and calves, vapor of vinegar (484) containing mill-dust (177), incinerated iron filings (289).

---

[285] See above, fn 26. A 'cord' was a combination of nerve-tendon-ligament, as defined in Tr. I, Ch. 1, fn 12 . See Galen, *De Usum Partium*, transl. by MT May, Cornell Univ. Press, 1968. Vol.II, pp 554-5 (LDR)

Emollient Compounds

These are graded according to the degrees of their contents. Galen (*Catageni*, part 3) said that they can be prepared in advance and that every surgeon should have on hand a weak and an exceedingly strong ointment with which he can make others to fit his needs when the case arises. No two cases have the same complexions and they differ in age, sex, professions, etc. He may use more of one or less of another or equal amounts. Galen (ibid. part 7) said that there is a wide gap between the weakest and the most potent medicaments in all the categories (ie maturatives, emollients, etc.) When he treats cases with varying degrees of seriousness the surgeon must adjust the potency of his medicines by adding strengtheners and weakeners to his topicals. Galen added (ibid.) that no simple remedy is suitable for all cases.

1st degree compounds: Galen described one (ibid.), the so-called Oil of Wax, made by mixing a little wax with oil. Many authors have modified the recipe by specifying a mix of four parts of old oil to one of wax. This softens the hard lumps in membranes, favoring their expulsion and resolution. It works by dilating the pores (ie an aperitive action) of a membrane. 2nd degree compounds; This is Galens recipe: Take 4 parts of of butter and 1 part each of Greek pitch (352) and fresh wax. It seems to be stronger than 1. The five 3rd degree compounds which follow here can be used for the 3rd degree indurations mentioned at the end of this list, the sort of induration which persists after a solid union of a fracture, and after dislocations when the reductions have been delayed and in cases where faulty reductions require repairs. 1. Mince 1 oz each of guimauve and bryony (67). Place them in a glass vial containing 4 oz. of oil of lilies. Boil until the watery juices of the plants have evaporated. This ointment can be spread on the indurated region. Spread fatty wool over it. 2. Take 3 oz. of unsalted pork-lard, 1 oz. each of fats from geese and hens. Melt all in an oven and filter through cloth. Add 1 oz. of wax. This is called the Ointment of Fats, and it is useful on dry and emaciated bodies without swellings, in which there are infiltrates of humors (ie edema) and which tend to spasms due to inanition. 3.

Add to 4: 1 oz. of tithymal (459), 1 oz. each of flours of fenugreek and flax-seeds; 1/2oz. each of bdellium, opoponax, mastic (272), and frankincense (188). Macerate the gums in wine and dissolve them in the fats, waxes and oils of recipe 2. Then add the flours while grinding all. This ointment is a potent softener by virtue of its terebinth and the frankincense and mastic are styptics. We call it the Theriac of Venus.; it softens the most resistant indurations. 4. 4 oz. of old pork-lard, 1 oz. each of lees of lily-oil (175), flax-seeds and bdellium; 1 oz. each styrax (447), calamint (72), galbanum, opoponax and gum ammoniac (23). Macerate the gums in wine, and when they are soft, grind them in a mortar and add the melted fats. This is an excellent emollient topical. 5. This ointment is described in *Catageni*, Part 7 where it is said to be of the 2nd deg. Make it with litharge-nutritum (441), oil and fresh pork-lard. Galen said it can be of the 4th deg. if all the contents are old and are not salted. We suppose that contents of that sort are stronger because they are older and that salt will diminish their potency.

III. Eight Explanations

1. Before applying ointments, emollients and other topicals, you should use fomentations on the affected part until it turns red; do no go beyond that limit. Use decoctions of flowers of camomille and melilot, seeds of fenugreek and flax, roots of guimauve and mauve and others. When fomenting, massage gently and move the joints to the fullest range possible. Do it gently and repeatedly and for a long time.

2. When the induration is of the 4th deg., heat to redness some fragments of mill-stone or iron filings and staunch them in some strong vinegar. Cover the region and fumigate it with vapors of the vinegar before applying the other topicals. Instead of the iron filings you may use marcassite (267) to fortify the vinegar for the fumigation. It will help, also, if you make a powder of marcassite and mix it with your resolutives. I once read a description of that remedy in one of Galen's books, but I no longer can remember where. Galen said that the remedy had a magical capacity to soften any induration.

3. You should know how difficult it is to assign an exact degree of complexion to the topicals you use in any case You must know the particular complexions of your patient and of the affected part; the complexion is idiosyncratic for every person. But if there truly is no single remedy appropriate for all cases, you make one from simples according to the needs and the degrees which you know, conforming to the general principles of medical science.

4. All of the simple and compound medications which you use to treat indurations of 2nd to 4th deg have been described in Chapters 2 and 11, Doct.1, Treat.II, and in the preliminary explanations for those chapters. However, we made no mention there of the degrees of their (ie of the medications themselves) complexions or their potencies, because those degrees, albeit they are necessary, were established by the Ancients. Because his real concerns are the treatments for the last three degrees of induration, the surgeon can be satisfied with a single medicament, or another that is very much like it in degree. Furthermore, as to the 1st degree of induration, the old-time and the modern surgeons agree about its unique signs and about the topicals to use for it.

5. When induration occurs after a nerve has been cut or a dislocation has been reduced, use emollients assiduously before the place has been hardened, solidified and dried by air. Make it easier to remove the damaging matter from the indurated nerves and the mushy scars. But, if you use them too soon, the firm reattachment of a dislocation and the solid union of a wound would be impaired.[286]

6. I refer you to Rhazes' (*Almansor*, VII, Ch. 2) who listed the causes for various indurations, as did Galen (*Simple Medicaments*) and Avicenna (Bk.IV, Fol. 5, Treat. 2, "Medicines that Soften Nerves", etc. and in Bk.IV. Fol. 4. Treat. 4, "On Indurated Nerves".)

---

[286] The emollioents may delay the agglutination of fractured or dislocated bones (LDR).

7. Afamous practitioner once told me that nerves which contract due to inanition can be cured by long daily tub baths using a decoction of roots of ferns (181a). Theodoric, Bishop of Cervia, in a little book of his secrets wrote that he had cured cases of contracted nerves by massaging the patients with watered human blood, distilled seven times and mixed with niter. The prescription has to compounded by a Master (ie apothecary or physician).

8. In regard of the foregoing, one may ask if all the topicals which are applied on nerves in their natural condition or which have been disturbed should be warm in their degrees of potency and action. It seems that should be so. The authors of handbooks and Hippocrates (*Aphorisms,* Part 5) state that cold is the enemy of nerves, bones and others. Therefore,we may conclude that warmth is favorable. Galen (*Catageni,* at the beginning of Part 3) tells of cases where he saw nerves suppurate in wounds amid warm and moist medicines. He then considered dressing them with contraries, that is, with cool and dry. Although he agreed that cold is harmful to nerves, in the end, he recognized and concluded that the best topicals for suppurating nerves should be dry and moderately warm. Haly came to an opposite conclusion from what Galen said in *Techni, Treatise on Causes,* Ch. 25. Haly wrote that warm things are helped by the warmer, (*Commentaries*) and that when you know the normal complexion of a body or a part of it and what the disturbances are, you try to preserve the normal state by using medications which resemble it in structure, and in the degrees of the vital elements of life. It follows then that we should use cold topicals when nerves suppurate. In his *Megatechni* (Bk. VI, Ch. 1), Galen wrote that warmth dissolves and liquefies everything that cold has coagulated and congealed, as happens with ice and fats. But he also (ibid.) remarked that nerves are made of cool materials and they are congealed by cold.[287] Consequently, we should not use warm things in treating

---

[287] Nerves are formed from warm humors that are condensed (congealed) by cooling (LDR).

them. And Avicenna (*Canon*, Bk. IV, Fol. 4, Tr. 4, "On Wounds of Nerves") wrote that tuthie (471) is the most acceptable topical medicament for nerves. According to him it is cool in the 1st deg., dry in the 2nd. He repeats that in Bk II. Serapion (*Agregations*) assures us that it also is styptic. Therefore, topicals applied on nerves should be warm.

The question is difficult and no answer is beyond doubt. Let us leave it to the Physicians to find the correct answer.[288] Added to what already has been said, we say that we are led to doubt that a complexion, be it normal or abnormal, is better treated with substances that are like its true self (ie ideosyncratic), and we are led to question whether or not it helps or harms. Now, inasmuch as cold medicines are harmful to nerves in whatever condition (ie normal or abnormal), one of two conclusions is apparent: either nerves are not cool or the natural complexion will not welcome anything that is not like it.

Finally, another doubtful question: Since the disturbed natural complexion can be restored to its normal condition by remedies which have the degrees and the forms of contrary properties. a fact established by the authority of Galen (*Techni, Treatise On Causes*) and since, on the other hand, cool topicals in action and potency are harmful to nerves that have have suffered accidental heating by whatever cause, how does it happen that a hand or a foot (ie notably cool-complected nerve-bearing regions) that is heated by an abscess or a wound or a burn does not suffer an incurable disturbance of its complexion when it is treated with moderately contrary medicines (ie cool). Thus we see an example of a contrary (ie the cool complected nervous organ) not corrected by a contrary (ie but by a similarly cool).[289]

---

[288] Mondeville's tongue-in-cheek wit (LDR)

[289] Mondeville's curious and tortured arguments are not explicit. He uses cool remedies to treat the 'accident', the hot abscess and wonders why it does not harm the cool complexion of the region. (LDR).

CHAPTER 9.

SYNONYMS AND EXPLANATIONS

For The Obscure Names of the Simples

This chapter has two parts:
I. Five reasons for writing the chapter:

1. All students do not have access to the synonyms (ie to be found in other books.)
2. Those who do have access, may have it only to synomyms for the legitimate names in current use.
3. The information obtained in 2. may not be clearly presented.
4. Because it is obscure, it is not clearly understood.
5. And, finally, this will absolve me of accusations of error, obscurity and corruption of the very doctrines that I teach. Galen *(Simple Medicaments,* Doct. 3, Ch. 7) said that Thessalus and his disciples had corrupted and made a muddy mix of the names and the meanings in their books, not having researched the subject nor thought it through.

In this Antidotary, I have designated many simples with names which are not recognized by many young practitioners. It will be worthwhile if I pre-chew for them the tough matter so it may be more easily absorbed. Let them not commit errors in their procedures that result from ignorance. To avoid such, I think it will be useful to explain the names and to use synonyms which, if not well known, can be found in other books. Where I use no synonyms, I give explanations and interpretations.

II.

In this section the synonyms are not listed alphabetically; they follow the names used in the text according to the chapters where they appear. First we give the synonyms for the repercussives, etc.

Whenever the medicament serves in two (ie or more) therapeutic roles, as, for examples, solathrum which is a repercussive and a maturative, or camomille which is both a resolutive and a maturative, or litharge which is both a repercussive and an emollient, and there are many more mentioned by pother autorities, they will be cited by me in every chapter, after a full description in only one. Do not be displeased if I am in error in assigning degrees of potency etc., because there is much disagreement among the authorities. Take quick-silver, for example and others. Serapion, himself, much to my astonishment, in his *Agregations* gives every medicament the same degree in all of its properties, heat, dryness, etc., active and passive. Yet we know for certain.that some may be hot in the 1st degree and dry in the 2nd or vice-versa. The same holds for all the properties.

THE SYNONYMS

Here Nicaise inserted a long note to explain his own supplement to Mondeville's list of synonyms. He had attempted to identify the many medicaments mentioned in the SURGERY by using botanical terms current in France in the late 19thC. He welcomed the assistance of Professor Saint-Lager, a respected botanist who had devoted several years to compiling and publishing a medieval medical pharmacopeia. Nicaise's additions extend Mondeville's text to double-length.

In this my edition I will not include the sections provided by Professor Saint-Lager. My reasons are first that botanical classifications have undergone many revisions during the century-plus since Nicaise's edition was published, and I have been troubled by the tautologies and variances. I know, too, that Mondeville himself had cited many items that he had culled from authors whose terms he accepted without precise identifications. And, second, I have included all the medicaments mentioned in the SURGERY in the Compendium Pharmacopeia in the Appendix I which follows this chapter. In that list I have used many of Professor Saint-Lagers synonyms as well as others from more recent sources, which I accredit there. I can make little claim to precision, having no one such as Professor Saint-Lager at my side.

I have not altered the French spelling of the Arabic terms as they are in Mondeville's chapter. (LDR)

CHAPTER 9 SUPPLEMENT:
SYNONYMS FOR THE MEDICAMENTS MENTIONED
IN THE TEXT[290]

SYNONYMS FOR THE REPERCUSSIVES

1. **Solathrum**: solanum, strychnos, morel, *Inab al tha, alab (camel)*, uva vulpa, uva canis, uva lupina, cuculus herba, herba salutaris, called such because it is dispersed immediately after being consumed (combusted) in the stomach. It is cool and dry in the 2nd deg. or cool in the 1st and dry in the 2nd deg.

   There are five types of solathrum: A. Common garden solanum (after Dioscorides). Mildly styptic and easily digested. B. Solanum ligneum, in France it is called dog's death; it grows in hedgerows. Its oval fruits resemble bedegar and eglantier, being red and fluffy within. C. Red Solanum, small, with red flowers and seeds. D. Black solanum, called deadly nightshade. It is large, has red flowers and black seeds. Called Marmot in France. E. Crazy solanum, unknown to me. The authors give many different names to each of the five, not counting the variety of common names used in different regions with different argots. The preceding list are the names commonly used in France, and by the herbalists.

2. **Crassula,** two types: A. The Large is called faba inversa. B. The small is called mamilla muris, vermicularis, tegularia, portulaca, andrachne (andrago), herba fatua, olus fatuum, *Bakla al-hamaka (bachal)*. It is cool in the 3rd deg. and moist in the 2nd deg.

3. **Virga pastoris**, after Dioscorides, centinodium, proserpinaca, polygonum masculum (polygonia), lingua passerina, sanguinaria, *As'a al Rai*. It is cool and moist in the 3rd deg. according to Serapion.

4. **Psyllium (psillium)** derives from the Greek word psylla which

---

[290] Arabic terms are in Italics. The terms in parentheses immediately after the title-word belong to the English translator). (LDR).

means flea, because its seed resembles one. It is cool and moist in the 2nd deg.

5. **Hyoscyamus** (jusquiamus), cassilago, symphoniaca, horse-tooth, caniculata. There are two types according to the seeds, white and black. It is cool and dry in the 3rd deg.

6. **Hedera**, in Greek it is cussus, or cissos or cyseos, according to Dioscorides. I think it is a type of convolvulus (volubilis) major, because Avicenna and Serapion speak of it as a type of ivy, in very obscure terms. If indeed it is conv. major then it is not the convolvulus used as a depilator and to kill lice. There are three varieties of **Ivy**. 1. Tree Ivy, well-known, in common use as cool and dry. 2. Poor-man's rope, so-called because the poor folk braid it into cords. It grows profusely among hedges, and bears small pointed leaves and a hairy flower which are wind-borne. We call it flammula because it burns like fire when rubbed on the skin. It is warm and dry in the 3rd deg. 3. Ground Ivy, also called ground-herb, because it sends its branches over the surface of the ground. It is said to be cool and moist.

7. **Acetosa** (acedula) is a variety of lapathum; some call it ribes, *Ribas*. It is a hearty medicine, cool and dry in the 3rd deg.

8. **Scariola** and **Endives** are kinds of lettuce, chicory. All are cool and moist.

9. **Nenufar,** *Nilufar,* aquatic horse-hoof, water-cabbage, dardana, farfara, water-poppy. Cool and moist in the 2nd deg.

10. **Plantago**, two types. A. Garden Plantains, large and small. The Large is called ram's tongue. The Small is called lanceola, quinquinervia, arnoglossa (lamb's tongue), *Lisan al-hamal,*. B. Wild Plantain has larger leaves than the cultivated type. It is hairy and deeply rooted, and is perennial. All plantains are cool and dry in the 3rd deg.

11. **Quercus (Oak),** *Dusberos,* has two species, Large Oak and Small, usually considering only the large among the medicaments.The small also is called Robur (in Fremch, Rouivre). The acorns of both are *Schah bullut,* or glands. Sometimes the small oak bears the styptic galls. They are cool and dry in the 2nd deg.

12. **Mespila (medlars)**, *Gubeira*, styptic, cool in the 1st deg. dry in the 2nd deg.

13. **Pyra (pear)**, *Kumathra (camentre)*, cool in the 1st deg.and dry in the 2nd deg.

14. **Cotoneum Malum (quince)**, cottanus, cydonium malum, *Safardjal (safaret)*. Styptic, cool and dry in the 2nd deg.

15. **Sorba**, sorbs, corms, are like the pears in shape and have thre cartilaginous seeds, hence are called tricoccum (triceon). In Arabic they are ˆZa'arur (Zacor), because they are appetizers; cool and dry.

16. **Cornus,**(cornel berry), cornea, the fruit of cornouille. Cool and dry.

17. **Rubus (blackberry)**, two types, one bears blackberries, the other bears a different fruit. A. The first, **Ronce**, is a small bushy growth called batos (baccus) in Greek. There are three varieties: 1. The large grows among hedges and is famous throughout the world (ie the wild blackberry). 2. The middle-size is cultivated in gardens and in France is called Framboise (raspberry). 3. The Small ronce grows wild in mossy areas. B. The second type has different kinds of fruit, and exists in three forms. 1. The small one is thorny and grows in the brush, called rhamnus, in French it is groseillier (currants). The large one is called bedegar, a small tree growing amid brush, larger than the rose bush which it resemble. It has small crisp leaves, small rose-like flowers, red and sweet-smelling. In France we call it eglantier. 3. The third ronce resembles the second except for its larger flat leaves and white flowers with little odor. We call them the dog-rose, the French boutonnier.

18. **Salix**, *Sif'saf (safsaf) and Khilaf (chulef)*, saule. It is cool and dry.

19. **Populus** (poplar), in Greek it is Aigeiros. Cool and dry.

20. **Hordeum (barley)**, orges *Schâr (scephair)*, cool and dry.

21. **Siligo (rye)**, *Sult (sulech)*.cool and dry.

22. **Avena (oats)**. Avicenna said that the Arabic term was *dausar (deuser)*, and Serapion called it avena and aegilops. However, I think that there are two varieties of avena. the common or

cultivated type which is the *Dausar* and the wild type, the aegilops, which in French is called haveron, which is useful in treating abscesses at the corner of the eye, hence aegil*ops*, or lacrimal cyst.

23. **Lolium, (darnel grass),** solium, zizania, *schajlam (sceilem).*

24. **Berberis (mountain grape),** fruit of a bush (*Zarschak (zacharach) or Amirberis (amirberberis).* In France it is vignette. Cool and dry in the 3rd deg. (see # 173)

25. **Sumach,** *Sumak.* Cool in the 2nd deg. and dry in the 3rd deg.

26. **Myrtilli (myrtle),** in Greek it is myrsine. Cool and dry.

27. **Uva acerba,** or green raisin-grapes, called agresta. A potion, cool in the 3rddeg. and dry in the 2nd deg.

28. **Gallae (oak galls)** the fruits of the small oak. cool, dry and styptic.

29. **Psidiae,** the peels of the pomegranate fruit.

30. **Balaustiae,** the flowers of the pomegranate.

31. **Antherae,** the yellow pollen-bearers of roses.

32. **Bolus armeniaca,** bol d'armenie, some wrongly claim that it is the same as Magra, which is a clay called sinapi, used by carpenters to mark trees (ie for cutting).

33. **Cachymia (cathimia),** *Kalimijja,* climia, cadmia, calamary stone, found with copper-minerals. Also, it is called fumus, the slag from the smelted ore.

34. **Lithargyrum** (litharge), merdensem, *Murtak,* burnt lead. It is the scum or lees (spume or faex) that surfaces on melted metal, es[ecially on melted silver. NB: Lithos is stone and Argyros is silver. Litharge of gold can be called climia. It tends to be cool and dry according to which metal gives rise to the scum.

35. **Merda ferri (scoria),** or iron filings, ferrugo, dross of smelted iron

36. **Corallus (coral),** *Basad (basad) and Mardjan (mergen),* either white or red. An aquatic plant, used when dried and hard. Moderately cool and dry. Some say in the 2nd deg.

37. **Antimonium,** in Greek it is stibum. A mineral used for ocular disorders. Warm and dry in the 4th deg.

38. **Cimolea.** A clay mineral not found in France. Here we use the name to describe filings and grindings found near mill-stones where metal tools are sharpened. The dust is cool and dry.

39. **Cerusa,** in Greek it is psimythion, *Asfidadj (aladheg)*. Cool and dry in 2nd deg.

40. **Argilla,** a strong clay. In Greek it is argillos. Cool and dry in 2nd deg.

41. **Acacia,** *Akakija (achacia),* the dried juice of unripe prunelles. Styptic, cool in 1st deg. and dry in 2nd deg.

42. **Opium**, the juice of the black poppy from overseas. Cool in the 4th deg. and dry in the 2nd deg.

43. **Atriplex**, *Kataf (cataf),* chrysolachanon (crisolocanna). a golden yellow vegetable, cool and moist.

44. **Mercurialis**, in Greek it is linozostis. A cool and moist herb.

45. **Malva (mauve),** in Greek it is malache (malochia), *Muluchijja, a tuber.* Mauve is cool and moist in the 2nd deg. Two varieties: Four types of the Cultivated Variety: 1.The common garden mauve. 2. Guimauve (bismalve), ebiscus, hibiscus, althea. Called bismalve because it is nearly twice as large as Mauve. It is moderately warm. 3. Termalva, nearly three times as large as mauve. It bears red roses called Spanish Roses in France. 4. Quadrimalva, a single, thick and tall stalk which has many branches, resembling a small cherry-tree.

46. **Rape**, the roots only, a type of navet. The seeds and the leaves are used to make oil. They and the spindly roots are inedible. They resemble a kind of white cabbage.

47. **Acetum,(vinegar)** *Khall (cal),* is cool in 1st deg, and dry in the 2nd deg.

48. **Viola,** *Banafsadj (beneffig),* is cool and moist in the 1st deg. There are four species of violets. 1. The common ones in France are saffron-colored flowers. 2.One usually grows on walls and is called *Keyri,* a purple flower with the odor of cloves, and is called clovette or caryophyllata. 3. One has a blue flower and is frequently used as in medicines. 4. One has a white flower. The last two named grow wild as well as in gardens.

49. **Mandragora**, *Iabru'h (iabroch)*. It is narcotic, cool and dry in the 3rd deg.

50. **Verbena**, hiera, botane, in Dioscorides it is Hierobrotanum, a word derived from the adjective hieros which means sacred; verminacula in Italian (vermiciata). It is a cool and dry herb.

51. **Hepatica** grows on rocks in rivers and in humid places. In France it is called Compierre.

52. **Corrigiola (polygonum)**, geniculata, in Fance it is cesune. It drags along on the ground at harvest time and is pulled off with metal rakes. You dry it as hay.

53. **Papaver (poppy)** has four types. 1. The white is cool and dry in the 2nd deg. We grow it in gardens and make an edible oil from its seeds. The other three are used in medicines rather than for foods. 2. The black is cool and dry in the 3rd deg., perhaps even in the 4th deg. 3. The red poppy has the same properties. 4. The hard (cornu) poppy is called the marine poppy. I think it is the same as memitte *Mamitha*. Wild Celidoine is also called memitte

54. **Semperviva,** sanguinary, jupiters' beard (joubarbe in France), smpervivum, crassula major, aizoon (aizon), succirum.

55. **Gramen**, *Nadjem (Nelgenteil and negil)*, an herb, one of a group of cereals which resembles a farmer's herb with nodular roots which creep over the ground. It has hard and pointed leaves like those of marsh reeds, but much smaller. Cattle browse it.

56. **Hypocystis**, hypoquistidos, *Hibukastidas (heifistidos)* is the juice of a mushroom. Cool and dry in the 2nd deg.

57. **Umbilicus Veneris**, cotyledon, cymbalaria, cool and dry in the 2nd deg.

58. **Capillus veneris (venus hair)**, *Scha'ar al-gul (algol)*, pig's hair, coriander-in-wells, *Barsijauschan (belchegnasten)*. Cool and dry in 2nd deg.

59. **Muscus aquae (moss)**, water lentil, *Ta'hleba (tahaleb)*, cool and moist in the 3rd deg. Also called pulmonaire because it is porous, light and spongy, resembling lung.

60. **Fabaria aquae (lovage)**, hydroselinon (hyposelina), is called

such because it is thought to be an aquatic bean. It grows on river-banks.

61. **Gratia Dei**, also grows alongside streams. It is about as long as a forearm (ie cubit), and has a round tap root which is soft and hollow, easily broken, not resistant. Its leaves resemble willow, but much softer, larger and less pointed. The flowers are small, bluish. It is very cool and moist.

62. **Spica** includes two species: 1. The Celtic and 2. **Spikenard**, or neridea. Some authors call them both spikenard.

63. **Absinthum** includes four varieties. 1.The commonly called garden-grown absinthe. 2. Wild gallic, teutonic, provençal, and aquatic. 3. Sweet absinthe or aneth. 4. Pontine, originally from the Pontus (Black Sea Region) having an astringent flavor.

64. **Caulis,** the Greek caulos, meaning a tree that bears thorns. The term itself denotes a wild cabbage, a very beautiful plant which at times grows on seaside cliffs. Macer said that caulis and brassica are synonyms. Cabbage is warm in the 1st deg. and dry in the 2nd deg.

65. **Fumus terraee (fumitory),** *Schahtarradj (secetherigi),* gingidium (sigidium), in France called white vulva, sparrow's beak. Cool in the 1st deg., dry in the 2nd deg.

66. **Marrubium,** in greek it is prassion, *Farasijun (parison),* linoscrofon. Warm in the 3rd deg.,dery in the 2nd deg.

67. **Polium,** *dja'ada (iahade),* two varieties: the small and the mountain. Both are warm in the 2nd deg. and dry in the 3rd deg.

68. **Stoechas (sticados),** *Astuhadus (astochados),* two varieties: arabian and citric. Both are warm in the 2nd deg. and dry in the 2nd deg.

69. **Myrrha (myrrh),** in Greek it is smyrna (spimyrrha), *Murr (mor),* warm and dry in the 2nd deg.

70. **Thus,** *Kundar (konder),* one of the three species of **Frankincense,** as previously described.

71. **Mastix (mastic),** *Ma'staka (mastica),* Roman gluten, meschelze. Warm and dry in the 2nd deg.

72. **Lupinus,** fava aegyptica, thermos (tarinus)in Greek. Warm in the 1st deg., dry in the 2nd deg.

73. **Alumen (alum)**, *Schabb (alafur and seb)*. Chemists distinguish several varieties of alum, but the physicians uses only three; 1. Rock alum (crystals), 2. Feather alum, and 3. Sugar alum. The first is humid and is called liparin. The second is called feather, jamen, fissile, flaky as asbestos (amentum), amer. The third is called sugar alum as it refers to sugar. It is sweet, leafy and round. Warm and dry in the 3rd deg.

74. **Sal (salt)** *Mil'h (malhi)*, The chemists desitinguish many varieties, only some of which are used by physicians, and seldom other than the common salt, our food-condiment, which is warm and dry in the 2nd deg., and as warm as it is bitter. We use it for clysters. lotions and plasters.

75. **Sulfur**, *Kibrit (cebrith)*, is better when it has not been heated, when it has a bright yellow color and shines when not lumpy. What has been melted by heat is weaker.

76. **Oleum rosarum (oil of roses)** is intermediate, according to Serapion (*Agregarions*) between oil and roses. And he added that the oil is used for its consistency whereas the roses are the virtual element, even in small amount. Roses are cool and dry in the 2nd deg. The oil dilutes them, such that the oil of roses is cool and dry in the 1st deg.

77. **Schoenanthum (squinantum), (sedge)**, camel-grass, hay, *Adskhar (adcher)*. Two varieties: one bears fruit, the other is infertile. Styptic, warm and dry in the 3rd deg.

78. **Abrotonon** (abrotanum), *Kei'sum (caisum)*, cultivated and wild. When not defined we mean the cultivated type, called camphorata. The meadow abrotonon is called *Schi'harmani (sichen amemunic)* in Arabic. It is warm in the 2nd deg. dry in the 1st deg, but according to Avicenna it is warm in the 1st deg. and dry in the 2nd deg.

79. **Centaurea,** called the Physician-centaur, has two types, the large and the small. Some call the large variety narce (narche). It grows in humid places and near streams, amid many other herbs which it resembles, such as the deer herb and others.

The herb-collectors do not know how to differentiate them and make mistakes. One is about two cubits high, has a slim stalk, is reddish, has leaves and flowers that resemble those of the small centaurea which are reddish and cluster at the tips like a crown. However, the floral stems are longer than those of the centaurea and form a larger crown The small centaurea is well-known and is called Earth-Bile. It grows in meadows and dry places, and is about one foot high.

80. **Aristolochia**, in Greek it is feralos or apiston, *Zarawand (zarapz)*. There are four species.: 1. The round, called Malum terrae (malum storacis) or female aristolochia. That is the usual connotation of the term aristolochia alone. 2. The long is called dactylitis (dactyla), saracenic or male. 3. Aristolochia clematis or tritica. 4. Pollion or polyrrhizon. The last two are little-known. All the aristolochias are warm and dry in the 2nd deg., but, according to Avicenna, they are warm in the 2nd deg. and dry in the 3rd deg.

81. **Terra sigillata,** *Tin maktum (ten machtum)*, lemnia, cool and dry in the 4th deg.

82 **Sandali (sandalwood)**, *Sandal*, three types: white, red and yellow. All of them are cool and dry in the 4th deg.

83. **Camphora**, *Kâfûr (hastor)*, the sublimated gum of a tree, exported to us. Cool and dry in the 3rd deg.

84. **Spodium**, *Tabâschîr (aheusir)*, we do not know it, except that it is made by burning the roots of marsh-reeds. It is cool in the 2nd deg. and dry in the 3rd deg.

85. **Cera (wax)**, in Greek it is propolis, *Scham'a (schamha)*. It mixes with any of the four qualities.

86. **Amidum (wheaten flour)**, amylum, in Greek it is amylos, unmilled. Cool and dry in the 1st deg.

87. **Crocus (saffron),** flower of carthamus (carcamus), *Za'afaran (zarafaran)* warm in the 1st deg and dry in the 2nd deg.

88. **Lana succida (lanolin)**, *Suf (sauf)*, its action is moderate in degree.

Synonyms For Resolutives

89. **Chamomilla** (camomilla), anthemus (antimus), leucanthemon (leukintimus), camomillum, camentilon. *Babundj (bebong)*. Dioscorides and Pliny said there were three varieties. However, based on other authorities since them I have come to believe that they wrongly included the two species of Cotula which lack fetid odor. Therefore, the single Camomille is warm and dry in the 1st deg.

90. **Melilotus** is a kind of meadow clover. Its name derives from mel (honey) and lotus, because is honey-colored Also called corona regia and sertula, from serta, a garland of flowers, melliton, *Iklil-al-Malik (almanik)*, which means royal crown. There are two melilots, white and yellow. Both are warm and dry in the 1st deg.

91. **Parietaria** (paritaria), absinthe, wind-herb. shadow-herb, glass-herb because it is very good for cleaning glass, in Greek it is perdicias (perdiciados), *Haschischat-al-Zadjadj (haschischalzadjadj)*, which means glass-herb in Arabic. The name parietory comes from its growth on walls.

92. **Anethum (aneth)**, sweet absinthe. Warm and dry in the 2nd deg.

93. **Urtica**, califax ignita, acalephe, *Hundjur (huniura)*. There are five species of **Orties (Nettles)**: 1. The seeds are contained in small, smooth and white globular capsules that resemble flaxseeds, but are much smaller. 2. The seeds are smaller and are not encapsulated. Thep dangle separately on twigs. These are called the large nettles. 3. Smaller than the preceding two they have small barbed leaves that sting badly. 4. The dead nettle which does not sting. It bears white flowers, has a sweet flavor and is called archangelica. 5. Black nettles (maura), ficaire, millemorbia, (milimorbia). scrophularia, because it cures scrofules amd is called quadrangula (castrangula), and wrongly called wild rue and peganon (piganon) by some. The stem and the leaves resemble those of the true nettles and it has many thick knotted roots, lumpy like scrofulas.

94. **Inula (enula or elna)**, helenium; aunée. Warm and dry in the 2nd deg.

95. **Lingua bovis, bugloss**, *Lisân-al-Thaur (lisan)*. Warm and moist in the 1st deg.

96. **Ebulus (elder tree)**, ysatis, chamaeactis (the smallest), acté, sureau. Warm and dry in the 3rd deg.

97. **Valerian** phu (fu), fistra, amentilla, potentilla. Two kinds: The cultivated valerian has thick and lumpy roots, growing on the surface, odorous. Some authors call it indiginous nard. The rural French call it Zedoara. Wild valerian grows in forests and sometimes on river banks. It has many long slim roots which emit a fetid odor. It is what the French call the true valerian. Warm in the 1st deg. and dry in the 2nd deg.

98. **Apium**,*Karafs (karfin)*, in Greek it is selinon (silmon). Warm in the 1st dfeg. and dry in the 2nd deg. Two forms: The cultivated type has two species. The first is the common form called **Parsley** or petroselinon or rock-parsley. It is our kitchen herb. The second, the wild type, includes seven varieties. 1.Water-cress, hydroselinon, hydor, hydros, seneçon. water-pepper, cyminum, cumin. Avicenna (*Canon*, Bk. II, Ch. on Apium etc.) speaks of all of the them. 2. Frog-apium, batrachion (natracheon). The herborists of Montpellier call it a type of crow's-foot with unstained leaves. The crow's-foot that has dark leaves is called patalupi (wolf-paw). 3. Berula, so-called pupil of the eye or eye of the pepper. 4. Macedonian parsley, well known. 5. Clover. 6. Laughing apium, batrachion, crowfoot herb. Many of our own folk call it cancer-herb because they think it can cure cancers. It is a small green plant that grows abundantly in forests and resembles self-heal, and has white flowers. 7. Apium hemorrhoideum is a small herb that grows in gardens and seems to sprout from a knot in the center of a root, like a spindle that leaves the wheel before it reaches the end of the traverse. People are confident that it relieves the pain of hemorrhoids.

99. **Foeniculum (fennel)**, in Greek it is marathron (maratrum).

Cultic=vated fennel is warm and dry in the 2nd deg., the wild type is in the 3rd deg.

100. **Furfur**, cantabrum, mill-dust. Bran is moderately warm and dry.

101. **Farina**, *Dakik (sanich)*, barley meal, is similar to wheat flour, and is used by surgeons in the same way. The wheaten meal is less cool than the barley, itself, and I think that *sanich* is crushed (ie in a mortar) barley rather than milled, called 'pileiche' (ie pounded) by the common folk. Some say that sanich is a broth of ground barley which then is thickened to become a porridge such that barley-meal and barley-porridge are different substances. Others agree with Avicenna (Bl. IV, Fol. 4, Tr. 2) that barley-meal, moist hyssop and sanich are purifiers and detergents of dead (ie stagnant) blood etc. Amidon (ie wheat meal) is cool, moist and viscous.

102. **Orobus (vetch)**, vicia, *Karsana (kessene)*, orobus in Greek, a plant fed to horses. It is warm in the 1st deg. and dry in the 2nd deg. Two types: Ers is cultivated and well-known. The wild type is called Ervum (herbus) and grows in grain fields, often to such an excess that it crowds out the crop. It also grows in the filedswhere horses are pastured.

103. **Faba**, *Bakila (kaskille)*, the unripe bean (fava) is very cool and moist. But the mature bean is very dry and cool.

104. **Pinguedo**, *Samn (semen(*, is different from adeps (*scha'hm or axa ham*). Pinguedo is the lard of such moist animals as pigs. **Adeps** is the **Lard** of such dry animals as beeves and sheep. Melted pinguedo is sagimen[291] melted adeps is sepum. These two lards when freshly prepared more or less warm according to their sources. When they are not fresh they become warmer by the day, and are more liquid and more resolutive in action.

105. **Medulla**, *Mukhkh, (moch alhadam)*. All marrows are warm and moist, more or less as for adeps and pinguedo. They soften the hardness of nerves and nerve-bearing regions by increasing the humidity of the tissues which they infiltrate.

---

[291] Not sagimen nitri. See Pharmacopeia (LDR)

106. **Ammoniacum**, *Wuschschak (vaxak)*, the gum or the sap of a tree. Warm in the 3rd deg. and dry in the 1st deg.

107. **Serapinum**Sarabina, *Sakbinadj (sarabezugo)*. Warm and dry in the 3rd deg.

108. **Opopoponax**, *Djarvschir (jausir)*,. the sap of the herb panax. Warm and dry in the 3rd deg.

109. **Asa,** laser (lasar) in Greek, has two varieties. One is sweet and malodorous and does not grow in France. Both are gums and warm in various degrees, according to the complexion of the source.

110. **Terebinthina**, gluten,[292] *Butin (albotin)*, it is the best gum after mastic. It is the Theriac (tyriac) for nerves

111. **Faex Cerae (beeswax)**, *Mum noir,* propolis in Greek. Warmer than wax; a more vigorouis attrtactant. Warm and moist in the 1st deg.

112. **Butyrum (butter)**, *Zubd (zebd)*, warm and moist, but mostly moist.

113. **Costus.** Two kinds of roots. One is bitter the other is sweet, neither of them is native in France. They are warm in the 2nd deg, and dry in the 2nd deg.

114. **Piper (pepper)**, same in Greek and Latin. Many species, including water-peppers or water-cress, ethiopian pepper, or acorus (ie sweet sedge), climbing peppers whose seeds are called *Fulful (fulfol)*, in Arabic. We know three species as the long, the white and the black peppers. The long are picked when the clusters darken like grapes before they are fully ripe. White peppers are harvested before they are ripe, when the their stems are separate. Black peppers are picked when they are fully ripe. All of them are warm and dry almost at the 3rd deg.

115. **Pyrethrum**, the pyrethron of Dioscorides, the piperetron of Pliny, *Akirkar'ha (marchicha chahara)*. The name comes from the Greek for fire, because the plant has a burning quality.

116. **Euphorbium**, euphorbion (euforbion) in Geek, *Furbium*

---

292 Not animal gluten. See Pharmacopeia (LDR)

(eufarbium), The words' root are eu meaning good, and formido, meaning fear, because one should be prudent and hesitant in regard of the violence of the medication. It is warm and dry in the 4th deg.

117. **Castoreum** (beaver musk), *Khu'sja (anchia)*, is more dilute (volatile) than any other medicine. It is warm between 3rd deg. and 4th deg. and dry in the 2nd deg.

118. **Lilium**, crinon and narcissos in Greek, *Nirdjis (narges) and asmandjuni (ansea)*, is warm and dry in the 2nd deg. There are two forms: the cultivated type has two varieties: a garden white annual and a perennial blue. The white **Lily** has a single flower whose petals are solid at their base but divided toward the tips, and they have a saffron-colored center (ie fruit). The blue lily is called iris, *Asmandjuni/ (arsenium)*, or arc of the goddess (ie of the rainbow) because the variety of its colors calls the goddess to mind. Some people, in error, call it flammula. Iris is warm in the 3rd deg. and dry in the 2nd deg. The wild lily is warmer and drier than the cultivated ones. There are four species: a. The Arabian *Susan (kumac)* has petals joined at the base like a cap. They are more often saffron-colored than white. The leaves and roots resemble those of the iris but they are smaller and the roots are about as long and thick as a finger. They grow in forests and shaded places. b. Is called *Casoras*, and c.is*Sa'afrani (safarani)*. d.is Xiphion in Greek. It has tiers of leaves, and the petals of the flowers are split. The root has two little separated nodules, resembling onions that arise from the central portion of the root.

119. **Foenum graecum (fenugreek)**, *Hulba (halbal)*, telis (cili) and buceros in Greek, de bus, boef, cerus (ceron), corne or aegoceros, de aïx (aeglon), chévre, in other words, bull-horn or stag-horn, because the hulls of its seeds resemble those animals' horns. Warm and dry in the 2nd deg.

120. **Semen Lini (flaxseed)**, used by surgeons in place of fenugreek and vice versa Warm in the 1st deg, and half-way into 2nd deg.of humidity.

121. **Balsamus** *Balasan*, opobalsamon (opobalsamos) in Greek, opos. The sap drips like tears from the tree, as a gum rather than as an oil.

122. **Carpobalsamum** is the fruit of the tree of 121. The Greek word carpos means a tree-fruit.

123. **Cassia,** *Salikha (salicha)*,. Many varieties, of which only two are medicinal. Warm and dry.

124. **Caryophilus, gariofilus**, *Karanful (karunfel)*, caryophyllon in Greek caryon from karyon, which means the nuts of trees, and phyllon, leaves, although the medicine is not from the leaves.Warm and dry in the 2nd deg.

125. **Cinnamomum**, *Dar'Sini*, includes several types, all similar. Warm and dry in the 3rd deg.

126a. **Bdellium,** a gum called proceron and malochia,*Mukl al-Iahud (molcal ieuz)*.

126b. **Ficus arboris (figs).** The cultivated fig is edible. When green it is slightly warm and moist, but when the fig dries it is warm and dry in the 2nd deg. The wild fig has the same characteristics, but grows in remote regions and in the mouintains. Another type is called Pharaohs Fig (ie Egyptian) *Djummaz (iumaz)*, *and Mais*, crazy fig, sycamore, (sicomoros). It is called crazy because the fruit is inedible, hard and dry. In France it is called sicamor, in Greek it is sycon. It gets its name from its resemblance to figs. Its leaves resemble the mulberry, moron in Greek. A fourth type of ficus is called caprificus. Its sap is used to make lycium (licium).

127. **Uva passa (sour grapes)**, *Zabid (zibid)*, In France we call them imported grapes, and the seeds are used, *Kischmisch (kesmes)*.

128. **Scilla (squills)**, in Greek it is spolia, because the bulb is covered with many thin layers, which are peeled. We call it the rat onion. Squills are warm and dry on the 2nd deg.

129. **Colla (Collagen Glue)** comes from the Greek word collesis, which means 'sticky', and is applied to any glutinous substance. Physicians and surgeons use only the collas made from animal hides, which are warm and dry in the 1st deg., or from fish-skin which is less warm but equally dry. 130. **Pix**

(pitch), *Pissa (zaft)*, Three types: Naval tar, used on ships; Colophane, Greek tar (colophonia); Cooked Resin is bremo in Greek,*Ratindj (rating)*, named for sap of the tree of its source and the place of origin. All tars are warm and dry in the 3rd deg.

Synonyms of Simple Maturatives

131. **Branca ursina**, resembles beets. Warm and moist in the 1st deg.
132. **Bryonia,** vitis alba, ampelos in Greek, *Faschira (fesire)*, has many synonyms, and is cited by Pliny. It is warm and dry in the 3rd deg. It grows in hedge-rows at the forest's edge.It has creeping vines, resembling grapes, and long thick roots that suggest a man's thighs. The white inner pulp is soft as in a turnip.
133. **Lapathum** (lapatium), *Humadh (humad)*, five varieties. 1. Cultivated, called the Royal by Serapion. It is edible and a source of oil. A common vegetable in France where it is called foreign-cabbage. Its leaves resemble round oxalis, but are larger, softer and darker. 2. Sour oxalis, so-called Lapathum acetosum, or acedula acetosum, in Greek it is oxylapathon (erroneously called ribes). In France it is called **Sorrel.** It is used to increase the acidity of agresta as a medicinal cordial. 3. Lapathum acutum, paratella, paradella in Italy, **Dock** in England, in France docque. It grows equally well in moist and dry places. 4. Round lapathum (Lap. magnum and aquatile), grows in water or on the banks. Its leaves are longer and more pointed than the others. 5. Lappago major and lapathum inversum, lappa inversa, bardana (lardana), personata (personata), scurfy-hat, philanthropos in Greek, which derives from philos meaning love and anthropos meaning man, because it loves man, named so because its fruits catch hold of his clothing as he brushes against them. It grows in dry places, has large leaves and round fruits that resemble small nuts, which adhere to

anything that they come against. All the lapathums are cool and dry in the 3rd deg.

134. **Baucia (carrot or parsnip),** pastinaca, *Alfaneria and Djazar (zezar),* staphylinon in Greek.Wild and cultivated. Both are warm close to 2nd deg. and moist in the 1st deg.

135. **Farina frumenti, (wheat flour),** is tempered and viscous. *Dekich.*

136. **Ova, eggs,** *Beidh (baidh),* they are midway between warm and dry, a bit moist and a bit cool, according to Serapion.

137. **Cepa (onions),** *Ba'sal (basal),* many kinds: dog-onion, narcissus, rat onion, squill-onion (bas'al al-zir, alzir), resembles squill. The edible onion appears in three forms: the common red, the sweet, large, .white onion called ciboule in France, the small onion of Ascalonia (Escalonia), named for a small Judean town Ascalon.

138. **Allium,** *Thaum,* scorodon in Greek. In France we include it as a **wild Garlic** that grows in fields. Its leaves resemble those of the small onion, a bit slimmer. rounder and tubular and more fragile. Allium and garlic are warm and dry in the 4th deg.\ 139. **Fermentum (yeast),** *Khamir (chamir),* Serapion (*Agregationes*) said that leaven is mildly warm and moist when it nitrous, and it is dry when its less nitrous or salty. In itself it exhibits contrary virtues; it is cool based on its acidity and warm based on its saltiness and its starch. It dilutes, extracts and resolves deeply situated matter, without any risks. It is incisive by means of its acidity, extractive by its warmth and maturative by its viscosity.

140. **Mel (honey),** *Asal (han).* In rosat (rose-honey) the roses dilute the heat and the dryness of the honey, which alone, according to Avicenna, is warm and dry in the 2rd deg., but according to Serapion it is warm and dry to the 3rd deg. It is less warm and dry by a full degree than is simple honey. As such it purifies and fortifies. We use it to advantage both inside (ie an open abscess or ulcer) and outside the warm dyscrasias.

141. **Aqua (water),** hydor in Greek, a word with which we form the

word hydrops. It is a simple element, cool and moist, a moderately dense and heavy fluid. When it is part of a matter (ie pus) it contributes coolness and wetness.

142. **Lac (milk)** *Laban (leben)*, the aqueous part of milk which we call serum (ie whey) is warm, the buttery part is tempered and verges on warmth. Sour milk is cool and dry.

143. **Oleum violaceum (oil of violets)**, in which the naturally cool and moist violets, in the 1ˢᵗ deg., dilute the complexion of the oil. In all the compounded oils, the complexion and the virtue of the simple oil are modified according to the medicaments which are mixed with it.

144. **Stercus (feces)** *Zibl (zebel)*, the droppings of pigeons is more combustive than that of other animals, but all of them are warm.

145. **Borax,** the solder of gold, *Tinchor*, is used by goldsmiths. It is a kind of salt, warm a dry when diluted.

146. **Sepum bovis (beef lard)** and of all the other horned animals, such as deer. The lard is warmer and dryer than that of animals without horns, such as pigs. Male lard is warmer and dryer than that of females. The same holds for non-castrated animals' fat in contrast with that of castrates, and for adult fat over that of youngsters.

Synonyms for Mondificatives (Detergents)

147. **Sarcocolla,** a gum. The name derives from sarco meaning flesh and colla meaning collagen, ergo, gluing the flesh. In Arabic *Anzarut*, in Hebrew angelot, in Greek, argemone (argimon). It is warm in the 2ⁿᵈ deg. and dry in the 1st deg.

148. **Vinum. New Wine** is warm and dry in the 1ˢᵗ deg. Aged for two or three years it is in the 2ⁿᵈ Deg. When older than six years it is warm and dry in the 3ʳᵈ deg. Those figures are approximates. One knows by experience that wine is harmful to nerves. It blunts the senses and aggravates maladies everywhere in the body. However, taken as a tonic, in small

amoints and diluted, it does no harm. I have not found Greek or Arabic words for wine.

149. **Myrrha** is a gum. *Set,* smyrna in Greek (spimirrha). Warm and dry in the 2nddeg.

150. **Aloe** is the sap of a plant, *Saber (sabar).* There are three kinds. Socotrin (cicotrin) is the best, the hepatic and the caballin. It is warm in the 2nd deg. and dry in the 1st deg.

151. **Serum (whey),** the liquid that remains after the cheese comes from the milk. *Mâ al-Djuban (mathon-nagna),* cleans when it is applied both from within and on the outside.

152. **Aqua hordei (barley-water),** ptisane, is less dry than barley. It cleanses from within and on the surface and produces many good effects which Hippocrates listed. (*Diet for acute illnesses,* Part I, Chs. 1-3, etc).

153. **Lexivium (lye).** Lessive from the ashes, after extracting the liquid and gaseous parts and leaving the earthen. The lessives vary with the sources of the ashes that have been rinsed. All lyes are abstersive except that from the fig-trees and other plants that have milky saps. Ashes are warm and dry in the 3rd deg., and the lessives are nearly the same.

154. **Urina,** *Baul.* The word derives from the Greek ouron (urich) from the Latin uro uris, meaning I burn, you burn. All urine is warm and dry, and strongly abstersive.

155. **Cyperus (a sedge),** cypeiros in Greek (cyperon), juncus, trangularis, *Sua'd (sahade),* Schoenos (squinum) according to Dioscorides. It grows near streams and along the banks. Its stem grows to about one and a half cubit in height, and bears at its tip a black fruit. Its leaves are dentate like the pear, sharp enough to cut the hands of those who gather them. The roots are long, round, dark and emit a delicate odor and a bitter flavor from may shoots. Sedge is styptic, desiccative but not tart. It is warm and dry in the 2nd deg.

156. **Chelidonium (celidonia),** *Hurd (hauroth) and Kurkum (curcuma),* in Greek it is Chelidonion (oromon, kilidon), which is to say, the swallow. One type is cultivated, and

includes two types. The large is the well known common one with saffron-colored flowers and a yellow sap. The smaller is called cantion in Greek, *Mâmirâm* in Arabic. The wild form is called memithe or memitha, and is the horned poppy. (see 175)

Synonyms for Regeneratives

157. **Chalcanthum** (Dragantum), vitriolum, *Zadj (zegi)*, colcotar *(kulkatar)*, couperose (according to only some authors). Couperose if not identical always is very close and similar.

158. **Vitriol (Couperose)** which has been burnt in order to reduce its irritating and corrosive qualities without diminishing its drying virtue. If you apply non-combusted vitriol on ulcers of the penis it will cause an erysipelous ulcer, as has been the experience of the practitioners. Many of the synonyms for the various kinds of vitriol now are out of fashion but can be found in the older treatises.

159. **Vernix**, sandaros and sandarac in Greek, but in the more recent translations from Arabic, *Sandaraca* is a clay-mineral called red orpiment. We know it as two kinds of vernis. The white is warm and dry gum in the 2nd deg. which is used by the scribes who powder their moist and soft parchments to dry and stiffen them. The red vernix is considered to be the same as the Arabic *Kahruba (karabe)* which is called amber in France (ie see #183).

160. **Sanguis draconis (sang dragon)**, *Dam al-alkhwein (alachiten and alakhuein)*, is an herbal sap, moderately warm, and dry in the 2nd deg.

Synonyms for Incarnatives

161. **Palma,** *Nakhl (nacla)*, the date-palm. It is styptic.

162. **Dactylus**, *Rutab (rotab)*, the **Date** itself, the branches, the leaves, and the bark of the tree. Cool and dry in the 2nd deg. The sweet date is cool and moist in the 4th deg.

163. **Malum granatum (pomegranate)**, pomum granatum, *Rumman* (*ruman*).

164. **Cypressus (cyperus)**, cyparissos in Geek, *Halharem*,. Its gall (ie cone) is *Sâru (saro)* in Arabic, is warm in the 1st deg. and dry in the 2nd deg. Serapion said that it is both warm and dry in the 1st deg.

165. **Quinquefolium (tormentilla)**, pentatomon (pentaston) in Greek, pentaphyllon, pentapetes, (pentarasson), callipetalon, (camelleston), pseudoselinum. Extremely dry without being corrosive.

166. **Seges silvestris (wild sedge)** resembles the forage hay that grows alongside narrow lanes where it stands upright bearing slim clusters that resemble wheat. It is cool and stypric.

167. **Porrum (garlic)**, *Kurath*, prason cephaloton in Greek (that is, porrum capitatum). Two forms: the garden porrum is well known, warm in the 3rd deg. and dry toward the 2nd deg. The wild forms include so many varieties that we do not know how to distinguish them.

168. **Farina volatica (mill-dust).** The powder that is found in mills, often settled on the walls. Cool, styptic and restrictive.

169. **Cauda equina.** A common herb that grows along streams. *Dsanab al-Kheil (deneb achil)*, herba caballina, hippuris and hippos in Greek, cheval, horse-tail. Cool in the 1st deg. and dry in the 2nd deg.

170. **Calx viva (quick-lime),** *Nura*, is the ash of limestone. Warm, dry, caustic.

Synonyms for Cicatrixatives

171. **Cortex pini (pine-tree bark).** Very styptic, warm and moist. The word pine derives from 'point', referring to its needles which are its leaves

172. **Os sepiae (squid bone),** the name of a fish, *Rubijjan (rubien or rubijan)*. The entire bone is removed. It is cool and dry. The topical is applied as a powder to consolidate flat ulcers which are slow to heal.

173. **Cortex bugiae (barberry)**, the bark resembles cinnamon (cannelle) but it is thinner and less yellow. Some say that it is the bark of the tree called berberis. See #24.

174 **Nux cypressi,** the fruit of the **Cypress**, warm and dry in the 1st deg., and styptic. See #164, above.

175. **Curcuma (turmeric)** (see #156). It is a saffron-colored root, used for dying cloth. According to Avicenna it is the root of the small chelidonium. Warm and dry in the 3rd deg.

176. **Fraxinus, (ash)**, dogwood tree, *Dardar (dirdar)*, some authors say they are the same, whereas others say that the oak and the elm are the true dogwood (or Stinking Trees), because many small worms are engendered in their twisted leaves. The Ash is cool and dry in the 1st deg. The inner bark quickly heals fresh wounds.

177. **Aes (bronze)**, a metal *Nu'has (nohas)*, ios (ion) in Greek. Warm and dry in the 3rd deg. There are five types. 1. The patina of bronze **(Vert d-arain)**, green bronze, *Zindjar (ziniar)*. 2. *Fleur d'arain*, chalcanthos in Greek, chalcos meaning bronze, and anthos meaning flower. It forms as fine scales on the surface of bronze that has been heated by a fire enhanced with bellows. 3. Battitura, *Tubal*, are flakes produced by hammering sheets of bronze. 4. Filings, limailles. 5. Burnt bronze, chalcophonon (chalkumenon) in Greek. So-called because it resounds when struck. You heat the (sheet) metal in an earthenware bowl placed in a furnace until it is calcined.

178. **Thanacetum (tansy)**, athanasia. Two types: Tansy is the well-known cultivated plant, called Saint Mary's Herb. Wild tansy grows in dry harvested fields where it lies close to the ground rarely lifting its leaves or stems. The inferior face of the leaves is white, resembling absinthe; the flowers have a pure golden color.

179. **Rubea (madder)**, rubia. Three varieties: 1. Large Garance, growing wild among hedge-rows, often profusely. 2. Spargule, the small garance resemble the large, but it differs in having sharp bristles and is much smaller. It grows in hedge-rows and in shaded places. When touched, the bristles prick the

hand and are painful to remove. They commonly are called 'goose-grass'. 3. The dyers' garance, *Fuwwa (faue)*. The roots are called 'the dyers' veins. It is cultivated, and well-known. All of the garances are warm and dry towards the 3rd deg.

180. **Lumbrici, earth-worms**, earth-pipes, *Kharâtin*, gês entera (gisenteria) in Greek.

181. **Axungia (veal-lard)** is heavier than pork-lard.Therefore, it is better for treating wounds and ulcers that attract humors, because it obstructs the inflow, as a defensive shield.

Synonyms for Corrosives

182. **Sapo (soap)**. Two types: Soft white soap, savon in French. Hard soap, gray soap, so-called saracenic, cut in slices to make suppositories. Both types are caustic and ulcerate, heat, putrefy and absterge.

183. **Arsenic** (Greek), from arsenago. It is **Orpiment**, *Zirnikh (zarnikh)*. Two types: the yellow and the red, called sandarac (see #159).Warm in the 3rd deg. and dry in the 2nd deg.

184. **Esula**, *Schabram (scebrum)*, an herb resembling cataputia or euphorbia, lathyris. It does not grow in France, and is not one of the six varieties of plants that have corrosive milky sap, all described by Dioscorides. The three that grow here are cataputia, anabula and marsilium. All these plants with milky saps belong to the species of tithymals, the name derived from the Greek tithyon (tithê) meaning breast, referring to the milky nature of the sap. There are other plants that have milky saps which are not corrosive, such as endives and others not classified as lacticinia. Finally, there are some without corrosive milky sap such as laureole, and crow-foot.

185. **Anabula,** a common plant belonging in the group with corrosive milky saps. In France it is called amblete. It grows profusely in sandy soils.

186. **Cantharis (spanish fly),** cantarica, *Dsarârîh or Dsurâ'h (adherira)*, is suppurative, dissolvent and ulcerative, warm and dry in the 3rd deg.

187. **Ruta (rue),** *Sadsâh (sadeb).* Two types: The cultivated rue is well-known. The wild rue has three species. 1. This is like the garden type, but grows in meadows and on mountains. It is applied on the surface of ulcers. It is sweet-smelling, warm and dry. 2. Harmala or moly as called by Dioscorides. I mention Peganum here, only because it is called rue in error (ie by common folk). 3. So-called scrophularia and quadrangula (castrangula), a kind of thornless nettle. The true garden— type rue is warm and dry in the 3rd deg. The sap of the wild rue is called *Thapsia (tefisie)* in Arabic. It is warm toward the 1st deg and dry in the 3rd deg., and it is slowly acting and corrosive (see #193).

188. **Anacardus (cashew)** seems to be a Greek term., *Balâdsûr.* The fruit of a tree called Elephant's Foot by some writers. Warm and dry in the 3rd deg.

189. **Spuma maris,** halcyonion (alcionum). It is spongy and includes, according to dioscorides and Pliny, five sorts which we do not know how to distinguish. It is warm and dry in the 3rd deg.

190. **Staphis agria,** *Habb al-Râs (habelraz),* mountain grapes or wild grapes, louse-herb because it kills lice, head-berry because it attracts a lartge amount of phlegm from the head. Warm and dry in the 3rd deg.

191. **Hermodactylus** (in Greek), Mercury's fingers, colchicon, *Surandjan (surungen),* a theriac for joints. Warm and dry in 3rd deg.

192. **Gentiana,** called myrrhica by some authors, genista by others. Warm and dry in the 3rd deg.

193. **Thapsia,** actors' herb because they use it to disguise the face and hands to mimic leprosy (ie it is not wild rue, #187).

194. **Pes (foot)** a commonly used name for many plants.

   a.   **Pes gallinaceus (hen's foot)** called thlaspi by Dioscorides, capnos and fumaria by Pliny.because it cures invisible lesions (caliginem et fumum oculorum)

   b.   **Pes pulli (portulaca),** Robert's herb, hawkweed, musk-needle

   c.   **Pes columbi (dove's foot).** spergule, gautier's herb, small mauve.

   d.   **Pes alaudae (larkspur),** delphinium, jonquarola, calisiana

   e.   **Pes corvi (crow's foot),** a ranunculus, wolf's paw.

   f.   **Pes locustae (locust's foot)** see #43. atriplex hortensis.

   g.   **Pes leporis (hare's foot)**, sanamunda (q.v.)

   h.   **Pes vituli (calf's foot)**, arum, Aaron's beard, serpentaria, dracunculus.

195. **Realgar**. A mineral imported from India, resembling orpiment. Strongly corrosive, dangerous and toxic. It should be applied only in small amounts and after being tempered as follows: grind it and then soak it in white vinegar for ten hours, Filter it through a double-fold of linen, and allow it dry in the sun. Repeat the maceration in vinegar, the filtration and the drying two or three times. Then you may safely apply small bits, about the size of a lentil, even on feeble patients. However, the old practitioners and those who continue to use it say that it will corrupt and destroy when applied to noble and non-fleshy parts such as the head, the penis, the lips and the nose. However, I believe that they refer to realgar wrongly applied in excessive amounts.

196. **Sal alkali (salsola)**, so-called *Kali*, a plant which grows at the sea shore, and is used in the manufacturte of glass.

197. **Sal gemma (rock salt)**, Cappadocian salt.

198. **Sagimen nitri,** is considered by the apothecaries to be as corrosive as couperose. I think it is spuma nitri, *Baurak (baurach)*, aphronitron in Greek.

199. **Tartarum (tartar),** *Durdj (tartar)*, the dry sediment of wine that adheres to the inside of containers. After it is combusted it becomes dissolvent, abstersive and corrosive of tissues. Warm and dry in the 2nd deg.

200. **Asphodelus**, hundred-heads. albucium, porrago because it resembles pears, sometimes called anthericon (aliteron), *Khanta (khunta)*,. A corrosive. Avicenna said that the plant called hundred-heads is not asphodel but is eryngium

(yringi) and is warm and dry. Those who claim it to be cool and moist are far from the truth.

201. **Nitrum,** *Zadjâdj or Zidjâdj (zugeg),* a mineral of two types: The usual white niter and the red, which is identical with borax. According to Serapion it is one of the most unusual of stones. It is analogous among the stones to the court-jesters among humans, because it can wear all colors and is used in all dyes. Warm in the 1st deg. and dry in the 2nd deg.

202. **Muscus arboris (tree moss)** is the woolly coat on branches of old trees in the forests. Avicenna called it usnée, *Uschna,.* He assures us that usné does not grow on styptic trees such as oak, walnut orpine. It is moderately cool, styptic and acvcording to some writers, warm in the 1st deg. and dry in the 2nd deg.

203. **Alleluia,** cuckoo-bread, a kind of clover with a sour flavor,. It grows among trees. In the springtime it is used to augment the sour quality of agresta.

204. **Gallitricum, centrum galli,** crista galli, sclarea in Lombardy. Two types: The edible, cultivated type is larger. The wild type has much smaller leaves. The seeds are blackish, and not quite round. Some practitioners introduce them into eyes where they are tolerated without discomfort, and they move about between the eye and the lids. When they are discharged from the eye without effort by the patient, they are swollen and coated with accumulated viscous matter. That gives them the name Christ's Eyes.

205. **Malum terrae(cyclamen),** chamaecissos (cassamus), swine-bread, ciclamen, *Buchûr Marjam (buchormarien),* cyclaminos of Dioscorides, tuber terrae of Pliny. Pliny also called the round aristolochium the malum terrae. In French it is commonly called gesnote. It grows in the same places and somewhat resembles fennel. It has a whistle-like stem about one cubit in length, bears at the top small white flowers. At the bottom are many edible roots projecting like small fingers.

206. **Acus muscata,** two varieties: 1. The large is **Robert's herb,** cancer herb, crispula, ziputa, sparrow-foot. Its flower is odorless. It grows in hedge-rows. Its stem and leaves usually

are red. 2. The **Geranium,** with its scented flower, resembles
the large variety, but rarely is red. It grows in sandy soil, along
roadways and walls.

207. **Rapistrum (horse-radish,** raifort), ambrathea, armorica,
raptusan, raphanus agrestis.

208. **Marsilium (a euphorbia),** wolf-bean because the shredded
hard ripe pods will repel wolves. A very corrosive plant. I saw
a deep corrosion produced in a peasant's body by some of
this plant crushed and wrapped in nine folds of a tough
cloth. It resembles black hellebore.

209. **Cataputia (a euphorbia),** cantharides (ie the plant. not the
insect). The large one is called pentadactylus, *Khirwa (kerwa),*
commonly called Christ's Palm in France, garden-guard in
Lombardy because it repels moles. The small cataputia is
common in France, and grows in gardens. It has a stem a
cubit in length which bears a large number of leaves that
resemble sage but are larger. It is a variety of tithymal and has
a corrosive milky sap. Some authors call it the little laureole.
It is a potent corrosive, warm and dry. Some call its fruit
coconidia.

210. **Cannabis** (canabus, canabs), **(Hemp),** *Schah-dânadj
(scehedenig).* Warm and dry in the 1st deg. Its seeds are potent
warmers and desiccatives.

211. **Ligusticum (lovage),** levisticus, called so because it grows
profusely in Liguria. Two types: The cultivated typoe is well-
known, and resembles the macedonian petroselinon. The
wild type grows in meadows and resembles the garden type.

212. **Scabiosa** is so common that it is not mentioned in the practica
or treatises. We are assured, however, that it is a species of
jacea, also largely unmentioned. Both are alike as antidotes
for poisons, a matter less affirmed by the professionals than
by the common folk. In France we have two varieties: The
large scabious is often used by me. It is a large, rough plant,
less good to look at than the small variety, and less potent.
The small plant is lovely, and recently has become popular as
the Scabious of Montpellier.

213. **Jacea** also has two well known species. The black is called macefelon in France. The white has much larger leaves and has the common name of Devil's Bite, because its main root is bitten off near the stem.

214. **Capitellum (a lye)** is the first drops of lye produced by soap-makers which they used in their manufacture. It is a strongly corrosive liquid. The recipe for it and the formulation is given in our chapter on the corrosives.

215. **Cinis (ashes)**, *Ra mâd*, all types of ashes which are warm and dry in the 4th deg. according to Serapion, especially the ashes of bean stalks and wine-tartar (ie potassium carbonate) used to make capitel lye.

Synonyms For Emollients

216. **Faex Olei (lees of olive oil)**, the residue recovered from empty containers. Moderately tempered. The lees of olives are called amurca and are watery. Serapion said that is warm and dry in the 2nd deg., and that it is very resolutive.

217. **Oesypus (ysopus) (lanolin)**, an ointment, so-called Hysop wax. It is described in the chapter on resolutives.

218. **Storax (styrax)**, *Ma'ja (mehahach)*, three types: the moist, the dry and calamite. All of them are chiefly warm maturatives.

219. **Medullae Vitulli et Cervi (bone marrow of calves and deer)**. These are known to be better than others as emollients and resolutives.

220. **Faex alveolarum apium**. *Mu*, two varieties: Pure **Beeswax** is the temperate inner hive where the bees deposit their eggs and honey. It is warmer and drier than wax alone. The less pure beeswax is from the outer hive, with open alveolae. It is dark or red. This variety makes a better medicine.

221. **Mola molendini (mill stone and metal dust)**. I cannot attest that this does any more than the vinegar vehicle in which you have quenched red-hot zinc. The vapor of the vinegar is very soothing (softening).

222. **Sicadis**, white bryony, wild courges, sicyos in Greek, described in # 132. It has many foreign names which are not used in France.

Here Ends Treatise V The Antidotary

FOLLOWED BY

APPENDIX I THECOMPENDIUM PHARMACIOPEIA

A LIST OF SIMPLES AND COMPOUNDS

THE FORMAT

This long list of medications is compiled from the English editions of the eight treatises by Roger Frugard, Roland of Parma, Bruno of Longoburgo, Theodoric (transl.by Campbell and Colton), William of Saliceto, Lanfranchi of Milan, Henri de Mondeville and Jehan Yperman. In the treatises of William, Lanfranchi, Henri (Treatise V only) and Jehan the substances are numbered so to more easily identify them in the pharmacopeias which I have appended to all seven of the works translated by me.[293]

The numbered medications are the major element in the list, and they include most of the items used by the authors in whose Englished texts numbers are not used. Most of the un-unumbered items are common names for popular herbs and medicinal substances, variant spellings and some terms that may lead the Reader to what she or he seeks among the numbered items.

Following each entry there are synonyms, one or more common names and variant spellings, and the names which modern translators have used in the Italian and Anglo-Norman French editions of Roger and Roland (**R** for both of them), the Italian of Bruno (**B**), The English of Theodoric (**T**), the French of William

---

[293] L.D. Rosenman, *A Compendium Surgical Pharmacopeia and Formulary, 1999. Awaiting publication* (LDR).

(**W**), Henri (**H**), and Yperman (**Y**) and the Middle-English of Lanfranchi (**L**). The initials in bold-face type placed at the end of most of the entries indicate the authors who mentioned the substances which I have culled from their texts. Many of the items are followed by "see", to refer the reader to other items for additional information.

Wherever I can, I identify the plants by botanical terms in an *Italic font*. I cannot claim accuracy in many cases, especially where the 19thC and 20thC botanists and herbalists have themselves confused me with variant nomenclatures. I have used Professor Saint-Lager's terminology in Henri de Mondeville's text and Mrs. Grieve's book as well as dictionaries and encyclopedias for assistance in those matters. The Alphita has helped in some cases.

As stated in my General Introduction, I believe that there are only a few items missing from the list, and I apologize for any such deficiencies. The list probably is a nearly complete *surgical* pharmacopeia for the century-and-half during which the eightmajor works were written.

A Supplementary Bibliography for This Pharmacopeia immediately follows the list of medications

A PREFACE FOR THE LIST OF MEDICATIONS

A Medieval Surgeon's materia medica for the most part were Topicals. The List that follows names the substances used by the surgeons as Simple Medicaments or as combined in various forms of Compounds, nearly all of them for external applications, that is, as Topicals. Many of the same substances, indeed many more than appear here, were used by the non-surgical Physicians who prescribed them as medicines to be taken by mouth or administered as clysters, etc. Aside from the very few of them that the surgeons were permitted to use, the laxatives, for example, and some potions, they are not in our list.

WEIGHTS AND MEASURES FOR MEDICINALS

The standards changed from epoch to epoch and from place to place, and the modern equivalents are approximations.

Baas (cf. p.263) cites The Antidotarium of Ncolaus Praepositus of Saler 1140):

1 scruple (scrupula)	20 grains (grana)
1 dram (drachma)	22 scruples
1 hexagium	1 1/2 dram
1 ounce (uncia)	6 hexagia
1 pound (libra)	12 ounces
1 sextarius	2 1/2 pounds

Hunt (cf p. 59) states that the Salernitan system was based on the grain (grana)

1 siliqua	1 grain of wheat
1 obol	3 grains
1 scruple	2 obols
1 dram	3 scruple
1 ounce	6 drams
1 pound	12 ounces

Other measures are stated in the texts. For examples:

A maniple (Yperman): somewhere between a large pinch and a small handful.

A liter (Yperman): a liquid amount somehow equivalent to a pound.

THE LIST

1. Absinthe   oil of wormwood, avrotonin, aloisne.
   Artemisia (sev.species). **R B W L H Y**

1a.	Abrotonin	aurone, southern-wood. see 35. **Y**
2.	Acacia	sap of *Acacia arabica,* Gum Arabic.sometimes dry prunella (368b) see 212. **B T W L H Y**
3.	Acanthus	Bear's breech. see 64. *Acanthus mollis.* **L**
4.	Ache apium.	see 29, 97. *Apium graveolum.* **R W L Y**
	Acedula	rumex, acetosa. see 395. **R**
4a.	Acetum	vinegar. see 484
5.	Acoris sweet	sedge. see 74, 75, 422, 434, 444. **W L**
	Acorns	glans quercus. see 203, 309a. **Y**
6.	Acus muscata	Robert's herb, crispula, geranium, Gratia Dei. see 209, 366.*Geranium Robertianum.* **H Y**
7.	Adiantum	delicate ferns, maiden hair, venus hair. see 362, 363, 476. *Adiantum capillus veneris.* **T W Y**
	Adeps	beef lard. see 206. **Y**
8.	Adracanth	tragacanth, diagradon. see 466. **H Y**
9.	Adustum	ashes. see 38. **L Y**
10.	Aes	chalcanthum, ios, bronze. see 65a, 482. **W L H**
	Affodil	garlic. see 43,198 and sometimes daffodil 301. **R**
11.	Agaric	a mushroom. *Polyporus igniarius.* **R T W H**
11a	Agnus castus	goat-weed. see 22, 205. also chaste-tree, *Vitex trifolia.* **T**
12.	Agresta	sour-grape (uva acerba) juice, verjus; (also memitte, a poppy). see 245a, 335, 478. *Vitri vinifera.* **T L H**
13.	Agrimony	the herb. see 172. *Agrimonia eupatorium.* **T H Y**
	Ailes	allium. see 16a, 43.
14.	Airaine	'flower' of bronze, areim, ereim, eneim. see 481. **R W L H**
	Alabaster	calcium carbonate, resembling gypsum. see 214. **Y**
	Albugasse	donkey's milk. see 286. **W**
	Alexander's Ivy	lovage. see 247, 256, 430. **T**

15. Alkanna   henna *Lawsonia inermis.* **T W H Y**

15a. Alkitron   cedar resin. See 96b. **H**

16. Alleluia   a clover, weed sorrel, oxalis, paniscuculi.
                      see 395 et al. **R L H**

16a. Allium   garlic. see 43, 198. **H Y**

17. Almonds   bitter or sweet. see 309. *Amygdalis communis.*
                      W H Y

18. Althea   guimauve, geumalve, marsh mallow,
                      bismalve. see 273. **R B T W L H Y**

19. Aloes   aloe vera, picra, pigré. Horse aloes are
                      unpurified. *Aloe socotrina.* **R B T W L H Y**

20. Alum (alun)   from wine lees (potassium tartrate) or from
                      crystals. Alun de roche is sugar-alum (
                      aluminum sulfate), alumen de pluma
                      (feather alum) is halotrichite (iron-
                      aluminum sulfate) etc. **R B T W L H**

20a. Amber   see 85. **T W H Y**

20b. Ambergris   from whales. here more likely ambrosia.
                      see 35. **Y**

21. Amidum   flour of various grains, amylum, usually
                      wheat, froment. see 177, 187, 395a, 436,
                      494. **T W L H Y**

22. Ammi   goat (or gout) weed, seeds. see 11a,
                      205.*Ammi visnaga or majus.* **W**

23. Ammoniacum   hore-hound. resin of *Dorema ammoniacum.* or
                      Bubon gummifer. **B W L H Y**

24. Amome   ginger family. see 201. **W**

24a. Amorantus   almarus, amens. see 42 *Amaranthus
                      hypochondria and blitus.* **T H**

24b. Anabulle   euphorbia. see 173. **H Y**

25. Anacardus   cashew nut, elephant-foot tree.*Semicarpus
                      anacordium.* **T W L H Y**

     Anagallis   pimpernel. see 292, 348. **Y**

25a. Andronium   a primrose. *Primula vulgare.* **T**

25b. Anemone   wind-flower and herb. *Anemone pulsatella.* **T**

26. Aneth   dill, or sweet anise. see 158. *Anethum
                      graveolens.* **R W L H Y**

26a.	Angelica	emperor's herb. see 167. *Angelica archangelica.* **W H**
27.	Anise	leaves and seeds. see 349. *Pimpinelle anisum.* **R W H Y**
27a.	A(r)naglossa	arnaglossus, plantain. see 93, 202, 353. **Y**
	Anthera	from roses. see 389. **H**
28.	Antimony	stibium, metal filings. **T W L H Y**
28a.	Ants	insects and eggs. **T L H**
28b.	Apis	honey bee. **H**
29.	Apium	many species. ache, batrachium, patalupi, wild celery, persil, wild parsley, petroselinum. see 4, 97, 344. *Apium graveolens.* **T W L H Y**
30.	Apostolicon	Apostles' Ointment from twelve ingredients: 23, 33, 53, 188, 194, 253, 299, 312, 321, 379, 482, 493 also called Black Ointment and Venus ointments. **R B T W L H Y**
30a.	Apricot	see 33a. *Prunus armenica.* **T**
	Apple	poma. see 363a.
30b.	Aqua forte	nitric acid. **W**
31.	Araignee	spider web. **R H**
31a.	Argentum Vivum	mercury. see 283. **W H Y**
31b.	Arancium	orange. see 119, 322. **W L H**
32.	Argilla terra s	igillata. clay. see 59, 119a, 460. **W L H Y**
33.	Aristolochium	snake-root, malum terrae, polyrrhyzon, clematis, birth-wort, serpentaria, see 58, 120, 142, 376. *Aristolochium rotunda.* **R T W L H Y**
33a.	Armoniacum	armenian apricots (not ammoniacum). Sal armoniacus see, 30a, 424. *Prunus armeniacus.* **R T W**
34.	Arsenic	auripigmentum, orpiment, sublimate of arsenic sulfate or oxide. see 43c, 327 (not 424). **R B T W L H Y**
35.	Artemisia	abrotonon, citronella, mugwort, armoise,

		aurone, ambrosia St. John's plant, absinthe. see 1a, 20b. *Artemisia vulgare.* **R W L H Y**
35a.	Arum	dracuntium minus, cuckoo-pint, dracunculus, calf's-foot.see 500. *Arum maculatum, Dracunculus vulgaris.* **T H Y**
35b.	Arundo	marsh-reeds. see 386. **W**
36.	Asa Dulcis	laser, benzoin, tapsii. see 41, 447, 461.*Styrax benzoin.* **T W H**
37.	Asafoetida	assa *Ferula foetida.* **B T W L H Y**
38.	Ashes	see 9, 60, 99, 111, 117. **R B T W H Y**
38a.	Ash tree	tamarix. see 189, 264, 453. **R H**
	Aspen tree	poplar. see 364a.
39.	Asparagus	asperge, asparagus roots or stems. *Asparagus officinalis.* **T W H Y**
40.	Asphodel	albutum, anthericon, centum capitum, porrago. roots of *Asphodelus albus.* **W L H**
41.	Assefan	styrax. see 36, 447. **W**
41a.	Assius	lapis a stone from Assa (the Troad). **W**
	Atrement	atrament. vitriol based. see 490 et al. **W Y**
42.	Atriplex hortense	orache, arroche, pes locustae, strawberry spinach, blite. see 24a *Atriplex hortensis.* **T W H**
43.	Aulx	ailes, garlic, allium, elinnium, affodile (wild), ramsome, porrum. see 16a, 43, 365a. *Allium sativum.* **R T W L**
43a.	Aunée	scabwort, inula. spikenard. see 437 et al. *Inula helenium.* **T W**
43b.	Aurea	aurum. a decoction of gold particles in water or vinegar. **T**
43c.	Auripigmentum.	orpiment. see 34, 327. **R T W L H**
	Avellana	hazel-nut. see 309.
	Avena oats,	haveron. see 310. **T H**
	Avrotonin	abrotinin. see 1a, 35. **T W L Y**
44.	Axonge	lard, lardon, oint, sui, axungia. see 206 et al. **R L H Y**
45.	Bacca populi	poplar berries, bourgeon. see 364a. **W**

46.	Baccar	asarum, asarabacca. *Asarum europaeum.* **W**
47.	Balaustium	wild pomegranate flowers. see 364. **R B W L H Y**
48.	Balsamdendron	balm of gilead, xylobalsamum. see 53, 90, 299. *Balsamodendron opobalsamum.* **T W H Y**
49.	Balsameta	St. Mary's herb, costmary, mace. see 454 *Tanacetum balsamita.* **H Y**
49a.	Barba	hircina salsify, goat's
49b.	Bardana	sorrel. see 236 and 395 ety al. (also Burdock). **R H**
50.	Barecha	melon. see 144a 280, 369. **W**
	Barley	see 323
50a.	Basilicon Oint.	53, 206, 312, 360, 379, 493. see 314. **H Y**
51.	Basilium	basilicon, herb basil, wall-thyme. *Ocymum basilicum.* **T W H**
	Battitura	chalcanthum. see 106 et al. **T**
52.	Baucia	pastinaca. parsnip or carrot. see 152. **L H**
53.	Bdellium	procerion. a balsamic resin. see 48. *Balsamodendron africanum.* **T W L H Y**
	Beans	fava. see 178.
	Bedegar	wild roses, eglantiere. see 53c. **R H**
53a.	Bees	**H Y**
	Beeswax	mu. see 176, 293, 329a, 397 329a.
	Belladonna	henbane. see 217 et al. **Y**
53aa.	Belliculus	blata bysantia, belliculli marini. purple and white marine snails, source of royal purple dye. **T W**
	Belsegensina	coriander. see 132.
53b.	Ben	been. *Moringa aptera.* **W**
53c.	Bendegard	bedegar. gall from stem of eglantine rose. see 389. **W Y**
53d.	Bennett	Herb Bennett. see 115, 116. **Y**
54.	Beta	Bleta, betta, beets. *Chenopodiacia.* **W H**
55.	Berberis	bartberry, cortex bugia. see 68. *Berberis vulgaris.* **L H Y**
55a.	Berula	an herb, a variety of apium. *Berula angustifolia.* **T H**

56.	Betonica	betony, scrofularia, figwort, vetony,. citonia. see 109, 155a, 185a *Betonica officinalis.* **R T H Y**
57.	Bile	melanchiron, fiel, fellis. see 183. **B T W L H Y**
	Bindweed	scammony. see 416. **H Y**
58.	Birthwort	asirinum, aristolochium, snake root. see 33 et al. **T Y**
	Black Ointment	see 30. **Y**
58a.	Blackberry	sometimes mulberry.see 294, 393. **H Y**
	Blood see 409.	
58b.	Blueberry	whortleberry, airelle. *Vaccinium* var. **H**
59.	Bol d'armenie	a clay containing iron oxide, various clays— German, Bohemian, etc.—sometimes fuller's earth, cimoleam. **R B T W L H Y**
60.	Bombacyna	ashes. see 107, 333. etc. **T W**
61.	Bone-marrow	medulla, midolle, meule. see 275. **T W H**
62.	Bones	of geese, hens, deer. etc. hooves, ivory, horns, antlers, squid etc., all burned and powdered. see 133. **T W L H Y**
63.	Borax	borax, sodium borate; see sal de nitre, also sagimen nitri, nitrum, boracis, bourach. see 402. **R W L H Y**
63a.	Borrago	borrage. *Borago officianalis.* see 69. **R T L H y**
	Bouillon	see oat, wheat.
63b.	Boxwood	dogwood, ammon's horn. *Cornus amonum and floridum.* **T H**
63c.	Bran	sou. see 21. **T H**
63d	Brandy	of any fruit. Eau de vie is from grapes **H**
64.	Branca Ursina	acanthus, brama, bear's brush. see 3. *Acanthus mollis or Heraclium spondylum.* **T L H Y**
64a.	Brassica	generic term. see 71, 377 et al.
65.	Bread	crumbs are mica panis, see 285. **T W L Y.** Panata is bread soup. **W Y** Opirus is whole-wheat bread. **T**

	Bresillet	caesalpina sappan (later called the Brazil tree) **H**
65a.	Bronze	aes. see 14, 481. **H**
66.	Broom	planta genesta. *Cytisus scoparius.* **H**
	Brown Ointment	a regenerative mentioned only by referrence to Nicholas.
66a.	Bruscus	butcher's broom. see 221, 226a. *Ruscus aculatis.* **T W Y**
67.	Bryonia	brionee, viticelle, root of bryony, also sicadis, labrusca. see 120, 144 182, 489. *Bryonia dioecia.* (White bryony, a poison, is *Momordica elaterum*, really a cucumber). **R T W L H Y**
68.	Bugia	root-bark of barberry bush, cortex bugia. see 55. **W H Y**
69.	Bugloss	borrago, blue weed, lingua bovis, lithospermum. see 63a. *Echium vulgare, Borrago officianalis or Anchusa officianalis.* **R T W L H Y**
	Bulbus	onions. see 319, 415.
	Burdock	sorrel. see 236 and 395. **R Y**
	Burith	soap. see 412, 430. **R**
70.	Butter	from cows, also sweet butter, May butter, bure de mai. **B T W L H**
71.	Cabbage	choux, chaux, caulis, cholet. *Brassica oleracea.* **R T W L H Y**
	Cacumia	climie. see 121.
71a.	Cadmia	cadmium sulfate, sory, climie. see 96a et al.
72.	Calamint	thyme. see 278. *Melissa calaminthus.* **T W H Y**
73.	Calamite	styrax. see 447. **T W**
74.	Calamus	aromaticus, squinanthus, sweet sedge, sweet flag. see 5, 75,422, 444. **B T W L Y**
	Calf's foot	arum. see 35a et al.
74a.	Calx viva	quick-lime. see 251. **L H Y**
	Camedreos	germander. see 200.

75.	Camel	grass calamus. see 74, 416a, 422, 444. *Andropogon schoenanthus.* **H**
76.	Camomille	chamomile (several varieties) cotula. *Anthemis nobilis and A. pyrethrum.* **B T W L H Y**
76a.	Campanula	rampion. *Campanula indicas.* **T**
77.	Camphor	champhore, caphura. *Laurus camphora.* **R T W L H Y**
78.	Cannabis	hemp seeds or leaves, chévenis. *Cannabis sativa.* **H Y**
	Canne	reeds. see 378a and 386.
79.	Cannelle	cinnamon. see 94, 117a *Laurus cinnamomum.* **R W H Y**
80.	Cantharides	spanish fly. Eloe vesicatorum. **B T W L H Y**
81.	Capers	capparis. fruits, shells, bark, oil. lonacera, a honeysuckle. *Capparis spinosa.* **R T W H**
	Capillus veneris	venus hair. see 7.
82.	Capital Powder	19, 33,188 228, 299, 325, 329, 408, 410. For head-wounds. **T H Y**
83.	Capitellum	capiteils, potash lye, lessive. see 254. **R L H Y**
	Capreoli	reeds. see 208.
84.	Caprifolium	caprifici, chevrefoile, honeysuckle, licium. a fig. see 260. *Lonicera caprifolium.* **R H Y**
84a	Caputpurge	head-purge. a mega-compound laxative containing marjoram, chelidonium, nasturtium, pyrethrum, staphisagre, nutmeg, long peppr, euphorbia, scammony and rose water. **H**
85.	Carabe	amber, possibly containing a scarab beetle. see 20a. **W H**
86.	Cardamom	amomie, granum paradisis, granum solis. *Amimum cardamomum.* **T W L H Y**
	Cardo	see 102. **H Y**
87.	Carmingella	an aromatic herb, unidentified **W**
88.	Carnis serpentum	snake meat. see 129. **H**

89.   Carolus   dry-rot wood of fallen trees. **W**

90.   Carpobalsama   mace. see 49. **W H**

      Carrot   see 52, 152. **T Y**

91.   Carthame   safflowers and seeds. see 397a. *Carthamus*
                          tinctorius. **R W**

92   Carvum   caraway seeds. *Carum carvi.* **T W**

93.   Caryophyla   garyophyla, giroflé, sanamunda, herb-
                          benedict, cloves, eugenia, arnagallus,
                          carnations, chickweed, stellaria, ipia,
                          horse-foot. see 122, 202, 279. *Caryophylus*
                          aromaticus or Geum urbanum. **B T W L H Y**

      Cashew   see 25.

94.   Cassia   'bastard' cinnamon. and many others. see
                          79, 427 *Laurus cassia.* **T W L H Y**

      Cassilago   henbane. see 217. **R H**

95.   Castor bean   ricinus, cataputia, cocconidium. Diacastor
                          is a laxative electuary. *Ricinus communis.* **B T**
                          W H Y

96.   Castoreum   beaver musk. **B H Y**

96a   Cathimia   cadmia, lapis calaminaris. see 71a, 121, 439.
                          T H

      Cauda equina   horse-tail, hippuris. see 170. **W H**

      Caul   cabbage. see 71. **W H**

96b. Cedar   tree of Lebanon.see 15a. *Cedrus libani.*

97.   Celery (wild)   see 4, 29. **R T W L Y**

98.   Celandine   chelidoine (probably the 'lesser'), figwort;.
                          see 56, 109. **T W H**

99.   Cendre de chêne   ashes of Turkish oak (Quercus cerris)
                          lexivium, see 117, 254. **W**

100. Centauria   jacea, narce, thistle, teazle. see 458. **T W L**
                          H Y

      Centinervia   rib-wort. see 428.

101. Cepa   oignun, ascalonia. onion. see 319. **W L H**

101a. Ceratonus   any of four legumes including carob. **W**

      Ceratum   beeswax. see 176, 329a

102. Cerdone   cardo, chardun, calendula, thistle, senecio.

		see 100, 458. *Centauria centaurium or Cardo benedictus.* **R H Y**
103.	Cerebrum galli	chicken brain. **H**
104.	Cerisier	cerice, cherry tree, bark, sap. *Prunus avium.* **R L Y**
105.	Ceruse	white lead, psimythion, minium. see 242, 253. **R T W L H Y**
	Ceterach	see 69, 417.
106.	Chalcanthum	copper sulfate and other metallic salts, battitura see 481, 482. **T W L H Y**
	Chalmedrys	kamedrys, germander. see 200.
107.	Charte de soit	bombacyna, papyrus. ashes of paper made from silk. see 38,60,117 333. **T W**
	Chaux	lime, calx. see 251 et al.
108.	Chebules kebulus	(from Kabul). see 298 (unripe). **T W L H Y**
109.	Chelidonium	celidoine salvage, celandine, mentelicum. see 98, 155a. *Chelidonium majus or Ficaria ranunculoidis.* **W L H Y**
109a.	Chenopodium	pigweed, dog-foot. *Chenopodium alba and rubra.* **H**
	Cherry	see 104. **Y**
110.	Chestnut	nux castanearum, chastaine. see 309. *Castanea sativa.* **R T L**
111.	Cheveux humaine	ashes from human hair, capillae humani. see 38 et al., 117. **T**
112.	Chick	peas cicer. *Cicer arietinum.* **T W L H Y**
	Chick-weed	stellaria. see 445a.
113.	Chicory	leaves. endives, scariola, groin du porc. *Cichorium intybus.* **W L Y**
	China root	galingale. see 193.
113a.	Cicada	incinerated insect. **T**
	Cicer	chick-pea. see 112. **H**
114.	Ciclamen	pome de terre, malum terrae, sowbread, cyclamen, earth-nut, panis porcinus. see 262. *Conopodium majus or Cyclamen hederifolium.* **L H Y**

115.	Cicuta	cowbane, water hemlock, cicuta virosa, benedicta. see 116.*Conium maculatum*. **R L H Y**
116.	Ciguë	hemlock seeds, herb bennett, beneite. see 115.*Conium maculatum*. **R W H Y**
116a.	Cimolea	chimolea, cymolia, terra cimolea. mud containing metallic and stoney bits accumulated under whet-stones. see 289. **H**
117.	Cinis	ashes.various woods, bones, crabs, mouse, rabbit, scorpion, sponge, hair, grapevine, oyster shell, sea-shells, seashell, hair, paper, etc. see 38 et al., 111. **T W L H Y**
	Cinnabar	mercuric sulfidesee 283. **W**
117a.	Cinnamon	see 79, 94, 117. **T W H Y**
	Cinq-foil	tormentilla. see 465.**T H**
118.	Cissus	cissus, hedera, Virginia creeper. *Vitis hederacea.* **H**
	Cistus	ladanum. see 234
	Citonia	and oil. see 56
119.	Citrons	various citrus fruits and melons. Venarum citrinum is citrus fruit pulp, citron pips were used alone. see 31b, 250. **B W H Y**
	Citrullus	gourds, melons. see 125, 280
119a.	Clay.	all types. see 32, 59, 460.
120.	Clematis	the flower, ground-ivy, liere, viticella, vitis petit vigne, hedera, flammula, aristolochium. see 33, 215b, 376. *Clematis vitalis and Hedera helix, etc.* **R T L H Y**
121.	Climie argenti	metallic oxides, especially silver, gold and zinc. also cacumia, cathimia, spode, tuthie, iron, couperose. see 253,441. **T W L H Y**
122.	Clove	see 93, 202.
122a	Clover	various trifoils. see 277, 344a. *Trifolium pratensa.* **T**
	Cochia	a laxative compound pill, hierarufinum. see 125, 219d. **T**
122b.	Coctana	dwarf figs. see 185. **W**

123.	Coing	quince. citonium, diacydomite. see 372. **W**
123a.	Cold Seeds	'cold-weather seeds'. The Greater are 137, 144, 191, 280. The Lesser are 245, 366, 392. see 423. **T W Y**
124.	Colle	gelatin ('gluten'). Vellis vaccini is cowhide as a source. see 204. **W H**
124a.	Colchicum	crocus. see 141, 219. **Y**
	Colcothar	vitriol. see 490. **T**
125.	Colocynth	"bitter apple' or bitter cucumber, cucurbite, citrullus. see 280, et al. *Cucumis colocynthis.* **W L H Y**
126.	Colophony	pitch. see 352. **R B T W L H Y**
127.	Columbine	culver-wort, sparagus, geranium molle, pigeon foot. *Geranium columbinum.* **B T**.
	Comfrey	consolida. see 128.
127a.	Condisicale	condes. Not identified, perhaps an ointment based on rye flour. see 395a. **T**
128.	Consolida	comfrey, consoude, greinure. *Symphytum officinalis (large) or Brunella vulgare (small).* **R B T W L H Y**
	Convolvulus	scammony. see 416. **Y**
129.	Cooked meats	beef, lamb, pork, veal, snake, frog and various domestic and wild fowls, and their organs. Sometimes the specifiied source was a castrated male animal. Drippings from roasts. **T W L Y**
130.	Copper	calchanthum, malachite, salts see 106, 482, etc. **T W L H Y**
131.	Corail	coral polyps(red and white), sponge stone. see 162. **W L H Y**
132.	Coriander herb	belsegensima. *Coriander sativum.* **T W H Y**
	Corigiola	see 361
	Cornel	the dogwood berry. see 63b
133.	Cornu	powdered horns of deer and goats, etc. see 62. **L W H**
134.	Cortex pini	bark of pine tree. see 350. *Pinus sylvestris.* **T L H**

135.	Costus	costmary. roots and oils. See 49, 454 *Costus arabicus.* **T W H Y**
	Cotonia	and cotonea malum. quince. see 372.
136a.	Coudrier.	hazel. see 215a. **Y**
136.	Cotyledon	umbilicus venus, wall pennyroyal. see 420. **L H**
136a.	Couperose	see 106, 482, 490. **L H Y**
137.	Courges	zucchini, watermelon et al. Juices are elacterina. Watermelon is cocomero see 144a, 280. *Cucurbita maxima.* **R W H Y**
	Cowslip primrose.	see 368a.
138.	Crab (river)	cancer fluvialis, granchia titrata, crab-meat and shells. **B W L Y**
138a.	Crayfish	flesh, juice and ashes. **H Y**
139.	Crassula	major and minor, sedum, stonecrop, orpine, andrachne, mamilla muris, tegularia, vermicularis. *Sedum purpurascens.* **R L H**
	Cress	plantain. see 353
	Creta	marina chalk. see 251.
140.	Crisomel	oil from orange seeds. **W**
140a	Crithimum	samphire, St. Peter's herb, cretani. *Crithimum maritimum.* **T**
141.	Crocus	saffron. see 124a, 219, 398. **W L H**
142.	Crows' feet	pes corvus, pes milvi. see 33, 120, 376. *Ranunculus sceleratus.* **T W L H**
143.	Cubebs	a pepper. *Piper cubeba.* **W L Y**
144.	Cucumber	cultivated or wild, momordique, cucumiscelle. *Cucumis var. Ecballium elaterium.* **R W L H Y**
144a	Cucurbite	melon. see 137, 144. see 280. Many species. **R W L Y**
145.	Cumin	comin, the herb *Cumin cyminum.* **R W Y**
146.	Curcuma	turmeric. see 503. *Curcuma longa.* **T L H Y**
147.	Currants	red, fresh or dried.see 210a, 393. *Rubis rubrum.* **W H**
148	Cuscuta	dodder; see 148, 169. **T W**
	Cuttle-fish bone	squid. see 62

	Cylotrum	an arsenical and lime corrosive. see 327. **Y**
148a.	Cynoglossus	cinoglossus, hound's-tongue herb, leaves and roots. *Cynoglossus officianale.* **H**
148b.	Cyperes succus.	sedge, papyrus. see 422. berries are from *Cyperus longus, C. rotundus.* **B W L H Y**
149.	Cyprés	the tree. nuts (seeds), bark, leaves. see 396, 479. The nuts are from *Cupressus sempervirem.* **B W L H Y**
150.	Dactylus	dates of palm *Phoenix dactilifera.* Diaphoenicon is a laxative electuary of dates. see 155c. **R T W H Y**
	Daffodil	roots and flowers. see 301. **T**
151.	Damascus plums	see 355 **W**
	Dates	see 150
152.	Daucus	carrot, baucia. see 52. *Daucius carota.* **T H Y**
	Delphinium	larkspur. see 445.
152a	Dentale	entale, lead-wort. *Plumbago europaea.* **T**
153.	Diachylon	Mesüe's (one of several versions, see 284a.) ointment made of: 18, 26, 76, 124, 181, 185, 206, 228, 252, 253, 311, 379, 443, 459, 493. **T W H Y**
153a	Diadragon	an electuary. see 8, 466. **H Y**
153b	Diagridum	an electuary. see 416 **W**
154.	Diamargaritum	rubis troscicata, troche based on 342 and 393. **W H Y**
155.	Diamoron	a syrup of 294. see 317a. **W H Y**
155a.	Dianthus	a betony (56) and an electuary based on 387. **T Y**
155b	Diapenidon	a potion based on barley sugar. see 323. **H**
155c.	Diaphoenicon	Diapalma. see 150
156.	Diarhodon	an astringent powder from 2, 27, 59, 116, 407(3kinds),460.**W**
156a.	Diateron piperion	see 341 **Y**
156b.	Diatesseron	see 462. **T H**
157.	Diazinziber	a purgative made from 79, 86, 93, 201, 276, 295. **W**

158.	Dill	leaves and seeds. see 26. **B**
159.	Dipsaccus	teazle thistle, carduus, cardo. see 100, 458, 487. *Cardo fullonum and Dipsaccus sylvestris.* **L H Y**
	Dock	sorrel. see 395.
	Dodder	cuscuta, Hell weed, (lesser dodder). see 148, 169
160.	Dogwood	box-wood, stink-weed, punaise. The oil the berry is cornel. see 63b. **T H**
160a.	Dove's Dung	star of Bethlehem. *Ornothogalum umbellatum.* **T**
160b.	Dove's Foot	sparagi, *Geranium Molle.* **R Y**
161.	Dragantum	iron peroxide, calcantum, colcothar chalcidis. see 490. sometimes confused with 466. **T L H Y**
161a.	Dragon-weed	tarragon. *Dracunculus vulgaris.* **T**
161b.	Dwarf elder	goat-weed, hieble. see 205. **H**
	Earthworms	see 258 and 499.
161c.	Ebulus elder tree.	see 165 and 406. **H Y**
162.	Écume de Mer	meerschaum, spuma maris, at least 5 varieties, including sponges, algae and corals, halcyon. see 131. **R W H L Y**
163.	Eggs	ova. whites (album), yolks (moel), shells, whole. **B R T W L H Y**
163a.	Eglantiere	see 389. **H**
164.	(H)Ellebore	eleborum, ellebre, Christmas rose. *Helleborus album and nigrum, Veratrum album.* **R T W H Y**
165.	Elder	sureau, sambucus, ebulus. juice and a tube of bark slipped off a twig to serve as a cannula. see 161c, 406. **B T W H Y**
165a.	Elm	slippery elm tree bark. *Ulmus fulva.* **R T**
166.	Emeralds	aluminum silicate gemstones, smaragdus, praze, prassium. **W H**
167.	Emperor's herb	an umbellifer similar to angelica. see 26a. *Angelica archangelica.* **W H**

167a.	Encaustrum	caustic red ink, or vernis. see 479, 490. *Terminalis vernis.* **T W**
168.	Encens	see 188, 316, 463. **R Y**
	Endive	pig-snout, groin du porc, chicory. see 113, 392. **T**
	Entale	dentale. see 152a
168a.	Enula	inula, elecampani. see 437. **W H Y**
169.	Epithyme	like cuscuta. see 160. *Cuscuta major and minor.* **W H**
170.	Equisetum	asperella, cauda equina, queue equina, horse-tail reed, shave grass. *Equisetum arvense.* **T W H**
	Ers vetch.	see 325, 483.
171.	Eruca	mustard weed, charlock, rocket-root, sinapus. see 297, 426. *Sinapus avensis.* **R T W Y**
171a	Esula	a spurge. see 173, 442. **T W H Y**
172.	Eupatorium	agrimony. see 13. **T W Y**
173.	Euphorbia	amblete, custos hortis, marsilium, manne, solsequium, esula, titimalle, cataputia. anabulle is the sap. see all named here. *Many varieties of spurges, Euphorbeacea.* **R B T W L H Y**
174.	Fabaria	lovage, water parsnip. see 247, 256, 430. **T H**
175.	Faex	precipitate at bottom of oil jugs, faex olei. **T W H Y**
176.	Faex	cerae beeswax, faex alvearum, eryngium. see 293, 329a.368aa, 397. **H**
177.	Farina de Moulin	amidon, far, amylum, farina volatica, mill-dust (fine wheaten flour) also found at bake-ovens, see 21,187,494. **R T W L H Y**
178.	Fava	feve, fabba, beans or their stalks, especially *Vicia faba.* **W L H**
179.	Feces	stercus, tordus; sheep, goat, deer, birds, horses, human etc. see 184. **R T W L Y**

180.	Fennel	fenoil, marathon. leaves. see 307. *Foeniculum dolce et vulgare, and Nigella sativa.* **R T W L H Y**
181.	Fenugrec	aegoceros, buceros, telis. seeds, Greek fennel, ferrugine. seeds usually powdered. see 180. *Trigonella foenum-graecum.* **R B T W L H Y**
	Fermentum	see 502. **W H**
181a.	Ferns	filix, beech fern, royal fern etc. see 7,362 363. var.*Osmundia* **H Y**
182.	Fesire	white bryony. see 67. **W**
	Fever-few	tansy. see 454.
	Ficus	fig. see 185.
183.	Fiel	bile, fel, felles, (dog, cow, bull, ox, felles avium, colre, choler etc.). see 57. **W L H Y**
184.	Fiente	fimus columbinus or equus. pigeon or horse droppings. see 179. **R T W L H**
185.	Figs	alos, coctana, citonia. caprificus is a wild fig. see 84, 122b, 260. *Ficus arboris.* **B T W L H**
185a.	Figwort	pigamon, quadrangula. see 56.
186.	Filipendula	spirea, dropwort, meadow-sweet. *Spirea filipendula.* **H**
186a	Fish	many varieties by name.All authors
	Fisticus	pistachio. see 351.
	Flammula	clematis. see 120.
	Flaura	an herb of various sorts: a clover, a fumitory, etc. see 122a, 191.**R**
	Flax	linen. see 252.
	Flies	house flies. **H**
186b.	Flos	'flower', battitura aeris, aloxan. various films (usually metallic salts) deposited on metals and fluids. see 481, 482, etc. **W L H Y**
187.	Flour	farina, furfur, samich, froment, pigle (coarse-ground). see 21, 177, 395a, 436, 494. **R T W L H Y**

188.   Frankincense   gum resin. see 168, 316, 463. *Boswellia cartorii.* **R B T W L H Y**

189.   Fraxinus   bark of ash tree, fiêne, stone-mint, manna. see 264, 389, 453. *Fraxinus excelsior.* **T W H Y**

        Frog   see 375. **B T**

190.   Fuchsia   the flower. see 334. *Parietaria officianalis.*

191.   Fumitory   fumiterre, gingidium, perhaps flama, lady's mantle, hen's foot. *Fumaria officinalis or Gingidium.* **R T W L H Y**

192.   Fusco   unguentum fuscum, Brown ointment of Nicolas see 306. **R H**

193.   Galanga   galingale, melingalata, China root. see 276. *Alpinia galanga.* **R T W H Y**

194.   Galbanum   a ferula resin. *Ferula galbaniflua.* **R B T W L H Y**

194a.   Galen's Ointment   72, 105, 206, 253, 312, 388, 439, 493. **T L H**

        Galigan   rue. see 394. **Y**

195.   Galls   oak galls. **B T W L H Y**

196.   Gallitricum   salvia, centrum galli, crista galli, oculi christis, sclaria. see 400, 405. **H Y**

197.   Garance   madder root, rubeau, spargula. see 392a et al., 434a, 474, 497. *Galium molugo, G. aparime and Rubia tinctorum.* **Y**

198.   Garlic   affodil, ailes, aulx, allium. see 16a, 43. **R T W**

        Garyophyla   see 93

199.   Gentian   genista. bitter roots of *Gentiana lutea.* **T W L H Y**

        Geranium   see 6, 209.

200.   Germander   scordium, polium, chalmedrys, camedrios, yva. see 346, 418. *Teucrium polium and montanum.* **R T W L H Y**

        Geum   Eugenia. see 93

201.   Ginger   zingiber, genevrier, abel. see 24. *Amomum zingiber.* **R B T W L H Y**

202.   Girofle   cloves, caryophyllus. see 93 et al. **R Y**

	Git	coriander. see 307
202a.	Gladiolus	sword-grass, yellow flag, see 228. *Iris pseudoacoris.* **H**
203.	Glans	acorns (quercus). see 309a. **H Y**
	Glue	see 368aa. **W L**
204.	Gluten	colle, gelatin from fish or domestic animals, mucilage, animal collagen (not from grains). see 124, 242a, 430bb. **R B T W L H**
	Glycyrrhyza	licorice. see 246, 381.
	Goat's beard	salsify. see 49a.
205.	Goatweed	gout-weed, another umbellifer, sometimes Bishops-wort, Agnus caste. see 11a, 22. *Ammi majus, Aegropodium podagraria.* **T W H**
	Gold	incinerated or decocted. see 43b, 121.
	Goudron navire	ship-tar. see 304 et al., 352, 359, 360. **Y**
206.	Graisse	animal fat—chicken, geese, pork, turtle, etc., pinguedo, sepum, ysopus, bovis, gras, sui, adeps. see 225, 235, 311. Grassede are the drippings from roasts. **R B T W L H Y**
207.	Gramen	sedges, couch grass. see 422 et al. *Agropyrum repens.* **L H**
	Grana paradisi	cardamon. see 86, 332
208.	Grape	uva. leaves and fruit, and skins. raisins, sapa michum (with honey), saramitum, capreoli are the tendrils. see 12. **R T W**
208a.	Grasses and Reeds	see 378a, and 386. **R T W H**
209.	Gratia dei	geranium. sec 6 *Gratiola officianalis.* **H Y**
210.	Green Oint.	basically 14 and 273, with supplements by various authors including 188, 256, 272, 329, 360, 424, 459. **R T W L H Y**
210a.	Groseille	currants or gooseberries. see 147 or *Ribes grossularia.* **W H Y**
211.	Gui.	mistletoe. leaves, berries, twigs. *Viscum album.* **T W**
	Guimauve	malva. see 18, 273 **Y**
212.	Gum Arabic	from acacias. **R T W L H Y**
213.	Gummi	sap of cherry, plum, other gums as gumme

d'eve, gum evry, Persian gum. see 466. **T W L H**

214.	Gypsum	cockel, selenite, alabaster, plaster of Paris, calcium sulfate. **T W L Y**
	Hair	see 111
215.	Handacote	septemnerviée. see 428. **W**
	Hare's Beard	see 395
	Hare's foot	pes leporis, see 344a
	Hawkweed	mouse-ear. see 219a, 347
215a.	Hawthorne	spina, hazel. fruits and flowers. *Crategus oxyacantha.* **B**
215b.	Hedera	clematis, funis pauperum. see 120. **R H Y**
216.	Hematite	rust, emathitis, pierre sanguinis, lapis sanguinis, red ochre. see 227, 228a. **T W L Y**
	Hemlock	see 116. **T Y**
	Hemp	see 78. **T Y**
217.	Henbane	jusquiame, hyoscyamus, solathrum, nightshade, belladonna, cassilago, chenille, fabe lupini. see 232, 290a, 431. **R T W H Y**
	Henna	alkanna. see 15.
	Hen's foot	pes gallinacia, fumitory. see 191
	Hepatica	liver-wort, marchantia. see 268.
218.	Hericium	sea urchin bristles. ericium, herisson, hircis, lupis iudaici. see 249. **W H**
219.	Hermodactyl	digitus hermetis, colchicum and other related tubers. see124a. *Hermodactylus tuberosus.* **R T W L H Y**
219a.	Hieracium	hawkweed. see 347.
219b.	Hieralogodon	Galen's laxative. see 344c. **H**
219c.	Hierapicra picra.	see 344c. **W H**
	Hieromandrea	mandrake. see 263
219d.	Hierarufinum	a laxative of colocynth, germander, asafoetida, wild parsley, aristolochia, pepper, cinnamon, saffron, polium, myrrh and honey. see cochia, and 125. **H Y**
220.	Honey	miel, various kinds. hydromel is a mixture

		with water. beehive honey contained some wax. see 330. **R B T W L H Y**
	Honeysuckle	also oculum. see 84. **R**
	Hordeum	barley. see 323.
	Horehound	marrubium. see 271.
	Horse's tail	equisetum. see 170.
221.	Houx	bruscus. holly. see 226a. *Riscus aculeatus.* **T W**
222.	Hyacinth	squill, blue-bell, jacinth. see 443. *Hyacinthus nonscriptus.* **T W H**
	Hydromel	see honey, 220.
	Hyoscyamus	see 217, 232, 431.
223.	Hypericon	St. John's Wort. see 287, 501. *Hypericium perforatum.* **R H**
224.	Hypocystus	hypoquistidos. fungus on roots of *Cistinus hypocystus.* **T W L H Y**
225.	Hysop humidus	arabic for lanolin. see 206, 235. **T W**
226.	Hyssop	herb hasca. *Hyssopus officinalis.* **W H**
226a.	Ilex	holly. berries, leaves, bark. see 221. *Ilex aquafolium.* **T**
	Ink	vitriol. see 167a, 479, 490
	Inula	enula. see 168a, 437 et al. **W H**
227.	Irundinum	iron. see 228a, et al.
228.	Iris	yreos, powdered orris root, eris ustis is burnt iris flowers, oil. see 202a. *Iris germanica, illyrica, florentinma, etc.* **T W L H Y**
228a	Iron	fer. yellow ochre. see 227, 249a, 267, 289, 419. **R T L H Y**
	Isis	ysis. see 502a. **T**
	Ivy	various. see 120
228b.	Jacea	scabious.dijaciton is a laxative. see 414. **H Y**
	Jacinth	hyacinth. see 222, 443.
	John	Saint. see 372a, 401a et al.
229.	Joubarbes Jove's	beard. house leeks, stonecrop, tettesuriz, poireaux, sticado. see 101, 229, 243, 319, 425. *Sempervivum tectorum.* **R T W H Y**

230.	Jujubes	fruit of the trees. *Zizyphus vulgaris and Z. saturna.* **T W Y**
231.	Juniper	savine, sabine. needles, cones, oil. see 479. *Juniperus sabina.* **R H Y**
232.	Jusquiaime	henbane, aconite, marsilium, faba luparia, luparis, chenille, morel, caniculata, dens caballinus, symphoniaca. see 217, 290a, 431. *Hyoscyamus albus and niger.* **R T W H Y**
	Kabiteji	lupin. see 259. **R**
233.	Kekenji	winter cherry. *Physalis alkekenji.* **H**
	Knot	grass polygonum. see 361.
234.	Labdanum	sometimes ladanum or laudanum. resin of cistus trees, especially *Cystus creticus.* **B T L H Y**
	Labrusca	bryony. see 67.
	Lac milk.	see 286. **H**
234a.	Lacca	lactea. a red resin from litmus. see 325. *Roccella tinctoria.*
	Lacertus	lizard. see 257, 446
	Lactucca	lettuce. see 245. **H**
	Lana succida	wool-fat. see 235, et al. **H**
	Lanceola	plantain. see 353. **H**
235.	Lanolin	suint, lana succida, hysop, ysopus, oesype. see 206, 311, 498. **W L**
236.	Lapathum	sorrel, rumex, burdock paradella, lappa, lapazio. see 49b, 395. *Lappatum acutum.* **R T W L H Y**
236a.	Lapis lazuli	powdered blue gem, ferrous sulfate. **T W**
	Lappa	sorrel, burdock, lapathum. see 236 and 395. **H Y**
	Larkspur	delphinium. see 445.
237.	Laterinum	oil of the small fish. **T**
	Laudanum	labdanum. see 234. **L**
238.	Lauriers	lorier, baccis lauri. berries, leaves of bay trees. Laurin is oil of bay leaves. *Laurus nobilis.* **R T W L H Y**

239.	Laureole	de-barked daphne twig, spurge laurel, medulla milici. *Daphne laureola.* **R W L H**
240.	Lavender	*Lavandula stoechas and L. officinalis.* **T W Y**
241.	Lead, filings	limailles, minium. See 249a. **R W L**
242.	Lead,	white lead sulfate is galena. see 105, 253. **R T W L**
242a.	Leather	especially shoe-soles. See 124, 204, 431a **H Y**
243.	Leeks	see 229, 319. Prassium (emerald) because of its green. see 166. **W H**
	Lemon	the peel. see 250. **H**
243a.	Lentigo	water-moss, probably sphagnum. *Sphagnum cymbifolium.* **T H Y**
244.	Lentils	lens, lentes. **R W L H Y**
244a.	Lessive	a strong soap, a weak lye. see 254, 412, 430c. **H Y**
245.	Lettuce	laitues, lactuca, lettue. *Lactuca sativa; Lactuca agresta* i is wild lettuce (escarolle, endive). **R T W L**
245a.	Levain	leaven. Bakers' yeast. **H**
	Levisticus	lovage. see 247. **R**
246.	Licorice	liquoritia, herb reglisse. see 381. *Glycyrrhiza glabra.* **T W L H**
247.	Ligusticum	levisticus, lovage. see 174, 256, 430. *Ligusticum levisticum.* **L H**
248.	Lily	lis, arcus daemoniacus, crinon. usually oil of bulbs and leaves. *Lilium candidum et al.* Lily often included iris, narcissus and gladiolus. **R B T W L H Y**
248a.	Lily of the Valley	mayblossom (? May butter). *Convallaria majalis.* **T**
249.	Limacons	limax, limazun. snails with shells. **R T W H Y**
249a	Limailles	metal filings including bronze, iron, lead. see 241, 289, 419. **R T H Y**
250.	Limes	limo. limes, lemons. see 119. *Citrus limonum.* **H W Y**
251.	Lime (stone)	quick lime, chaux vive, calx, creta, chauz.

		fresh lime, powdered, used in cylotrum. See 74a. **R B T W L H Y**
252.	Lin	linois, semence, semen lini, flax seeds and oil or meal. *Linum usitatissimum.* **R B T W L H Y**
	Linaria	penny-wort. see 420. **R H Y**
253.	Litharge	litargerie, yellow lead oxide, scum of melted lead or silver. see 121, 242. for litharge nutritum (spumie argentum) see 441. **R T W L H Y**
253a	Liver wort	hepatica, marchantia *Peltigera canina or Marchantia polymorphi.* **H**
254.	Lixivium	aqua cineris, lessive, lye, also very strong soap. see 83, 240a et al. **B T W L H Y**
	Lizard	see 257, 446. **R B T**
254a.	Lodestone	magnetite, ainant, black iron oxide. **H**
255.	Lolium	panicium, zizania. darnel grass. *Lolium temulentum and perenne.* **R T H**
256.	Lovage	"Alexander's Medicine". see 174, 247. Black lovage is *Smyrnum dusatrum,* **T W L Y**
257.	Lucertoli	lacertus. lizard. see 446. **B L**
257a.	Lucius	magna the pike fish, Esox lucius. **W**
258.	Lumbrici	ges entera. earthworms (or maggots). see 499. **R L H Y**
258a.	Lungwort	water lentil, palma Marina, muscus aquae. *Lemna minor.* **R H**
259.	Lupin	kabitegi, flowers of faba lupina, usually powdered. *Lupinus album.* **R T W H**
260.	Lycium	licium. made from 84. see 185. **W H**
	Lye	aqua cineris. see 83, 244, 254. **H**
261.	Mace	myristica. *Myristica fragrans.* **W H Y**
	Madder	garance. see 197, 392a, 474.
	Magnet	see 254a.
	Maiden hair	venus hair. see 7, 363.
	Malum punicum	pomegranate. see 364. **H**
262.	Malum terrae	cyclamen, malot, earth-nut, etc. see 114. **R L H Y**

	Malva	mauve. see 18, 273. **B H**
263.	Mandragora	mandrake, belladonna. Hieromandrea is a potion based on mandragora. *Mandragora officinarium.* **R T W L H Y**
264.	Manna	tamarix, ash-tree. see 189, 453. **T H Y**
265.	Manne	euphorbia, esula resin. see 173 et al. **W Y**
266.	Manuchriston	like diamargariton, with added honey. **W**
	Marathrum	saxifrage. See 349, 413
266a.	Marble	powdered. **L**
267.	Marcassita	iron pyrites (sulfites). **T H**
268.	Marchantia	liver wort, hepatica. *Marchantia polymorphia and M. conica.* **T L**
269.	Margarita	pearls or daisies. see 342. **R H**
269a.	Marigolds	various. *Calendula officinalis.* **H**
269b.	Marine	pumace a coral fish. **T**
270.	Marjoram	marjolaine, oregano, hortensa, amoracus, maiorama. sweet marjoram. see 324. *Origanum marjorana and O.vulgare.* **T W H Y**
	Marrow	bone. see 275. **R T W**
271.	Marrubium	maruil neir, samsucus, linoscrofon. horehound. *Marrubium vulgare.* **B T W H**
	Marsilium	wolf-bean. **H**
271a.	Martiaton	soldier's ointment. 1, 51, 72, 238, 270, 312, 394, 400, 493, 496. **T L**
	Mary	Saint. See 361, 401a.
272.	Mastic	resin and oil. *Pistachia lentiscus.* **R W L H Y**
273.	Mauve	guimauve, mallow (various), althea, sanaticula, malve, ebiscus malaviscum, cubes, dialthea, diante, mallachee. see 18. *Althea officianalis.* **R B T W L H Y**
	Meadow-sweet	see filipendula, see186
273a.	Meats	carnes, including snakes and moles. see 88, 129, 375. 451. **T L H Y**
274.	Medlar	nespole, mespilus, the fruit. *Mespilus germanica.* **L H Y**

275. Medulla   moelle, meule, midolle, bone marrow,
                           specific animals and bones are named. see
                           61. **B T W L H**

      Meerschaum   spuma maris. see 162.
      Mel   also miel. honey. see 220, 330
276. Meligalata   galanga. see 193. **W**
277. Melilot   corona regia. sweet clover. *Melilotus leucothea
                           and arvensis.* **T W L H**
278. Melissa   sweet balm, calamint. see 72. *Melissa
                           officinalis.* **T W H Y**
279. Mellicrate   pomegranate fruit, or a sweet honey-water
                           beverage. See 364 **R**
280. Melon   barecha, citrullus, pumpkin, melo, citri,
                           pepo, squash. See 50, 144a, 492. **R T W H Y**
280a. Memitte   yellow-horn poppy (confused with 109)
                           Glaucum flavum. **R T H**
281. Menthastrum   wild mint, aquatic mint, horse mint,
                           sysimbro. See 288a. *Menthastrum sylvestris.*
                           R T H
282. Mercuriale   linozostis, mercurelle. common weed.
                           Mercurialis annua. **T W H**
283. Mercury   argentum vivum. quick-silver. see 186b. **T
                           W H Y**
284. Mesûe's Fetid Pills a laxative containing 11, 19, 79, 125, 220,
                           228, 271, 272, 299, 398, 410, 469. **W Y**
284a. Mesûe's Oint.   19, 65 (old white), 253, 278, 299, 321, 388
                           (or 300), 411, 496. see153. several were
                           attributed to Mesúe. **R B T L H**
285. Mie a pain   mica panis. bread crumbs. see 65. **W L Y**
286. Milk   mother's is lactus muliebris. cow's milk is
                           lait de scroppha, albugasse is donkey's milk,
                           also goat-milk-whey and cheese-water. **T W
                           L H Y**
      Mill Dust   see 177
287. Millefuilles   yarrow, St. John's herb. see 501. *Achillea
                           millefolium, Hypericum perforatum et al.* **B T H**

288.	Millet	milium, granum. *Panicum miliaceum.* **T W Y**
	Minium	lead filings. see 241. **Y**
288a.	Mint	menthe. see 72, 281, 367. **R T H Y**
	Mistletoe gui.	see 211.
288b.	Mithradates	penny cress. the Mithradaticon was a theriac *Thrapsus arvense.* **T H Y**
	Moles	see 273a. **H**
	Moly	rue. see 394
289.	Molo	molendini stercus irundini, cimolia, molybdenum. bits of iron or powder from mill grind-stones, limailles. see 116a, 228a, 249a. **R T L H Y**
290.	Money wart	wood pimpernel, serpentaria, nummularia. see 292, 348. *Lysimachus nummularia.* **Y**
290a	Morel	(not the mushroom) henbane. see 232, 431. *Morel officianalis* **H Y**
291.	Moschus	deer musk. see 96,446. *Moschus moschifer.* **H Y**
291a.	Moss	muscus, musceline, muscus aquae, sphagnum, etc. see 258a, 296. *Sphagnum cymbifolium and Usnea barbata. etc.* **T H**
291b.	Moss Ointment	27, 29, 35, 51, 72, 128, 180, 201, 226, 238, 270, 271a, 290a, 309 (oil), 367, 396, 411. **T**
292.	Mourons	scarlet pimpernel, anagallis. See 290, 348.
293.	Mu	beeswax. see 176, 329a, 397. **W H**
294.	Mulberries	mûres, mora, omorusia. see 155, 317a. *Morus nigra.* **W L H Y**
294a	Mullein	see 395, etc. **H Y**
295.	Mummy	momie, mumie, flecks of desiccated cadavers collected from tombs and catacombs. **R B W L Y**
	Muriate	of soda salt. see 404.
	Musa	plantain. see 353.
295a.	Muscade	nutmeg, muscat. see 308. **W H Y**
296.	Muscus arboris	tree-moss, lichen. see 291a. *Usnea barbata.*

	Mushrooms	various. see 11.
	Musk	deer, beaver. see 96,291, 446. **T W Y**
297.	Mustard	see 171,426. *Synapus alba.* **R B T W L H Y**
297a.	Mustard	Garlic *Sysimbrium allaria.* **R**
298.	Myrobalans	Indian, or yellow. unripe are chebules. emblicus, belliricus. see 108 *Myrobalans indica.* **T W L H Y**
299.	Myrrh	mirre, musa, smyrna, resins of commiphora plants. *Balsamodendron myrra.* **R B T W L H Y**
300.	Myrtle	seeds, leaves, berries, oil, wood, water. *Myrtis communis.* **R B T W L H Y**
301.	Narcissus	daffodils, affodil, asphodel. *Narcissus pseudo-narcissus.* **R B T L H**
302.	Nard	spikenard, spic, and oil. see 437, 475. *Valeriana jatamansi* and *Nardostachys jatamansi.* **R B T W L H Y**
303.	Nasturtium	cresson, water-cress, senationes, garden crew.*Nasturtium officinalis.* **B T W H**
304.	Navale	goudron de navire, ship tar. see 352, 455. **B T W**
304a	Nenufar	water-lily, farfar. see 491. **L H Y**
	Nespole	medlar. see 274. **H**
305.	Nettles	ortie, califex ignita, castrangula, millemorbia. seeds. Quadrangular, ficaria see 326. *Urtica urens.* **T L H Y**
306.	Nicolas' Ointment	see 365. Ung.Fuscum (Brown Ointment), see 192. **T W H**
307.	Nigella	Roman coriander, ciminum, nielle, gith. see 180. *Nigella sativa.* **B T W H Y**
	Nightshade	henbane. see 217. **T**
	Nitre	nitrum. see 402. **R H**
	Nummulare	money-wort. see 290, 292, 348 **Y**
308.	Nutmeg	muscade, nux muscata, noiz muscate, centrum galle. See 295a *Myristica fragrans.* **R W L H Y**
309.	Nux	meats shells and oils of nuts. chestnuts are

		castana, hazel nuts are avellanum, nual is walnut, brou is walnut shells. See17, 110, 309a **R T W L H Y**
	Oak fern	see 362
309a	Oak tree	quercus, robur, chene, glans. tan bark is cortex stypticus. See 203 **W L H Y**
310.	Oats	avoine, aegilope. *Avena sativa.* **T W L H**
311.	Oesypus	Ysopus, lanolin. see 206, 235. **T W L H**
312.	Oil	usually from mature olives, whereas onfacium (318) was made from the thin juice of green olives and was not classed as a real oil. see 248, 313, 317, 344b, 389, 394, 486. **R B T W L H Y**
313.	Oil of Deben	see 53b. from seeds of *Moringa aptera.* **W**
314.	Ointments	Apostles', Basilicon, Black, Brown, Diachylon Galen's, Green, Martiaton, Mesües, Moss, Mummy Nicolas', Palm, Populeum, Rhaze's White, Saracenic,William Somer's, Aurgheons, Yellow (citrin). **R B T W L H Y**
315.	Oleander	shrub. *Nerium oleander.* **T H**
316.	Olibanum	thus, frankincense, cortex olibanum. (thus masculinum from Lebanon). See 188, 463. **R B T W L H Y**
317.	Olives	oil of green or ripe fruits. wood of tree. *Olea sativa.* **R B T W H**
317a.	Omorusia	mulberry. see 294. *Morus nigra.* **T**
318.	Onfacium	omphacus, infantium. thin oil of green olives. see 312. **W H Y**
319.	Onions	cepa, as distinct from leeks, allium, poirium. see 101, 415. **R T W H**
	Opirus	bread. see 65
320.	Opium	pavot, philonium. seeds and pods of poppy. see 337. *Papaver somniferum.* **R B T W L H Y**
321.	Opoponax	epoponac, panax. like myrrh a commiphora resin. *Opoponax chironem.* **R T W L H Y**
	Orache	blite. see 24a, 42.

322.	Oranges	fruit or peel, arantium. see 31b, 119. **T W**
323.	Orges	barley, hordeum. penidium is barley-sugar cake, ptisan and vitis alba are barley broth. see 155. *Hordeum vulgare.* **T W L H Y**
324.	Origanum	oregano, wild marjoram. see 270. **T W H**
325.	Orobe	horobus, bitter vetch, ers, vicia, lacca is the gum. see 234a. *Ervium ervilia and E. lens.* **W L H**
326.	Ortie	nettles. see 305. *Urtie pilulifera et al. including Scrophularium nodosum.* **T W H Y**
327.	Orpiment	orphimentum, auripigmentum. yellow arsenious oxide, used in cylotrum. see 34, 43c. **R T W H Y**
	Ossa combusta	bone-ash. see 62
	Ossisacara	and ozzizacara, see 330. **B T Y**
	Ova	eggs. see 163. **H**
328.	Ova formicarum	ant eggs. **T L H**
329.	Oxalis	rumex, lapathum, wood-sorrel, alleluia, trefoil, oseille. see 395 et al., and 433. **T L H**
329a.	Oxycroceum	beeswax. see mu, 176, 293. **H Y**
330.	Oxymel	oxysaccharum. a honey-vinegar laxative mixture, ozzizacara, osisatum, oxylaxativum. **R T W L H Y**
	Oyster	shells incinerated and powdered. see 117.
330a.	Palm Ointment	made with red clay (32), 105, 121 (silver), 253, 289, 419, 441, 460. **W L H**
330b.	Palma marina	lungwort. see 258a. **R H**
331.	Palme vert	heart of palm or date palm. see 150. **W L H**
	Panada	bread soup. see 65.
	Panax	opopanax. see 321.
	Panicium	darnel grass, see 255.
	Papaver	opium. see 320. **H**
332.	Paprika	cardamon. see 86.
333.	Papyrus	paper, usually burnt. see 107. **T L**
334.	Parietoria	pellitory, lichwort, paritarie. see 190 *Parietoria erecta and P. diffusa.* **R L H**
334a.	Parsnip	wild. *Pastinacia sativa.* **T H**

335.	Passula	agresta. see 12, 478. **W L**
336.	Patience	wild dock, rumex, oxalis. see 395.
337.	Pavot	white and black papaver, poppy (seeds). see 320. **W Y**
338.	Peach	persica. fruit, seeds or leaves. *Persica vulgaris or Prunus persica.* **R T W H Y**
339.	Pear	fruit, flowers, leaves, wine. *Pirus communis.* **T W L H**
340.	Peas	pisum, any type, or beans. see 358.
340a.	Peganum	piganum, wild rue (not a rue) probably a scrofularia. **H**
	Penidium	barley sugar. see 323. **H**
	Pennyroyal	see 367
340b.	Peony	flower of *Nus paionia.* **R**
341.	Peppers	piper, serpyllum, various regions. Diateron piperion was made of three kinds of peppers. see 156a. *Polygonum hydropiper.* **W L H Y**
342.	Perles	pearls or marguerite flowers. leaves or oyster-pearls. see 269. **R T W**
343.	Persicaria	water-pepper, smart weed, cul-rag. *Polygonum varieties and Persicaria hydropiper.* **R H**
344.	Persil	parsley, selinum is the seed of apium. *Carum petroselinum.* **R T W H Y**
	Pes corvinus	crow's foot. see 142. **W**
344a.	Pes leporis	Haresfoot clover, sanamunda. see 122a. *Trifolium arvensis.* **T H** 344b.Petroleum oil of. Benite, Holy Oil. **R**
	Petroselinium	see 29, 247. **L H**
	Philonium	opium. see 320
	Pierre de lanternes	unidentified. probably crumbled sandstone. **Y**
	Pigle	wheat. see 494.
344c.	Pigra	picra, hierapicra and Galen's hierologodon. a laxative pill. see, 199, 79, 125, used with or

as alternative to cochia. Contains 19, 58, 79, 220, 272, 398, 437. **T W H Y**

Pig-snout endive, groin du porc. see 113, 392

Pigweed   dog-foot. see 109a **H**

345. Pili leporis   mullein. see 356. **B T L Y**

346. Pilium   a germander. see 200, 418. **W**

347. Piloselle   hieracium, hawkweed. see 219a. *Hieracium pilosella.* **T L H Y**

348. Pimpernels   anagallis, scarlet pimpernel, ipia, wood pimpernel is moneywort, a lysimachia. see 290, 292. *Anagallis arvensis or Stellaria media.* **R T W H Y**

349. Pimpinelle   pimprinelle, saxifrage, lesser burnet, anise. see27, 413. *Sanguisorbe officinalis.* **R T H Y**

350. Pine   stone pine and others. tree bark, seeds etc. *Pinus pinus.* see 134. **H**

Pinguedo   pork lard. see 206.

Piper   pepper, serpyllum. see 341.

Pira and   pyra. see 339.

351. Pistachia   fisticus. see 459 *Pistacia vera.* **L H**

Pisum   peas. see 340, 358. **H**

352. Pix   peize, pix alba and nigra, poix, pissa, resin, turbentyne, colophony. see 126, 304, 359, 360, 379, 469a. **R B T W L H Y**

353. Plantain   plantago, many varieties including watercress, psyllium, rib-wort, lanceola, lancelette, policaria, quinque nervicium, arnoglossa, waybread, yva. see 428. *Plantago psyllium, P. cynops, etc.* **R T W L H Y**

354. Plomb   brûle alanauch, plumbum ustum, yellow oxide of lead. see 242, 253. **W L H Y**

355. Plum   prune, prunella, plums, tree sap (see gummi), seneste, sebeste Damascus, etc. Diaprunum is an electuary of plums. *Prunus domestica,* also *P. spinosum,* the blackthorn, sloe. **T W L H Y**

356. Poil de lievre   pili leporis, hare's-beard (Great Mullein). see 395, 433, 457. *Verbascum thapsus.* **R T W H Y**

357. Poireux   garlic. see 43. *Allium porrum.* **W**

358. Pois   pisum, peas of all kinds. *Pisum sativum.* **H**

359. Poix   greque probably hemlock tree resin, see pix. **W**

360. Poix (noir)   shoe-makers' pitch or wax. **W Y**

360a. Polycaria   pulicaria. inula. see 168a. **T Y**

361. Polygonum   knot grass, corrigiola, cesune, geniculata, St. Mary's Herb, Solomon's seal. see 49, 361, 454, *Polygonum aviculare.* **T L H Y**

362. Polypode   oak fern, beech fern. ee 181a. *Polypodium vulgare or Gymnocorpium dryopteria.* **R T W L H Y**

363. Polytric   beech fern, hair-cap moss, golden maiden hair fern. *Polytrichium juniperium.* **R T W**

363a. Poma   Apples, sour apples pears, etc. fruit, wood, leaves, bark. **T L H**

364. Pomegranate   malum punicum, pomme gernette, fruit, leaves, flowers or bark (ecorce, cortica), mellicrate, psidia (the fruit peel), wine, water.

     Punica   *granatum.* see 47, 279. **R B T W L H Y**

     Pome de terre   cyclamen. see 114

364a. Poplar   aspen. leaves. buds are oculi populi and bourgeons, bacca, aigeros. see 364a et al. *Populus tremuloides.* **T L H Y**

     Poppy   opium. see 320. **R B T W L H Y**

365. Populeum Oint.   also populeon (contained poplar tree-buds—bourgeone de peuplier) also called Nicolas' Ointment, one of several recipes included. 44, 49, 217, 263, 336, 348. **R T W L H Y**

365a. Porrum   garlic, ptasion. see 43. **R Y**

366. Portulaca   wild and domestic. purslane, chicken-feet,

		olus fatuum, Herb Robert. see 368. *Lepidum campestre and L.ruderale.* **T L H Y**
	Potentilla	tormentilla. see 465
367.	Pouliot	pulegium, pennyroyal, polial roial, Dragontea used mint. *Menthe pulegium.* see 288a. **R T W Y**
368.	Pourpier	purslane. see 366. *Portulaca oleracea.* **H**
	Prassium	emeralds or leeks. see166, 243. **W H**
368a.	Primavera	primrose, primula, cowslip. see 25a. *Primula vera.* **H**
368aa.	Propolis	bee glue and wax. see 293, 329a. **T H**
368b.	Prunella	self-heal, fruit of blackthorn sloe, sanicula *Prunella vulgaris.* see 424a. **T H Y**
	Prunum	plums. see 355. **H Y**
	Psidia	pomegranate peel. See 364. **T H**
368bb.	Psyllium	plantain. see 353. **T L H Y**
	Pulegium	penny-royal. see 367.
368c.	Pumice	spuma maris. see 162. **T H**
369.	Pumpkin	citrouille, zucca see 50, 280. **R W H**
	Purslane	portulaca. see 368.
370.	Pyrecanthum	lycium. **T H**
371.	Pyrethrum	peretrum, pellitory. flowers of fever-few, chrysanthemum. *Anthrum pyrethrum.* **B T W H Y**
	Quercus	oak. see 309a.
	Quicklime	see 251
372.	Quince	coing, cotonia, cydonium malum, citonium malum. pulp and seeds. see 123. *Pyrus cydonia.* **T W L H**
	Radish	see 373
372a	Ragweed	St. John's wort. *Senecio jacobea.* **T**
373.	Raifort	radish, raiz, horse-radish, rapistrum, raphanus, rave. *Cochlearia armoracia, Raphanus sativus and other Raphani.* **R T W L H Y**
	Raisins	uva passa. see 473. **T H**

374.	Ramich	an Arabian compound 19, 30 (berries), 93, 195, 299, 308, 318, 264, 407, 433. **H**
375.	Rana	scortica frog-meat. see 273a. **B T**
376.	Ranunculus	clematis, pes corvus, crow's feet. See 33, 120, 142. *Flammula jovis.* **H**
377.	Rape	a mustard. *Brassica rapa and napus.* see 64a. **W L H Y**
377a.	Rave	turnip or radish. see 373 and 377.
377b.	Red Powder	Roger's suture powder. 59, 126, 128, 272, 316, 408. **R Y**
378a.	Reeds	canes, marsh grass, canne, panicium, darnel etc. see 386. **R T W L H**
379.	Resin	gumma pini. pine pitch rosin. see 15a, 352. **R T W L H Y**
380.	Realgar	red arsenic ore. see 34 and 327. **B W L H Y.**
381.	Reglisse	licorice. see 246. **W**
381a	Rhazes' Oint.	Ung. album, contained 17, 77, 105, 163(whites), 253, 388,493 **Y**
382.	Rhubarb	*Rhababarum of many species.* **W Y**
	Rib-wort	plantain. see 353.
	Robert's Herb	portulaca. see 366. **Y**
383.	Roche	alum. see 20.
	Rocket-root	eruca. see 171. **T W**
384.	Rognons	(oil from) castrated testes,—calves, sheep, etc. **T**
	Ronces	rubis,blackberry. see 393.
385.	Rosat	roset, rosamel, honey with crushed rose petals. see 390. **R W Y**
386.	Roseau	marsh reeds, canne. see 378a. *Arundo donax.* **W.**
387.	Rosemary	herb. see 155a. *Rosemarinus officinalis.* **W Y**
388.	Rose-oil	oleum rosarum, oil of petals. **R W L H Y**
389.	Rose	powder of petals or whole flowers including the anthers. Eglantiers are wild roses. see 53c. *Rosacea, var. species.* **T W H Y**
390.	Rose-syrup	rosamel, rosat syrup of petals with honey

or sugar, see 385. **W H Y** 391. Rose-water water of petals, eau rosat. **W H Y**

392. Rostrum — porcinum endive, groin du porc. see 113. *Cichorium endivia.* **W H Y**

392a. Rubia — madder. see 197, 434a, 474, 497 *Rubia tinctoris.* **T H Y**

393. Rubis — blackberry, bramble, framboisier moron. ronces rouge are unripe. see 154. *Ronces nemorosus et al.* **R T W H Y**

394. Rue — herb ruta, moly, galigan. oil. *Ruta angustifolia and R. graveolans. A. montana.* **R T W H Y**

395. Rumex — patience, sorrel, oxalis, lappa, burdoch, dock, acedula, great mullein, shepherd's crook, bouillon, hare's-beard. see 16, 49b, 236, 294a, 329, 356, 433, 457. *Rumex acetosa et al.* **T W H Y**

395a. Rye — flour. siligo. *Secale cereale.* **T**

396. Sabina — cypress, or juniper. the resin is sandarac. see 149, 479. **T**

397. Saccharum — beeswax. see 220, 293.

397a. Safflower — carthamus. (not saffron). see 91. *Catrhamus tinctorius.* **R W**

398. Saffron — safron, colchicum, zafranatis, crocus. see 91, 141, 219. *Crocus sativus.* **R T W L H Y**

399. Sagapenum — another ferula plant resin. see 37, 429 *Ferula persica.* **R B T Y**

400. Sage — many varieties. herb salvia, sange, lungwort, palma marina, pulmonaria, gallitricum. see 196, 258a, 405. **R T W L H Y**

401. Sagimen — aphroniton, spuma nitri. precipitate of potassium nitrate. **H Y**

401a Saints' Herbs — Saint Bennett's, see 116. **T**. Saint John's, see 13, 35, 287, 287a, 372a,**T**, Saint Mary's, see 49, 361, 454 Saint Peter's, see 140a.

Sal armoniac — see 424. **W L H Y**

Sal baurachi — borax. see 63, 402. **R W L H Y**

402	Sal de nitre	nitrum, borax, saltpeter. see 63,402. **R W L H Y**
402a	Salex	salix. see 495. **H Y**
403.	Saliva	spittle. **R T W L**
403a.	Salsola	samphire, glass-wort, sal alkali. see 140a, 406a. *Salsola kali.* **H**
404.	Salt	sal. common or rock, sel gemma, brine, aloxan (brine 'flower'), muriate of soda. **R B T W L H Y**
405.	Salvia	sage (many varieties), centrum galli, also darnel, cockle, clary, ieble, salge-damasche, salge-savage. see 196, 400. *Salvia officinalis.* **R H Y**
406.	Sambucus	elder tree, sureau, seu, ebulus. see 165. *Sambucus nigra.* **R L H Y**
406a.	Samphire	saphira, glass-wort, marine crest. see 140a, 403a, 487, 501. **T H**
407.	Sandalwood	sandalus. *Santalum album and rubium.* **R T W L H**
	Sanamunda	hare's foot, caryophylla. see, 93 et al. **H**
	Sandarac	the resin. see 396. **H**
408.	Sang	dragon sang de dragunt, sedge. resin of *Calamus draco.* **R B T W H Y**
	Sanguinaria	see 140a, 406a, 487, 501. **H**
409.	Sanguis	blood: goat, sheep, deer, tortoise, bat, frog, snake, ox, dove hare, menstrual etc. **B T H**
	Sapa	grape syrup. see 208, and sappa michum, a syrup with honey.
409a.	Saracenic ointment	102, 105, 173, 206, 242, 242, 286, 368c. **R T**
410.	Sarcocolla	argemone. resin of *Pinea mucronata or Astragalus fasciculoformis.* **R B T W L H Y**
411.	Saturieia	herb savory. *Satureia hortensis.* **T H**
411a.	Satyrion	an aphrodysiac orchid. see 472. *Satyrion hircinum.* **W**
412.	Savon	French soap, sapo, soapwort, saponaria, burith. soap. see 244a, 430c **R T W L H**

412a.	Savory	many varieties. *Satureia hortensis.* **H**
413.	Saxifrage	marathrum. flowers or leaves. see 27, 349. **T W H Y**
414.	Scabious	devil's bit, jacea, knautia arvensis, morsus diabole. see 228b. *Centurea scabiosa.* **R T W L H Y**
415.	Scalllions	bulbus, onions. see 101, 319. **W**
416.	Scammony	bindweed, convolvulus, anabula, diagridum. see 152b. *Scammonaciae, several varieties.* **T W L H Y**
	Scariola	chicory. see 113. **T H**
416a.	Schoenanthum	palea camelorum, juncus. see 444. **H**
417.	Scoloprendre	ceterach, hart's-tongue fern, cow-tongue, bugloss, blue weed. see 69. **W H**
418.	Scordium	wood sage, a germander. see 200, 346. *Eucrium scordium.* **W**
418a.	Scorpion	incinerated. **L H**
419.	Scoria	ferrugo, iron filings and rust, merda ferri and cimolea, limailles. the dross of melted iron. see 216.**W H Y**
420.	Scrophularia	pennywort, centumcellie, toad-flax, umbilicus venus, linaria, cymbalaire, quadrangular, many varieties of *Scrophularaciae nodosum, aquativca, etc. Linaria vulgaris.* **R T L H Y**
421.	Sebestes	sebesten plums, cordia myxa. see 355. **T Y**
422.	Sedge	cyperes. see 74, 148b, 207, 434, 444. **W L**
422a.	Sedum	stonecrop. see 446b. **R Y**
423.	Seeds	common seeds, including lettuce, endive, purslain, chicory, melons. see 123a. **R T W**
424.	Sel armoniac	sal ammoniac, ammonium chlorhydrate, (not from 23) (commonly called"arsenic"). **B W L Y**
424a	Self-heal	sanicula, prunella. see 368 et al. **H Y**
	Selinum	apium seeds. see 29, 344. **L**
425.	Sempervivum	crassula minor, house leek, leeks, joubarbe, sticado. see 229 et al. **R T W L**

426. Senape   synapus, senevé, mustard. see 171, 297. **R
                     B T W L Y**

     Senecio   also senacio. groundsel see 102. **T Y**

427. Senna   many cassia varieties.Diasunna is a potent
                     laxative. see 94.*Cassia fistula* **T W Y**

428. Septemnervée   handacotte, septfoil centinervia, ribwort,
                     quinquenervia, tormentila see 215, 353.
                     Plantago **R T W**

429. Serapinas serapias, sarapinas, sagapenum. see 37, 399. *Ferula*
                     persica. **R W L H Y**

430. Sesali   lovage, white gentian. See 174, 247, 256.
                     Laerpitum siler. **W**

     Shepherd's Crook rumex. see 395.

     Sicadis   white bryony. see 67. **H**

     Sifula   sider, aneth, sweet absinthe see 26. **H**

430a. Silex   flint. **Y**

     Siligo   rye. see 187 (white), 395a. **T**

430b. Silver   argentum. flos, ashes. **L**

     Sisymbro   horse-mint, menthastrum. see 281. **R**

430bb. Skin   cow and sheep, for making collagen,
                     parchment, pellis. see 204. **W**

     Snails   see 249. **T H Y**

     Snake-root   serpentaria. see 33.

     Snakes   various. see 273a. **H**

430c. Soap   usually a strong soap, lessive, perhaps
                     French soap, Saracenic soap etc. see 244a,
                     412. **T W L H Y**

     Soapwort   see 412. **R**

     Socotrin   aloes, succatrensis. see 19.

431. Solathrum   nightshade, henbane, solanum, morele,
                     camel, mors canis, egg-plant see 217, 232,
                     290a. *Solanum nigrum et al.* **R T W L H Y**

431a Solea   leather, soglia. see 124, 204, 242a, 430bb.
                     R H

431b. Soot   from ovens, chimneys etc. see 38, etc. **H**

432. Sorba   cormes. fruits of the mountain ash. *Sorbus*
                     domestica, and Sorbus ancuparia. **T L H**

433.	Sorrel	oxalis, oseille, rumex, lapathum, burdock see 329, 395. **T W Y**
434.	Souchet	a sedge, see 5, 74, 75, 422, 444.
434a.	Spergula spargula,	madder. see 392a. **H**
435.	Spathula foetida	stinking iris. *Iris foetidissima.* **R T H**
436.	Spelt	epeautre, hard wheat. see 21, etc. **W**
436a.	Sperm	goat. **T**
437.	Spic	spikenard, nard. See 43a, 168a, 302, 475. *Valeriana officinalis, Inula conyza et al.* **B T W L H Y**
438.	Spider web	toile d'araignée, tela aranea. cobweb. see 31. **B W H**
438a.	Spig	any lichen, especially *Lichen gyratis.* **W**
	Spina	hawthorn. see 215a. **B**
438b.	Spinach	*Spinacea oleracia.* **H**
438c.	Spleen-wort	a scaly fern. *Asplenum ceterach.* **T**
439.	Spode	zinc oxide, tuthie, pompholyx, cathimia, calamine see 121. **W L Y**
440.	Spodium	calcined arrowroot. *Maranta arundinacia.* **L H**
	Sponge	whole or ashes. see 117. **T H**
441.	Spuma d'argent	écume d'argent, flower of silver, litharge. see 121, 253. **R T W**
441a	Spuma maris	magnesium silicate, meerschaum. see 162. **W L H**
442.	Spurge	resin of a euphorbium. See 171, 173. *Euphorbia lathyris.* **R T L**
442a.	Squid	os sepiae, seiche, sepia. burnt bone. see 62. **T H**
443.	Squill	wild hyacinth. See 222. *Scilla maritima.* **R T W L H Y**
444.	Squinanthus	sinancie, another calamus, schoenanthum. see 5, 74, 75, 416a 422, 434. *Andropogon schoenanthus.* **R T W L H**
444a.	Stag's horn	any fern of genus *Platycerum.* **T**
445.	Staphisagre	larkspur, delphinium, cheif d'espurge, polycaria, pes alauda, jonquarola.

Delphinium staphisagre or inula policaria. **R T W H Y**

445a. Stellaria — chick-weed. morsus gallinae. *Stellaria media.* **H**

446. Stellion — gaulus, a musk-like excrement of lizards. see 257.**W**

Stercus — feces. see 179.

Sticado — joubarbs. see 229, 446a. also *Stoechas citrinus and Graphalum stoechas.*

446a Stoechas — many varieties. see sticado and 240. also *Stoechas arabica.* **W H Y**

446b. Stonecrop — vermicularis. see 422a. *Sedum acre.* **R Y**

446c Straw — reeds, grasses, etc. see 378a, 386. **H**

446d. Strawberry — *Fragaria vesca.* **Y**

447. Styrax (storax) — assefan, liquid amber, benzoin. calamite is inferior grade storax. see 36, 41, 73. *Liquidambar orientale (a tree).* **R B T W L H**

447a. Sudor — sweat, animal or human. **R**

448. Sugar — alun de sucre, zuccharum, nabete (powdered) sugar candy, penedis is a droplet of sugar, sugar of violets etc. **R B T W L H Y**

Sureau — elder. see 165. **H Y**

449. Sumach — the poisonous toxicodendron. *Rhus coriaria* as well as non-poisonous *R. aromaticum.* **B T L Y**

450. Sulfur — often stated as 'live', being fresh from the mine. **R T W L H**

450a. Swallow-wort — milk-weed. *Cynanthum vincitoxicum.* **T**

450b. Sycomore — sap of *Ficus sycomorus.* sometimes the Egyptian mulberry. **Y**

451. Talpa — the mole. see 273a. **H**

452. Tamarind — tamarind fruit. *Tamarindus indica.* **T W L Y**

453. Tamariscus — sap of manna, ash tree. see 189, 264. *Fraxinus ornus, Mysicaria germanica.* **R T W H**

454. Tansy   athanasia, St. Mary's Herb, fever-
                       few. *Tanacetum vulgare.* **T H Y**
455. Tar   naval, piotch, pix, Greek. see 304, 352, 359,
                       360. **R B T H Y**
      Tarragon   dracunculus. see 161a.
456. Tartar   potassium bitartrate from wine lees. **R T W
                       L H Y**
457. Tassus Barbatus tassebarbatus. great mullein (bouillon).
                       see 356. *Verbascum thapsus.* **R T W H Y**
458. Teazle   chardun thistle, dipsacus, cardo. see 100,
                       102, 159, 487. **T Y**
      Tenacetum   tansy. see 454
459. Terebinth   also olibanum, xylobalsamum. closely
                       related to mastic. alkitron is a distillate.
                       see 272, 351. also *Pistacia terebinthus.* **R W L
                       H Y**
      Terpentine   pitch. see 352.
460. Terra sigillata an astringent trochee of baobab fruit,
                       Adansonia digitata, or a reddish clay of
                       Lemnos, fashioned like an Egyptian seal.
                       chimolia or cymolea. **R T W L H Y**
461. Thapsus   tapsie, another scrofularia umbellifer. see
                       36. *Thapsia villosa and Th. garganilla.* **R T
                       H**
462. Theriac   many formulas through the centuries,
                       Galen's Greater Treacle. The diatesseron
                       variety contained 58, 199, 220 and
                       238(berries). Recently called treacle. see
                       156b. **T W L H Y**
      Thistle   teazle. see 458 et al.
463. Thus   tus, cortex thuris. a thick frankincense, see
                       168, 188, 316. *Boswellia thurifera.* **B W L H**
464. Thyme   calamint. *Thymus capitatus.* **T W H**
      Titimalle   euphorbia. see 173. **R T H Y**
      Toad flax   scrophularia. see 420.
      Tongue   see 69

464a. Tonnina   Mediterranean tuna. **T**

465. Tormentilla   sarsaparilla, quinquefolum, cinqfoil, potentilli, geranium maculatum, cranesbill or doves-foot, pie de colomb, pseudoselinon, callipetalon. *Potentilla reptans.* **T H Y**

466. Tragacanth   dragacanth, adracanth. see 8, 153a, 213 *Astragalus gummifer.* **B T W L Y**

      Tremula   poplar. see 364a (quaking aspen). **T**

467. Tribulus   water thistle, water chestnut or Burra Gukaroo. *Tribulus terrestris or T. aqautica.* **H**

      Triticum   wheat. see 494.

468. Tryphére   an electuary containing truffles and various sweets. **W**

469. Turbith   turpeth, diatesseron laxative, roots of *Operculum turpathum.* **B T L Y**

      Turmeric   curcuma. see 146 and 503.

469a Turpentine   pitch. see 352.**T W L**

470. Tussilage   colt's foot. *Ungula caballina.* **H Y**

471. Tuthie (tutty)   see spode. see 121, 439. **T W L H Y**

      Umbilicus venus   cymbalaria. see 136, 420. **R L H**

472. Unguis caprae   goat slipper. see 411a. **H**

472a Urine   animal and human. **B T H Y**

      Urtica   nettles. see 305, 326.

      Uva   grapes. see 208.

473. Uva passa   raisins. **T W H**

474. Valania   madder. see 197, 392a. *Rubia tinctorium.* **W**

475. Valerian   phu, amatilla, fistra. spikenard. see 302, 437. **T H Y**

476. Venus hair   maidens-hair fern, capillus venus, bed-straw. see 7, 36, 181a. *Adiantum capillus veneris.* **T H Y**

      Venus ointment   apostolicon. see 30

      Verbascum   tassus barbatus, see 457. **Y**

477. Verbena   vervaine, hiera botane, verminacula. *Verbena officinalis.* **R T W L H Y**

478.	Verjus	agresta, vinum acerbum, a potion from sour grapes or other sour fruits. see 12, 335, 485.
478a.	Vermicularis	stonecrop, sedum. see 446b
	Vermis	worms. see 258, 499. **H**
479.	Vernis	juniper tree sap, encaustrum. see 167a *Thuia articulata.* **T W H Y**
480.	Vespa	wasp. **H**
481.	Vert d'Airain	vert d'araim, ziniar, fleur d'arain, flos aeris, viride aes, bronze flower, similar to 482. **R T L H Y**
482.	Vert de Gris	ziniar, copper 'flower', copper acetate, chloride or sulfate. See 10, 65a, 106, 130, 481. **R T W L H Y**
483.	Vetch	ers, orobe. see 325. **L**
484.	Vinegar	acetum, aisil. **R B T W L H Y**
485.	Vinum goretum	raw styptic wine. see 12, 478. **W**
486.	Violets	oil of or water of many varieties of *Violaria.* **R B T W L H Y**
487.	Virgo pastoris	shepherd's purse, another dipsacus, sanguinary, centinodium, passerinus, proserpinaia. see 501 et al. *D. sylvestris.* **T W L H**
488.	Virgo cervi	deer's penis. **H**
489.	Viticella	white bryony, vitis. see 67, 120 etc. *Clematis flammula.* **R L H**
490.	Vitriol rosa	red iron sulfate, couperose, atrament, chalcantum, colcothar, ink, Roman vitriol. **R T W L H Y**
490a.	Water chestnut	*Trapa natans.* **T**
490b.	Water cress	plantain. see 353. *Nasturtium officianale, Lepidium sativum.* **R Y**
491.	Water lilies	nenuphar, dardana, fafara. see 304a. *Nymphea alba.* **T W L H Y**
492.	Watermelon	fruit, wild and cultivated. see 280. **Y**
492a	Water mint	*Mentha aquatica.* **T**

492b. Waters   sea, rain, spring, containing drownings etc.
                         see 286, 300, 364, 391.

493.   Wax   white and red (cera alba and rossa), cere,
                         sire, oxycroceum. see 329a. **R B T W L H Y**

494.   Wheat   froment, triticum. pigle is coarsely ground.
                         see 21 et al. **T W L Y**

494a. Wm. Somer's Oint. 379 and 485. **T L**

       Whey   serum. especially goat's. see 286. **Y**

495.   Willow   salex. withe. tree bark, and flowers *Salix
                         alba and nigra.* **W** L

496.   Wines   many varieties, named by color, region,
                         potency, acidity, thickness. see 485. **R B T
                         W L H Y**

       Winter seeds   the four cool-weather seeds. 137,144, 280,
                         492. see 123a. **T**

497.   Woad   pastel, the dye. See 197, 392a. *Isatis
                         tinctoria.* **Y**

497a. Wood   various kinds, including Brazil-wood. see
                         11a, etc **H**

498.   Wool   laine muste, lana succida, unwashed
                         fleece. see 235. **R B W Y**

499.   Worms   lumbrici, ver, verm, maggots. see 258. **T L
                         Y**

       Wormwood   absinthe. see 1.

       Xylobalsamum   terebinth. see 459 **W**

500.   Yari   cuckoopint arum, se. see 35a. *Arum
                         maculata.*

501.   Yarrow   mille feuille, sanguinary. see 223, 287. **B
                         W**

502.   Yeast   fermentum. leaven. **W H**

       Yellow Oint.   citrin. contains 181, 188, 312, 379, 493. **H Y**

502a   Yerasimum   erysimus, ? Saint Simon's herb, or
                         simissome. Yperman used it as a mild
                         laxative for children. see 458. *Erysimum
                         cheiranthoides* Y

       Yreos   iris. see 228.

	Ysopus	see 206, 311.
502b.	Ysis	isis. any lichen of genus *Isidium.* **T**
	Yva	gum eve, iva arthretica, a germander. see 353. *Teucrium chamaedrys.* **T**
503.	Zedoaria	turmeric. see 146. *Curcuma longa et al.* **H Y**
503a.	Zegi	ink, vitriol. **W**
	Ziniar	verdigris. see 482.
	Zinziber	ginger. see 24, 201. **H**
	Zizania	lolium. see 255
	Zucca	pumpkin. see 369 **R**

BIBLIOGRAPHY
FOR THE COMPENDIUM PHARMACOPEIA

Roger Frugard (1170): Chirurgia; in Vol. I of Anglo-Norman Medicine, edited by Tony Hunt. 1994, Cambridge, UK; D.S. Brewer

Roger Frugard: Chirurgia. An Italian Transl. by L. Stroppiana and D. Spallone. Rome, Istituto di Storia della Medicina della Universita di Roma, 1957. Engl. Transl. by LD Rosenman. Philadelphia, Xlibris Co., 2002

Roland of Parma : The Surgery;. An Italian Transl. by M. Tabanelli, in his La Chirurgia Italiana Nell'Alto Mediovo. Florence. Leo S. Olschki Editore, 1965. Engl.Transl. by LD Rosenman. Philadelphia, Xlibris Co., 2002

Bruno da Longoburgo (1252): An Italian Surgeon of the 13thC;(an anotated transl. of his text); M Tabanelli, 1970, Florence, Leo S. Olschki, Engl. Transl. by LD Rosenman, Pittsburgh, Dorrance Opublishing Co., 2002

Theodoric (1265): The Surgery of Theodoric Engl. Transl. by E Campbell and J Colton; 1955-1960, New York, Appleton-Century-Crofts, Inc.

William of Saliceto (1275): Chirurgie (1275), French. Transl. by P.Pifteau; 1898, Toulouse, Imprim. Saint Cyprien. Engl. Transl. by LD Rosenman, Philadelphia, Xlibris Co, 2002

Lanfranchi of Milan, Chirurgia Magna, Engl. Transl. by LD Rosenman, Philadelphia, Xlibis Corp., 2003

Lanfranchi of Milan: Chirurgia Parva (1295)[294]: see footnote below

Henri de Mondeville (1315): The Surgery of H.de M.: French Transl.

by E Nicaise; 1893, Paris. Germer-Balliere. Engl. Transl. in 2 Vols. by LD Rosenman, Philadelphia, Xlbris Co, 2003

Jehan Yperman (1320): The Surgery of J.Y. French Transl. by Doctor A. DeMets; 1936, Paris, Editions Hippoc. Italian Transl. by M Tabanelli, Engl. Transl. by LD Rosenman, Philadelphia, Xlibris Corp. 2003

Mrs. M. Grieve: A Modern Herbal in Two Volumes; 1931, reprinted in 1971; New York, Dover Publications, Inc.

Saint-Lager: Genre Grammatical des Noms Generique, etc.: JB Saint-Lager, 1897, Paris, Germer-Balliere

The Alphita: A Glossary and Herbal of French Origin, in Latin, probably of the late 14th C. now in the Collectio Salernitana. Edited by Salvatore de Renzi. 1853, Bologna, Forni Editore. Vol I, pp 271-322.

---

[294] A transcript of the title page which introduces it follows:"*A most excellent and learned work of chirurgerie called Chirurgia Parva Lanfranci. Lanfranke of Mylan, his brief reduced from diverse translations to our vulgar or usual frase, and now first published in Englyshe prynte by JOHN HALLE, chirurgien who has necessarily annexed a Table as wel of the names of diseases and simples, etc. Imprinted at London in Flete Streate, nyghe unto Saint Dunstones churche by Thomas Marshe. AN 1565. (LDR).*